Functional Polyimide Dielectrics

Functional Polyimide Dielectrics

Structure, Properties, and Applications

Edited by Jun-Wei Zha

WILEY-VCH

Editor

Prof. Jun-Wei Zha
University of Science and
Technology Beijing
No. 30 Xueyuan Road
Haidian District
Beijing 100083
China

Cover Images: © Eugene Mymrin/Getty Images, Courtesy of Jun-Wei Zha

Library of Congress Card No.: applied for

British Library Cataloguing-in-Publication Data
A catalogue record for this book is available from the British Library.

Bibliographic information published by the Deutsche Nationalbibliothek
The Deutsche Nationalbibliothek lists this publication in the Deutsche Nationalbibliografie; detailed bibliographic data are available on the Internet at http://dnb.d-nb.de.

Print ISBN: 978-3-527-35433-7
ePDF ISBN: 978-3-527-84902-4
ePub ISBN: 978-3-527-84901-7
oBook ISBN: 978-3-527-84903-1

Typesetting Straive, Chennai, India

Contents

1

Overview of Polyimide Dielectrics

Mengyu Xiao and Jun-Wei Zha

University of Science and Technology Beijing, School of Chemistry and Biological Engineering, Department of Chemistry, 30 Xueyuan Road, Haidian District, Beijing 100083, P. R. China

1.1 Introduction

The advancements in aerospace and new energy technologies have accelerated the demand for high-temperature-resistant and high-performance polymer dielectric materials. Researchers have developed a range of specialty engineering plastics such as polyphenylene sulfide (PPS), polytetrafluoroethylene (PTFE), poly(ether ether ketone) (PEEK), polyethersulfone (PES), and polyimide (PI) [1]. PI is a class of polymer that contains an imide structure in the molecular backbone. The rigid imide ring gives PI excellent high-temperature resistance; thus, only cyclic polyimides are of practical application. PI was first reported in 1908, [2] and DuPont commercialized Kapton films in the mid-1960s. Since then, PI has entered an era of booming growth.

PI has been fully developed as a promising polymer, especially in the field of insulating and functional materials. The reason for the considerable interest in PI, compared to other high-temperature-resistant polymers, is due to its high structural designability, numerous synthesis and processing methods, as well as its outstanding comprehensive performance and wide range of applications.

This chapter provides a concise overview of the design, fabrication, and application of PIs. Firstly, the PI is introduced from a structural perspective in order to provide an overview of the general relationship between structure and function, which will then be used to guide the subsequent design. Then, the synthesis of PI and the preparation of different types of PI dielectric materials are discussed. Subsequently, the various properties of PI and the corresponding application areas are summarized. Finally, the key points of PI dielectrics are summed up and the future development is prospected.

1.2 Structure Design of Polyimide

The process of designing a polyimide usually involves careful planning of its molecular structure to ensure that the resulting material has the desired properties.

Functional Polyimide Dielectrics: Structure, Properties, and Applications, First Edition.
Edited by Jun-Wei Zha.

PI consists of alternating electron donors (diamines) and electron acceptors (dianhydrides). Therefore, designing PI can be realized by changing the structure of diamine and dianhydride. More than 1000 diamines and more than 400 dianhydrides are currently used to synthesize PIs, resulting in thousands of PIs with different structures. Here, only some common structure–property relationships are briefly described, such as thermal, mechanical, and electrical properties.

For PI, during aromatic heterocyclic polymerization, the glass transition temperature (T_g) is mainly related to the relevant chain length of the macromolecule and the intermolecular forces. PI has a higher T_g and thermal stability than the corresponding polyether or polyester due to intermolecular forces other than van der Waals forces. Intramolecular or intermolecular charge transfer interactions are induced between the electron acceptor and electron donor of the PI, as shown in Figure 1.1. The charge transfer effect is related to the electron affinity of the dianhydride (or the hole affinity of the diamine) and the conformation of the molecular chain. It can be qualitatively argued that increasing the electron affinity of the dianhydride enhances interchain interactions, i.e., increases T_g. Intramolecular charge transfer is greatest when the electron donor and the electron acceptor are co-planar, but least when they are perpendicular. The planar structure between the electron donor and acceptor can be disrupted by making neighboring substitutions to the C—N bond. For example, with neighboring methyl substitution of diamines, the site-blocking effect disrupts the coplanar structure and increases the rigidity of the molecular chain, resulting in an improved T_g.

The mechanical properties of polymers are mainly influenced by the molecular structure, which determines the intramolecular chemical bonding forces and intermolecular forces. For example, increasing the polarity of the PI or creating hydrogen bonds can increase the chemical bonding and intermolecular forces of the main chain, resulting in increased tensile strength. In addition to the molecular structure, the mechanical properties of PI depend on the synthesis method and processing conditions. The thermal history during the formation of PI affects its aggregate state structure, which ultimately has an impact on the mechanical properties.

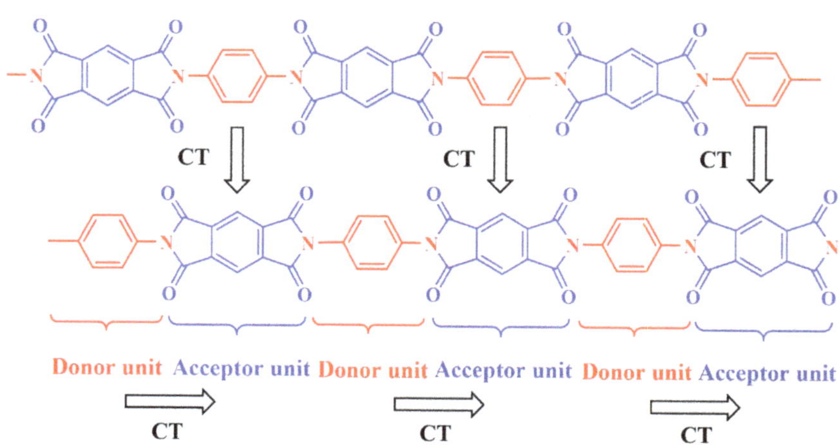

Figure 1.1 Intermolecular and intramolecular charge transfer (CT) in PI.
Source: Ref. [3]/Springer Nature.

The electrical properties of polymers refer to the behavior of polymers under the action of an applied voltage or electric field and the various physical phenomena they exhibit. PI is a typical linear dielectric with excellent dielectric and insulating and properties. The dielectric permittivity of PI is mainly contributed by dipolar polarization of their polar groups. Therefore, the introduction of polar groups into the PI molecular chain, or an appropriate increase in free volume can improve the dielectric permittivity of PI. For instance, Liu et al. [4] introduced polar groups -COOH and -CO-NH- into the PI main chain by controlling the degree of imidization of PAA (Figure 1.2), resulting in a dielectric permittivity up to 4.59@1 kHz. However, the promotion strategies of dielectric permittivity usually lead to a higher dielectric loss. Thus, systematic consideration of the molecular/structure design according to the polarization mechanisms is necessary to improve the dielectric loss of PIs. The insulating properties of PIs depend mainly on the bandgap and carrier traps. Wang et al. [5] found that the high-temperature insulation performance would experience diminishing marginal utility as the bandgap increases beyond a critical point (~3.3 eV) through the study of some series of PI derivatives. Therefore, it is essential to ensure a wide bandgap while constructing deep traps to enhance the insulation performance of PI [6]. There are three strategies for constructing carrier traps: adjusting the intrinsic structure of polymers, preparing inorganic/polymer composites and all organic polymer composites.

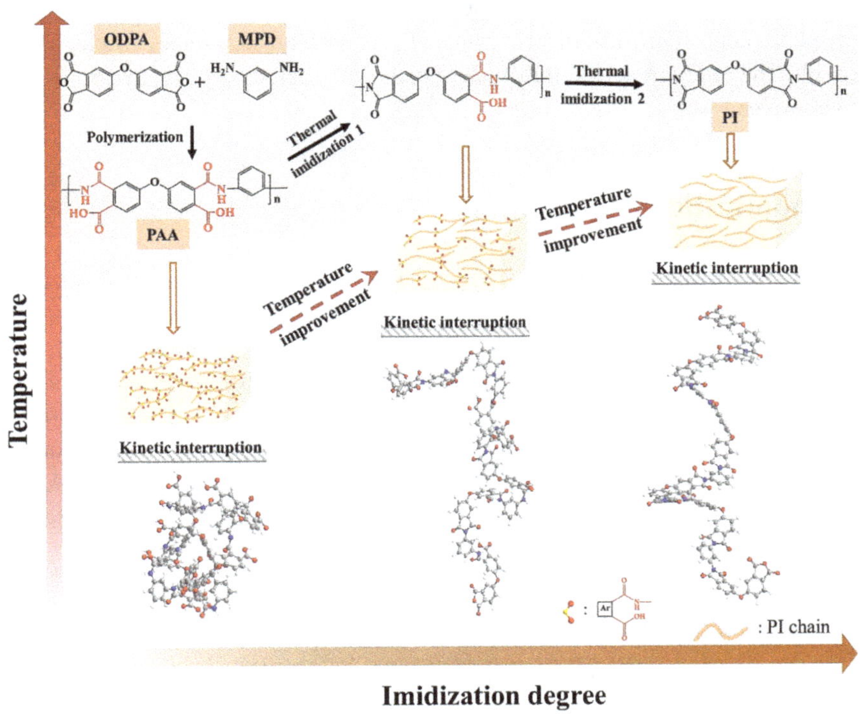

Figure 1.2 Schematic diagram of the thermal imidization process of PAA. Source: Adapted from Ref. [4].

1.3 Fabrication of Polyimide

PI boasts of numerous synthetic pathways and can be tailored for diverse applications, making it unparalleled among other polymers [7]. PIs are generally synthesized by a two-step process. Firstly, polyamic acid (PAA) is obtained by low-temperature polycondensation of dianhydride and diamine in a polar solvent (N, N-dimethylacetamide, N, N-dimethylformamide, and N-methylpyrrolidone). Subsequently PAA can be thermally or chemically dehydrated to obtain PI. PI can also be synthesized in one step, where the dianhydride and diamine are polycondensed by heating in a high boiling solvent (phenolics). The monomer ratio, dosing sequence, and reaction temperature are the core parameters affecting the PI polymerization reaction. For example, the ring-opening polymerization of dianhydride and diamine is an equilibrium reaction and has a high equilibrium constant, with the reaction heavily biased in favor of the product [8]. The polymerization process is significantly exothermic, and a proper lowering of the reaction temperature is more conducive to a positive reaction.

After decades of development, PI can be processed by methods suitable for most polymers. For example, PAA solutions can be utilized for cast film formation, spin coating, and spinning. Because inorganic salts are not generally produced during the synthesis of PI, there is no need for an additional purification step, which is very favorable for the preparation of insulating materials. PI can also be thermo-compressed, extruded, and injection molded by melt processing. Moreover, it is possible to utilize the easy sublimation of dianhydride and diamine for vapor-phase deposition. The diverse range of processing technologies employed by PI enables the production of a multitude of materials, including films, fibers, foams, and adhesives.

1.4 Applications of Polyimide

PI, a high-performance polymer, enjoys widespread utilization across various industries owing to its exceptional structural properties, facile synthesis process, and adaptability to diverse processing techniques. Its unique chemical and physical characteristics have allowed it to demonstrate outstanding performance in a broad range of applications, as shown in Figure 1.3 [9]. Each of these applications, from aerospace and automotive components to medical implants and semiconductor devices, has benefited from PI's ability to deliver superior performance in challenging environments. Consequently, PI continues to be a highly sought-after material for a diverse range of applications worldwide.

1.4.1 Capacitive Energy Storage

PI has a very broad application prospect in the field of dielectric energy storage. There is now a high demand for polymer dielectrics with outstanding high-temperature capacitance performance [10]. Because in many areas, such as

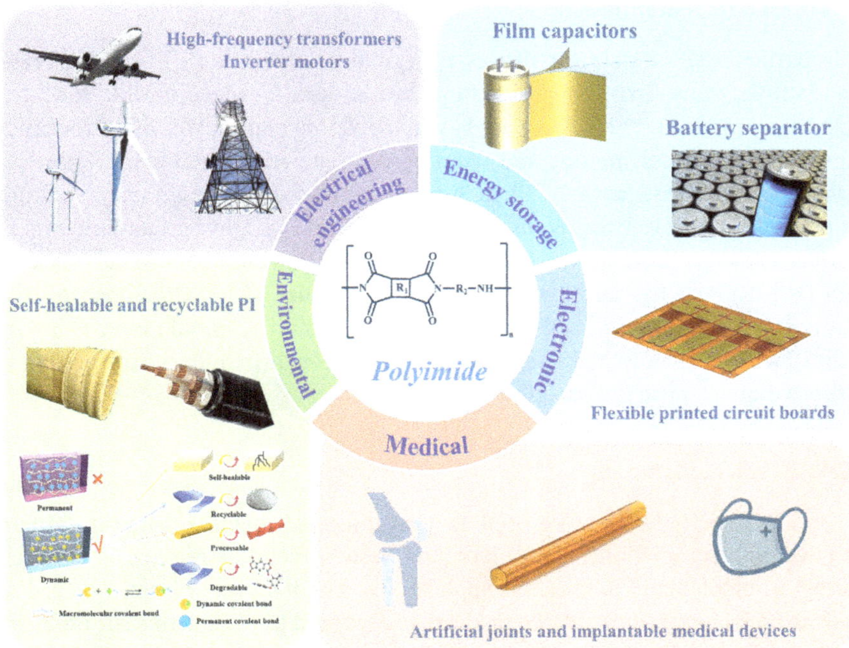

Figure 1.3 Applications of polyimide. Source: Ref. [9]/John Wiley & Sons.

hybrid electric vehicles, underground oil/gas exploration, and aircraft electrification, film capacitors typically operate at high temperatures (>150 °C) [11]. However, the maximum operating temperature of commercially available biaxial-oriented polypropylene (BOPP) should not exceed 105 °C [12]. PI is considered to be the most promising material for high-temperature energy storage. Therefore, many researchers have carried out studies in order to improve the usage of PI in high-temperature energy storage applications [13, 14]. In addition to physical energy storage, PI can be used in the new energy field to manufacture high-performance battery separators and fuel-cell bipolar plates.

1.4.2 Electronic Devices

As an important interlayer dielectric material for flexible copper-clad laminate, PI film plays a crucial role as the structural and functional basis for flexible printed circuit boards. However, the dielectric permittivity of PI is about 3.4, which is no longer sufficient to meet the needs of integrated circuits process development. In order to reduce the dielectric permittivity of PI, researchers have carried out a lot of modification work on PI from the Clausius–Mosotti equation. The dielectric permittivity of PIs can be effectively reduced by introducing fluorine atoms, bulky side groups, or porous structures [15]. Moreover, designing dielectrics with low permittivity and high thermal conductivity is crucial to improve the signal transmission quality of smart terminals and effective heat dissipation of devices [16].

1.4.3 Electrical Engineering

PI is particularly suitable for use as a flexible printed circuit board substrate. For example, transparent polyimide films can be used to make flexible solar cell substrates. Moreover, PI is widely used as an insulating material for high-frequency transformers, inverter motors, and wind power generation equipment. However, the poor corona resistance of PI at high-frequency voltages seriously shortens the service life of electrical equipment. The addition of appropriate amounts of nanoparticles has been shown to be effective in improving the corona resistance of PI [17]. DuPont has developed a sandwich structure of 100 CR films (top and bottom layers are Al_2O_3-doped PI composite films and the middle layer is a pure PI film), which provides higher corona resistance at high-frequency voltages, up to 10 times that of a pure PI film (100 HN) [18].

1.4.4 Medical Materials

PI has low toxicity and can be used as a multifaceted medical material. Certain PIs have high compatibility with blood and tissues, which is the reason for their interest in biocompatible material applications. For instance, PI can be used to make artificial joints, implantable medical devices, and biodegradable sutures.

1.4.5 Environmental Protection

The heat and organic solvent resistance of PI is particularly important in the separation of organic liquids and gases. High-temperature filtration materials prepared from PI fibers are widely used for exhaust gas treatment in cement production, steel, and other industries. In addition, PI has excellent stability and durability and maintains its original performance even in extreme environments, so it can be used for long periods of time without much maintenance or replacement. However, high T_g PI molecular chains are difficult to move at room temperature and cannot be actively recycled and self-healed after destruction. Self-healable and recyclable PIs have great practical engineering application value and environmental significance, promoting the development of a sustainable society. In recent years, a small amount of studies have been published to modulate the dynamic behavior of PI such as self-healable, recyclable, and degradable abilities by adjusting the structure of molecular chains and the composition of monomers [19, 20].

1.5 Summary and Outlook

In summary, PI can be used in many applications through the choice of structural design and synthetic processing methods. These areas include high-temperature energy storage, low dielectric, corona resistance, thermal conduction, electromagnetic shielding, optical transparency, corrosion resistance, and dynamics.

The primary reason that the PI has not become a more prevalent breed over the past half-century is its high cost. But from a chemical standpoint, the cost of PI can be reduced. There are two main ways to reduce the cost, one is to develop new monomer synthesis and polymerization methods, and the other is to use PI to modify other polymers. As preparation technology continues to evolve and costs are reduced, PI is poised to assume a more prominent role in the field of materials in the future.

It can be predicted that the future development direction of PI will mainly focus on high performance, multifunctionality, environmental protection and sustainability, and the expansion of application areas. This will bring new opportunities and challenges for the development of polyimide materials, and at the same time will promote the progress and development of related fields.

References

1 Feng, Q.K., Zhong, S.L., Pei, J.Y. et al. (2022). Recent progress and future prospects on all-organic polymer dielectrics for energy storage capacitors. *Chemical Reviews* 122: 3820–3878.

2 Bogert, M.T. and Renshaw, R.R. (1908). 4-amino-0-phthalic acid and some of its derivatives. *Journal of the American Chemical Society* 30: 1135–1144.

3 Zuo, H.T., Gan, F., Dong, J. et al. (2021). Highly transparent and colorless polyimide film with low dielectric constant by introducing meta-substituted structure and trifluoromethyl groups. *Chinese Journal of Polymer Science* 39: 455–464.

4 Liu, X.J., Zheng, M.S., Wang, G. et al. (2022). High energy density of polyimide films employing an imidization reaction kinetics strategy at elevated temperature. *Journal of Materials Chemistry A* 10: 10950–10959.

5 Wang, R., Zhu, Y., Fu, J. et al. (2023). Designing tailored combinations of structural units in polymer dielectrics for high-temperature capacitive energy storage. *Nature Communications* 14: 2406.

6 Zha, J.W., Xiao, M., Wan, B. et al. (2023). Polymer dielectrics for high-temperature energy storage: constructing carrier traps. *Progress in Materials Science* 140: 101208.

7 Liaw, D.J., Wang, K.L., Huang, Y.C. et al. (2012). Advanced polyimide materials: syntheses, physical properties and applications. *Progress in Polymer Science* 37: 907–974.

8 Ardashnikov, A.Y., Kardash, I.Y., and Pravednikov, A.N. (1971). The nature of the equilibrium in the reaction of aromatic anhydrides with aromatic amines and its role in synthesis of polyimides. *Polymer Science USSR* 13: 2092–2100.

9 Wan, B. and Zha, J.W. (2024). Dynamic polyimide dielectric film: a concept and application perspective. *Advanced Functional Materials* 34: 2312829.

10 Chen, J., Zhou, Y., Huang, X. et al. (2023). Ladderphane copolymers for high-temperature capacitive energy storage. *Nature* 615: 62–66.

11 Li, H., Zhou, Y., Liu, Y. et al. (2021). Dielectric polymers for high-temperature capacitive energy storage. *Chemical Society Reviews* 50: 6369–6400.

12 Yang, M., Guo, M., Xu, E. et al. (2024). Polymer nanocomposite dielectrics for capacitive energy storage. *Nature Nanotechnology* 19: 588–603.

13 Liu, X.J., Zheng, M.S., Chen, G. et al. (2022). High-temperature polyimide dielectric materials for energy storage: theory, design, preparation and properties. *Energy & Environmental Science* 15: 56–81.

14 Zha, J.W., Tian, Y., Zheng, M.S. et al. (2023). High-temperature energy storage polyimide dielectric materials: polymer multiple-structure design. *Materials Today Energy* 31: 101217.

15 Hu, R., Chen, Y., Zhang, C. et al. (2024). Porous monoliths from polyimide: synthesis, modifications and applications. *Progress in Materials Science* 144: 101284.

16 Dong, X., Wan, B., Zheng, M.S. et al. (2024). Dual-effect coupling for superior dielectric and thermal conductivity of polyimide composite films featuring "crystal-like phase" structure. *Advanced Materials* 36: 2307804.

17 Zha, J.W., Song, H.T., Dang, Z.M. et al. (2008). Mechanism analysis of improved corona-resistant characteristic in polyimide/TiO_2 nanohybrid films. *Applied Physics Letters* 93: 192911.

18 Meloni, P.A. (2004). High temperature polymeric materials containing corona resistant composite filler, and preparing methods thereof. US20040249041 A1.

19 Wan, B., Dong, X., Yang, X. et al. (2023). Rising of dynamic polyimide materials: a versatile dielectric for electrical and electronic applications. *Advanced Materials* 35: 2301185.

20 Wan, B., Yang, X., Dong, X. et al. (2023). Dynamic sustainable polyimide film combining hardness with softness via a "mimosa-like" bionic strategy. *Advanced Materials* 35: 2207451.

2

Structures and Properties of Polyimide Dielectric Materials

Lei Zhai[1], Yi-Kai Wang[1,2], and Lin Fan[1,2]

[1] *Key Laboratory of Science and Technology on High-tech Polymer Materials, Institute of Chemistry, Chinese Academy of Sciences, Zhongguancun North First Street 2, Beijing 100190, PR China*
[2] *University of Chinese Academy of Sciences, School of Chemical Sciences, No.1 Yanqihu East Rd, Huairou District, Beijing 101408, PR China*

Film capacitor technology has rapidly developed and is widely applied in various fields, such as power electronic equipment, high-voltage transportation, oil and gas deep-well drilling, high-frequency coupling circuits, and military direct-energy weapons [1–5]. As the key material for thin-film capacitors, polymer dielectric films possessing many advantages, including good dielectric properties and versatile capacitor designs, face increasingly stringent performance requirements. Biaxially oriented polypropylene (BOPP) film currently used in energy storage displays high breakdown strength up to 700 MV/m and low dielectric loss, but its temperature resistance and energy density are poor, making it unable to meet the increasing demand for high-temperature applications [6–10]. Polymer dielectric materials that can operate under high temperatures, such as polycarbonate (PC), polyvinylidene fluoride (PVDF), polyphenylsulfide (PPS), polyetheretherketone (PEEK), polytetrafluoroethylene (PTFE), poly-(oxa)-fluoronorbornene (POFNB), and polyimide (PI) [11–19], have attracted great attention from domestic and international researchers.

Among the above polymer materials, PI dielectric films typically have become a research hotspot and focus in recent years due to their excellent thermal stability, outstanding mechanical properties, rich structural designability, and high feasibility to scale up for industrial production, which is considered the preferred polymer dielectric choice for the development of high-temperature film capacitors [20–24]. Polyetherimide (PEI), a commercial PI film product developed by GE company (currently owned by Sabic), is known as a high-temperature polymer dielectric film, and a number of attempts have been made to validate such film for application in the film capacitor. The molecular chain of PEI is composed of alternating ether bonds and imide rings, which is polymerized by aromatic monomers using 2,2-bis[4-(3,4-dicarboxyphenoxy)phenyl]propane dianhydride (BPADA) as the dianhydride and *m*-phenylenediamine (*m*-PDA) as the diamine. This structure endows PEI with outstanding comprehensive properties. The PEI-based capacitor

Functional Polyimide Dielectrics: Structure, Properties, and Applications, First Edition.
Edited by Jun-Wei Zha.

was also found to have excellent thermal stability and insulation resistance up to 150 °C or even higher [21]. In addition, compared with the BOPP-based capacitor, it is proved that PEI film shows 50% higher dielectric constant and can yield a higher volumetric advantage. In general, such PEI film have relatively lower dielectric loss of about 0.003, good dielectric strength (500–650 kV/mm), and energy density above 3.5 J/cm^3. Yang et al. prepared PEI films using Ultem 1000 resin as the raw material supplied by GE company. The resultant PEI film energy density is around 3.2 with efficiency of 70% at 150 °C [25]. On the other hand, Zeng et al. also prepared this PEI film with same molecular structure through laboratory synthesis, but better experimental results were obtained [26]. The energy storage density can reach to 3.8 at 150 °C, and the charge–discharge efficiency is increased to 87%. It is expected that further process improvement will lead to mass production of high-temperature capacitor-grade PEI films to meet upcoming market demand.

Although it has shown great potential for application, the energy storage characteristics of traditional PI including PEI films still face many challenges, making it difficult to meet practical application requirements. Previously developed PI materials are not strictly designed for the field of energy storage; thus, it is necessary to reconsider the structure design of PI dielectrics for high-temperature energy storage. Researchers have paid a lot of attention to designing novel PI dielectric films with different molecular structures, and mainly improve the dielectric constant and energy band gap (E_g) of materials by introducing polar groups, non-conjugated main chain, non-coplanar unit, and chemical crosslinking into the molecular structure. In this work, the recent research and development status of PI dielectric films is systematically summarized, with a focus on the relationship between their structural design and dielectric performance. By elucidating the influence of polymer multi-level structures on energy storage characteristics, we hope to provide guidance for the development of high-temperature polymer dielectric materials and accelerate the application of PI dielectrics in film capacitors.

2.1 Introducing Polar Groups

On the basis of maintaining the aromatic structure units of PI main chain to achieve good heat resistance and thermal stability, introducing polar functional groups into the molecular main chain or side chain can effectively increase the molecular polarization, which helps to improve the dielectric constant of PI film and in the meantime enhance its energy storage performance. However, the improvement of polarizability is often accompanied by more polarization loss, leading to an obvious decrease in charging and discharging efficiency. Thus, how to balance the energy storage density and efficiency is the main challenge, especially for the evaluation at high temperatures.

2.1.1 Nitrile Group

Nitrile groups (—C≡N) have a large dipole moment, and the introduction of polar nitrile groups can significantly enhance the polarizability of polymers, thereby increasing their dielectric constant. Wang et al. synthesized an asymmetric diamine

(3,4-APBN)-containing nitrile group and polymerized it with five commercial dianhydrides including 6FDA, DSDA, OPDA, BTDA, and PMDA to obtain a series of aromatic PIs with structures shown in Figure 2.1 [27]. Compared with symmetrical diamines such as 3,3-APBN and 4,4-APBN, the solution viscosity of poly(amic acid) (PAA) synthesized using 3,4-APBN is much lower. This is mainly due to the increased conformational entropy of the asymmetric chain, making it difficult for PAA chains to stack with the reduction of interaction between polymer chains. Through the wide-angle X-ray diffraction (WAXD) analysis, it was found that asymmetric PI derived from 3,4-APBN/PMDA was mostly amorphous, while symmetric PI based on 3,3-APBN/PMDA and 4,4-APBN/PMDA exhibits semi-crystallinity. As displayed in Figure 2.1, the dielectric properties of these nitrile-containing PI films were investigated. By introducing nitrile groups, the dielectric constants of PI films range from 3.08 to 3.62 at 10 kHz, and the results are higher than those of PI without benzonitrile groups in the structure, which shows a dielectric constant of 2.92. For asymmetric PI synthesized through 3,4-APBN, their dielectric constants are higher than those of symmetric series, due to higher polarity of adjacent nitrile groups caused by asymmetric repeating units. In addition, the polarization response near glass transition temperature (T_g) of PI with asymmetric structure is significantly higher than that of symmetric PI, indicating that under high-temperature conditions, the enhancement effect of asymmetric structure on dielectric constant becomes more remarkable.

Wang et al. further synthesized three special diamines (*m,m*-3BCN, *p,p*-3BCN, and *p,m*-3BCN) containing three benzonitrile groups, which were connected by meta-substituted ether bonds [28]. A series of PI films were prepared from such diamines and four aromatic dianhydrides (6FDA, OPDA, BTDA, and PMDA), and these highly polar groups were incorporated into the molecular structure in order to increase the dielectric constant and thereby improve its energy storage performance. As listed in Table 2.1, the dielectric constant results of PIs containing three adjacent nitrile groups can be significantly enhanced that are measured by *D-E* loops at the temperatures of 23 °C, 100 °C and 190 °C, respectively, while maintaining a lower dielectric loss than conventional PI film products such as Kapton and Ultem. In addition, the test results of broadband dielectric spectroscopy indicate that molecular chain motion around T_g, especially the dipole motion of polar groups assigned to β transition, can effectively increase the dielectric constant of PIs with the maximum up to 4.9 while retaining relatively low dielectric loss.

Treufeld et al. reported twelve kinds of PIs with one or three nitrile groups in the structures and studied the relationship between dielectric properties and energy storage performance [29]. The results show that introducing highly polar nitrile groups to PI structure can indeed increase the dielectric constant, resulting in an obvious improvement of energy density. The more the content of nitrile groups, the higher the density of energy storage. It is further found that the polar groups have different effects on improving the dielectric constant with the structure change of dianhydrides. For diamines containing the same nitrile group, as the polarity of dianhydride moiety increases, the improvement of dielectric constant gradually decreases, indicating that the dipole moments ascribed to dianhydride and diamine moieties may cancel each other out at the molecular level. The dielectric constant of

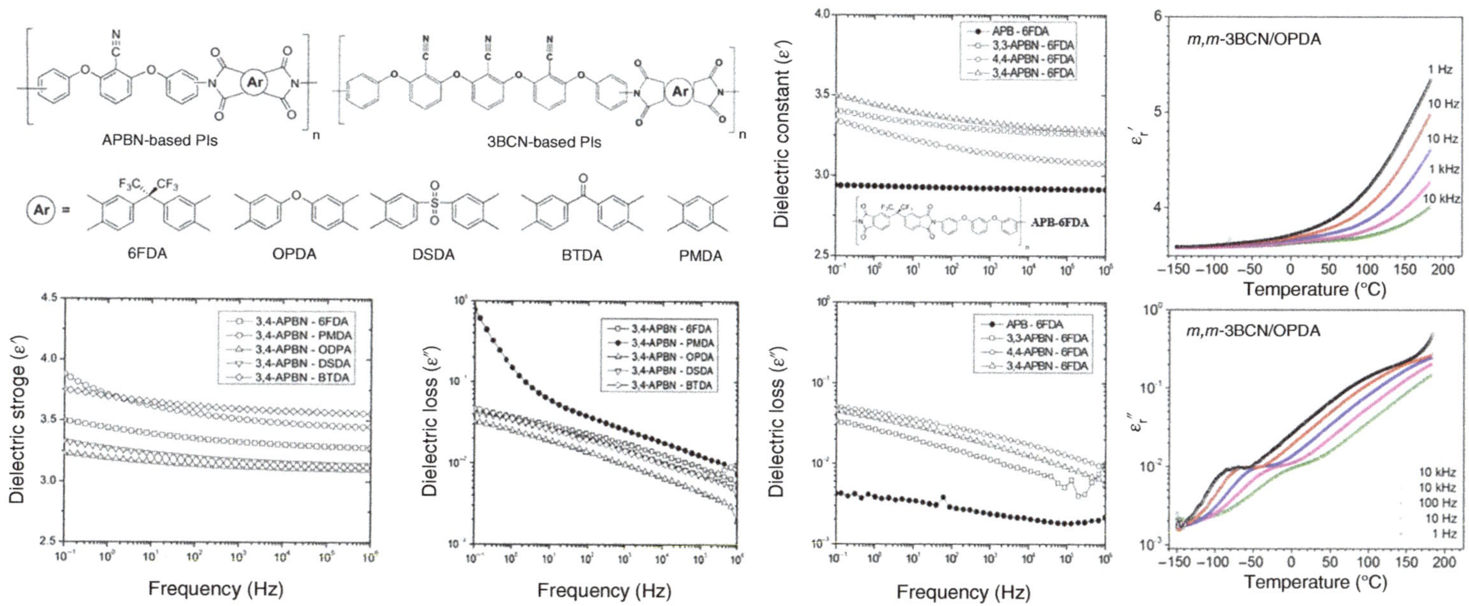

Figure 2.1 Structures of benzonitrile-containing PIs, dielectric properties of APBN-based PIs, and temperature-scan dielectric spectra of *m,m*-3BCN/OPDA including the real (ε_r') and the imaginary (ε_r'') parts of relative permittivity [27, 28].

Table 2.1 Permittivity and polarization response of benzonitrile-containing PIs and relevant structures.

	Apparent ε_r from *D-E* Loops			ε_r' from BDS		
Sample ID[a]	23 (°C), 10 (Hz)	100 (°C), 10 (Hz)	190 (°C), 1 (kHz)	−150 (°C), 1 (kHz)	190 (°C), 1 (kHz)	$\Delta\varepsilon_r'$[b]
m,m-3BCN/6FDA	3.7	3.7	4.1	3.13	3.62	0.49
m,m-3BCN/OPDA	4.0	4.2	4.5	3.58	4.34	0.76
m,m-3BCN/BTDA	4.0	4.3	4.3	3.42	3.85	0.43
m,m-3BCN/PMDA	3.8	4.8	4.6	3.42	4.06	0.64
p,p-3BCN/6FDA	4.3	4.5	4.9	3.45	4.30	0.85
p,p-3BCN/OPDA	4.0	4.1	4.5	3.33	4.47	1.14
p,p-3BCN/BTDA	3.9	4.0	4.3	3.56	4.14	0.58
p,p-3BCN/PMDA	3.7	4.3	4.6	3.42	4.79	1.37
p,m-3BCN/6FDA	4.1	5.2	4.6	3.25	4.17	0.92
p,m-3BCN/OPDA	4.0	4.1	4.3	3.63	4.48	0.85
3,4-APBN/OPDA	4.2	4.5	4.7	3.42	4.06	0.64
3,3-APBN/OPDA	3.8	4.0	4.0	3.30	3.71	0.41
CP2	3.4	3.1	2.9	2.93	2.98	0.05
Kapton	3.2	3.2	3.1	3.15	3.10	−0.05
Ultem	3.6	3.4	3.0	3.19	3.00	−0.19

a) Chemical strucrures of PIs with benzonitrile are shown in Figure 2.1.
b) $\Delta\varepsilon_r' = \varepsilon_r'(190\,°C) - \varepsilon_r'(-150\,°C)$ at 1 kHz.
Source: Adapted from Ref. [28] with permission of Wiley.

PI obtained by reaction with PMDA increased the most, while that of PI derived from BTDA changed the least, exhibiting the order of PMDA>ODPA>6FDA>BTDA. In the meantime, PIs based on para-substituted diamines are more prone to rotation, and this connection method has a higher polarization rate than the meta-substituted series. According to the results of *D-E* loops, it can be seen that such PIs combined with PMDA and para-substituted diamine with nitrile groups typically exhibit the highest discharge energy density and low losses. However, for most of other PIs containing nitrile groups, it is difficult to avoid an increase in dielectric loss while the dielectric constant is improved, which is not the desired result of structure design.

Zhu et al. designed and prepared PIs with high dielectric constant and low loss by directly introducing two ortho-substituted nitrile groups into the rigid skeleton, based on two diamines (2CN, 0CN as a contrast) and five dianhydrides (PMDA, BPDA, BTDA, ODPA, and 6FDA) [30]. Compared with diamine 0CN without nitrile groups (1.75 D), the two nitrile groups ortho-substituted in the diamine 2CN ensure the same dipole moment direction (8.60 D). The direction of dipole moment is ensured to be consistent during polarization, which is beneficial for increasing the overall dielectric constant of PI. In the meantime, the steric hindrance of ortho-substituted nitrile groups greatly limits its polarization motion

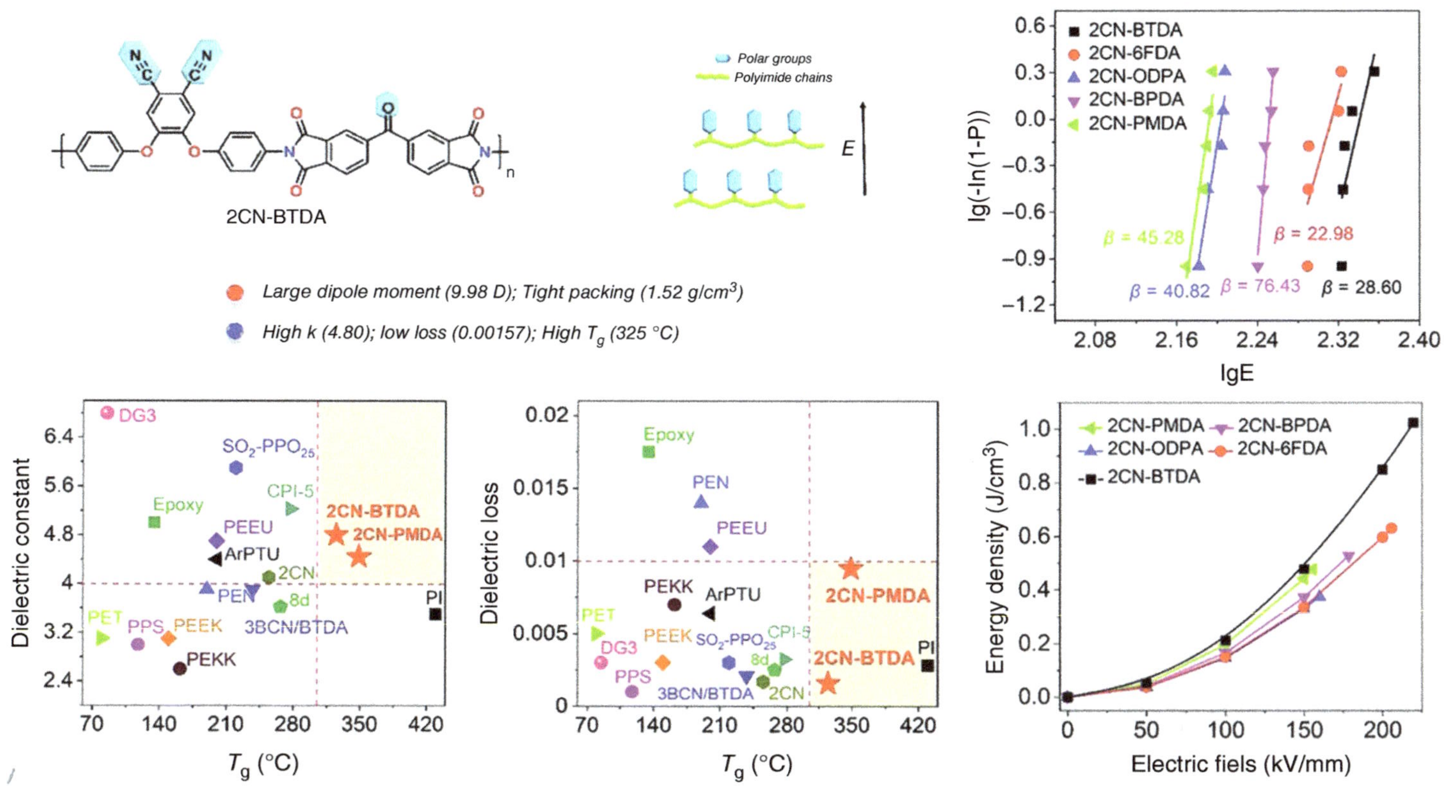

Figure 2.2 Structure of 2CN-BTDA, relevant T_g values and dielectric properties, Weibull breakdown strength, energy density, and charge–discharge efficiency at room temperature. Source: Ref. [30]/Royal Society of Chemistry.

at room temperature, thereby reducing the dielectric loss. More importantly, the rigid structure of molecular chain makes it more difficult for the glass transition movement, with T_g close to 325 °C. The rigid structure also facilitates the tight stacking of PI chains and contributes to the increasing density of repeating units, which are beneficial for the further promotion of dielectric performance. As shown in Figure 2.2, the dielectric constant of PI (BTDA/2CN) reached 4.80 and the loss factor was only 0.00157 that is measured at 1 kHz and 25 °C while maintaining high thermal resistance. Moreover, its breakdown strength is 219.4 kV/mm and the maximum discharge energy density reaches 1.023 J/cm^3.

Tian et al. synthesized a new diamine containing both nitrile groups and imine bond, i.e. bis(2-cyano-4-aminophenyl)amine (BCA) shown in Figure 2.3, and prepared a series of PI films by polymerizing with three dianhydrides (BPADA, ODPA, and PMDA) [31]. It is proved that the introduction of polar nitrile groups can dramatically improve the dielectric constant, mainly due to the high polarity and the formation of intra-molecular hydrogen bonds, both of which promote dipole polarization and thus increase the dielectric constant. These prepared films exhibited excellent dielectric properties and thermal stability. CPI-3 (PMDA/BCA) exhibited a substantially improved dielectric constant as high as 5.5 at 10^2 Hz, which is 2.5 times more than that of commercial BOPP film. Researchers found that the length of repeating units can affect both the density of nitrile groups and the rigidity of molecular chains. CPI-1 (BPADA/BCA) with longer repeating units shows higher breakdown strength, while CPI-3 with shorter repeating units has higher dielectric constant. Theoretical calculation results clarified that CPI-1 has the largest bandgap and trap energy level, which provides a higher charge injection barrier and effectively captures the charge carrier. Thus, its conductivity and dielectric loss can be reduced, and energy storage performance is greatly improved. The CPI-1 film exhibits excellent breakdown strength of 433 MV/m even at 250 °C and still maintains an energy density up to 2.5 J/cm^3, which is one of the highest values reported in this harsh environment.

2.1.2 Sulfonyl Group

Sulfonyl group (-SO_2-) is another functional group with a relatively high dipole moment of approximately 4.30 D. The double bond can increase the density of electron cloud and expand the conjugation effect, making the electrons more easily polarizable. To develop the high-performance PI materials for dielectric applications, it is an effective method to improve the polarizability and dielectric constant by introducing strong polar sulfonyl groups into molecular structure.

Tong et al. reported a series of sulfonyl-containing polyimide (SPI) films combined with different flexible linkages between benzene units in the molecular chains, and their chemical structures are provided in Figure 2.4(a) [32]. These SPIs exhibit high thermal stability, with T_g ranging from 244 °C to 304 °C and decomposition temperatures exceeding 500 °C. The relationship between molecular structure of SPIs and dielectric properties was explained, and the results show that introducing sulfonyl groups into the polyimide backbone can significantly increase the dipole

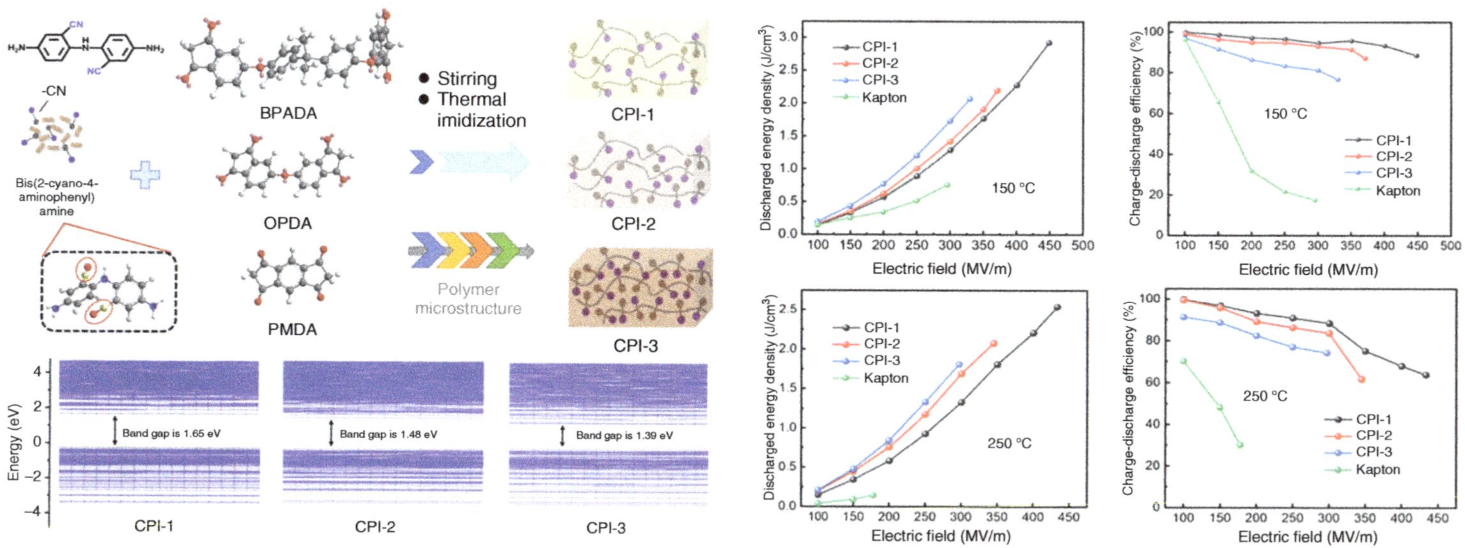

Figure 2.3 Structures of nitrile-containing diamine BCA and relevant PIs, energy level distributions, discharged energy density, and charge–discharge efficiency at 150 °C and 250 °C. Source: Ref. [31]/Royal Society of Chemistry.

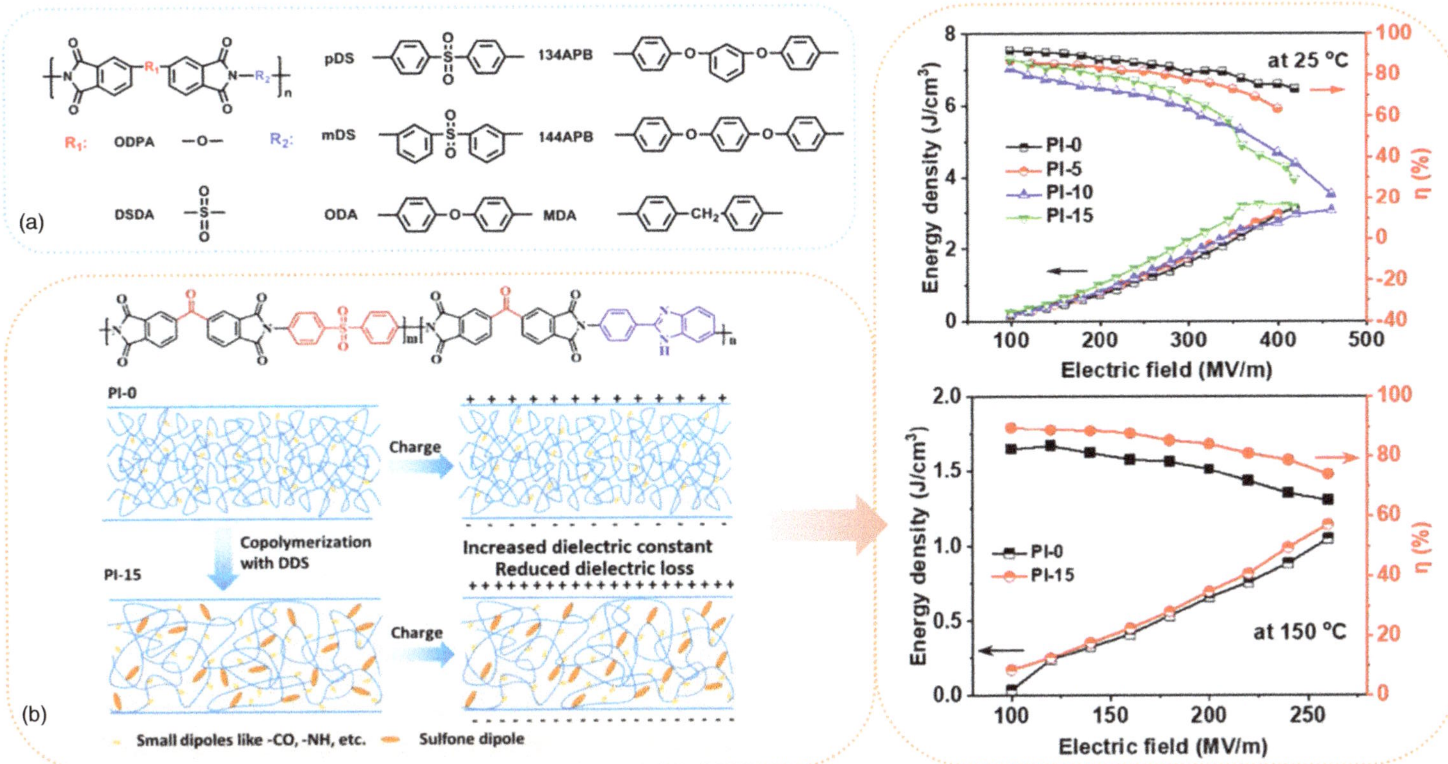

Figure 2.4 (a) Structures of sulfonyl-containing PIs based on ODPA and DSDA. Source: Ref. [32]/John Wiley & Sons; (b) structure of sulfonyl-containing PI (BTDA/BIA/4,4′-DDS), schematic illustration under charging state, energy density, and charge–discharge efficiency at 25 °C and 150 °C. Source: Ref. [33]/with permission of Elsevier.

density, thereby enhancing the dielectric constant. However, it is worth noting that the connected position of sulfonyl group can affect the rotation and orientation of the dipoles, thus affecting their effectiveness. Sulfonyl groups connected to the benzene rings in diamine (4,4′-DDS and 3,3′-DDS) have greater mobility and are more effective than those connected to the benzene-imide rings in dianhydride (DSDA). Moreover, flexible chains can promote the rotation and orientation of dipoles, resulting in the decrease of dielectric loss factor. For example, SPI-1 (ODPA/4,4′-DDS) with para-para-linked sulfonyl group in diamine moiety displays higher dielectric constant than the meta-meta-linked SPI-2 (ODPA/3,3′-DDS), mainly due to the more symmetrical structure of SPI-1 that makes the energy barrier of free rotation lower and also contributes to greater alignment of dipoles in alternating electric field. SPI-1 maintains stable dielectric properties up to 150 °C, with the discharged energy density as high as 7.04 J/cm^3 and charge–discharge efficiency of 91.3% at 500 MV/m, respectively. Furthermore, SPI-3 (DSDA/ODA) with a shorter repeating unit shows higher dielectric constant than other SPIs with similar molecular structures because of the higher sulfonyl group density. It proved that high dielectric constant and low loss factor can be achieved simultaneously without sacrificing heat resistance by introducing high polar sulfonyl groups, together with maintaining high dipole density and moderate flexible structure.

Wu et al. prepared a new kind of sulfonyl-containing PI dielectric film by copolymerization of BTDA dianhydride with benzimidazole diamine (BIA) and 4,4′-DDS, and the chemical structures and schematic illustration under charging state are given in Figure 2.4(b) [33]. As the content of sulfonyl-containing diamine increases, the dielectric constant of film is improved to 4.45 while maintaining a low dielectric loss factor below 0.005. According to the results of molecular dynamics simulation, the introduction of sulfonyl groups leads to an increase in the spacing between molecular chains, which is accompanied with more free volume. It is undoubtedly beneficial for the polarization process, resulting in a synergistic increase in dielectric constant while reducing dielectric loss. In addition, it is found that energy storage density can be increased when the amount of sulfonyl groups is appropriate, but an excess addition of sulfonyl-containing diamine will reduce the charge–discharge efficiency of film. PI-15 (BTDA/BIA/4,4′-DDS) with a mount of 15 mol% 4,4′-DDS exhibits high energy density of 3.27 J/cm^3 at 380 MV/m measured at room temperature and also has good dielectric properties at a high temperature. When the test temperature is raised to 150 °C, this film can maintain a low dielectric loss of 0.026 at 1 kHz and a high breakdown strength, with an energy density of 1.14 J/cm^3 at 260 MV/m and the efficiency around 74% as shown in Figure 2.4(b). The above results demonstrate that the sulfonyl-containing PI dielectrics are possible to be used as one of promising candidates for next-generation film capacitors, particularly in harsh environmental applications.

2.1.3 Carboxylic Group

Research indicates that PIs containing a certain amount of amide acid can play the same role as the high polar groups due to the carboxylic groups (-COOH), enhancing the dielectric constant as well as energy storage characteristics. In addition to introducing carboxylic groups through special monomers, the PI film containing carboxylic acid groups can be directly obtained by controlling the film preparation process. PI films are typically prepared from the precursor PAA through thermal imidization method, where cyclization and dehydration occur during the conversion process. Considering the beneficial effect of polar carboxylic groups on achieving a higher dielectric constant and improved energy density, it is possible to retain a portion of amide acid groups in the PI film by controlling the incomplete imidization degree of PAA through drying temperature and time.

Dai et al. prepared PI-PAA copolymer films with different composition ratios by adjusting the thermochemical process, using PMDA as the dianhydride and *m*-BAPS as the diamine with the chemical structures shown in Figure 2.5(a) [34]. The decrease of PAA content in the copolymer is conducive to enhancing the breakdown strength and energy density, but leads to a reduction in the dielectric constant and loss factor. In the meantime, the leakage current density is also reduced and the bandgap becomes narrower as the PAA content decreases. It is mainly because of the transformation from the carboxylic groups in PAA to the imide rings in PI, and low PAA content means less polar groups that are retained in the finally prepared film. Among various PI-PAA copolymers, the 0.87PI-0.13PAA film exhibits both the highest breakdown strength of 616 MV/m and the highest energy density up to 8.9 J/cm^3 at room temperature. This 0.87PI-0.13PAA copolymer is further improved by introducing 0.1 vol% boron nitride nanosheets (BNNS), and demonstrates a high discharged energy density of 1.38 J/cm^3 and high efficiency over 96% under the test conditions of 150 °C and 200 MV/m, which is much better than the properties of commercial BOPP. Furthermore, the results of 20,000 cycles of charge–discharge test and 35 days high-temperature endurance test are provided, the charge–discharge performance is proved to be robust, and the energy density or efficiency at 150 °C shows no sign of degradation.

Liu et al. prepared a series of PI-PAA films with different degrees of imidization (71%, 90%, and 100%) by polymerizing dianhydride ODPA with sulfonyl-containing diamine 4,4′-DDS [35]. In addition to introducing polar sulfone groups in the diamine moiety, the polarization rate and dielectric constant of film are further increased by retaining amide acid groups in the partially imidized PI-PAA structure. Both polar sulfonyl groups and carboxylic groups can act as traps to suppress the migration of charge carriers, thereby enhancing the breakdown strength and energy storage efficiency of film. However, it is also found that an excess of polar functional groups can lead to trap overlap, which in turn reduces the energy barrier. As presented in Figure 2.5(b), PI film with imidization degree of 90% named SO_2PI-2 exhibits the best comprehensive performance, and achieves a discharged energy density of 5.14 J/cm^3 and a charge–discharge efficiency of 90% at 25 °C and 535 MV/m. Especially, the discharged energy density of this film can reach 3.29 J/cm^3 under conditions of 150 °C and 400 MV/m, with a charge–discharge efficiency of 80%. Based on the above research, it is demonstrated that PI-PAA copolymer with suitable PAA contents can achieve superior energy

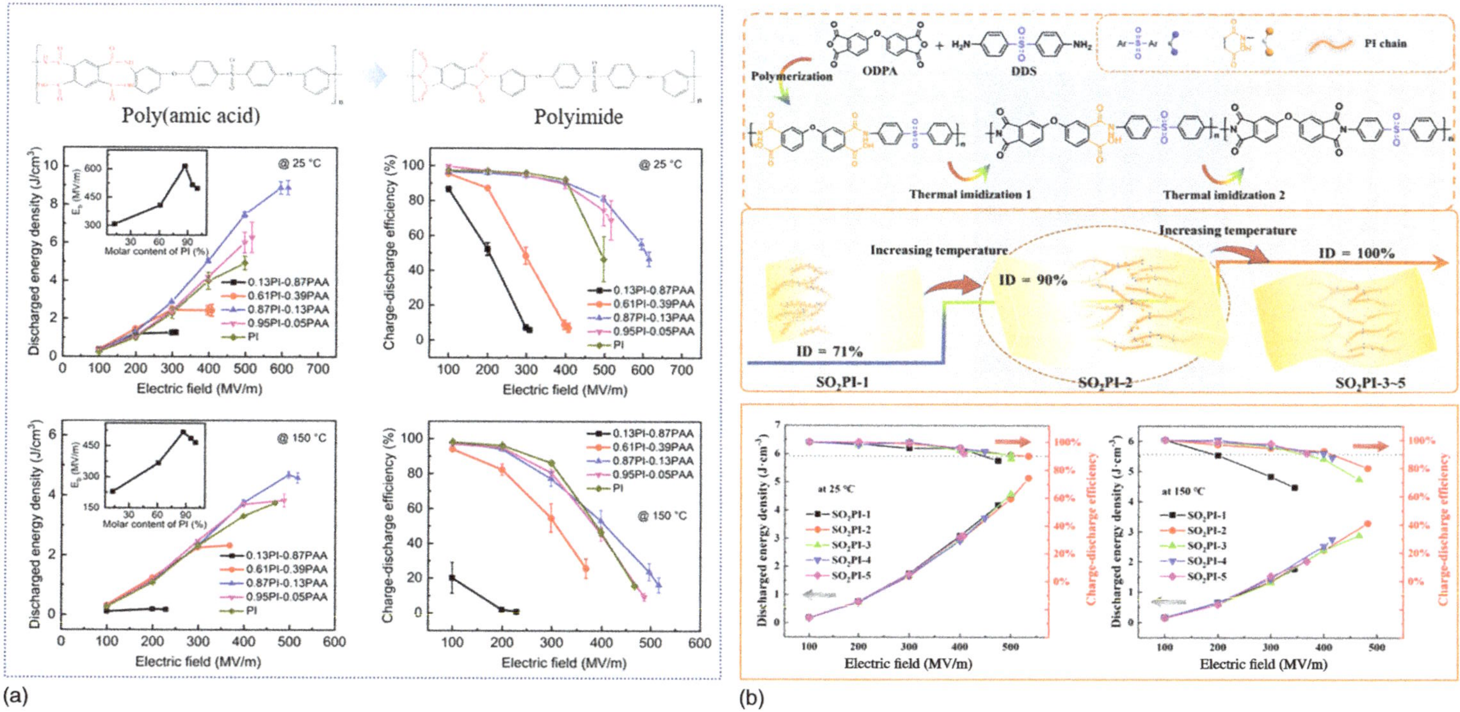

Figure 2.5 (a) Structures of PAA and PI based on PMDA/*m*-BAPS, discharged energy density, and charge–discharge efficiency at 25 °C and 150 °C. Source: Ref. [34]/John Wiley & Sons; (b) structures of PAA, PI-PAA, and PI based on ODPA/4,4′-DDS, discharged energy density, and charge–discharge efficiency at 25 °C and 150 °C. Source: Ref. [35]/AIP Publishing LLC.

storage performance. In view of the simple fabrication process, this special pathway showed the potential for constructing high-temperature capacitors.

2.1.4 Pyridine and Urea Group

Peng et al. utilized a newly synthesized diamine (BPBPA) and several commercial dianhydrides (BPDA, PMDA, BTDA, and ODPA) to prepare a series of PIs containing bipyridine units which are provided in Figure 2.6(a) [36]. These PI films exhibit high thermal resistance with T_g ranging from 275 °C to 320 °C and high mechanical properties with tensile strengths between 175 MPa and 221 MPa. Among them, PIs based on BTDA and BPDA show higher dielectric constants compared to those films based on PMDA and ODPA. The PI derived from BTDA/BPBPA has long-range conjugated units and polar carbonyl groups in structure, resulting in a dielectric constant as high as 7.2 and a dielectric loss factor of 0.038, with the minimal energy loss under high-frequency operation. This is attributed to the enhanced electronic and dipole polarization within the bipyridine units, where each bipyridine unit has two lone electron pairs in the conjugated system. Under external electric field, the high electron mobility increases the electronic polarization, and the strong polarity of nitrogen atoms further improves the dipolar polarization. The synergistic effect results in a significant increase in dielectric constant, and polarizability followed the decreasing order: benzophenone structure > diphenyl structure > diphenyl ether structure > phenyl structure. The energy density of BTDA/BPBPA film can reach up to 2.77 J/cm^3 at room temperature, suggesting a good energy storage performance.

Tang et al. prepared different novel PIs containing functional pendant groups in the side chains, including urea groups (BU-PI), carbamate groups (BC-PI), and sulfonyl groups (BS-PI) with structures illustrated in Figure 2.6(b) [37]. The dielectric properties of PIs were compared, and the influence of different dipolar groups on dielectric behaviors was investigated comprehensively. The results show that high dielectric constant and low dielectric loss can be achieved by both the enhanced orientational polarization and the suppressed dipole–dipole interactions of dipolar groups. It is worth noting that the free rotation of polar urea groups can significantly improve the oriented polarization, shown by the strong γ transitions with low activation energies, with the result that BU-PI has the highest dielectric constant of 6.14 and a relatively low dielectric loss of 0.0097. Moreover, BU-PI exhibits a stable dielectric constant over a wide temperature range, attributed to the presence of intermolecular hydrogen bonds. BU-PI shows the discharged energy density of 6.92 J/cm^3 with charge–discharge efficiency over 83% at 500 MV/m. Compared with the most commonly used polar sulfonyl groups, this research proves that urea groups as the dipolar pendant group can endow PIs with much better dielectric properties.

2.2 Introducing Non-Conjugated Structures

Heat-resistant polymers generally benefit from the conjugated benzene rings in their molecular chains, which ensure good thermal stability. However, the π-π

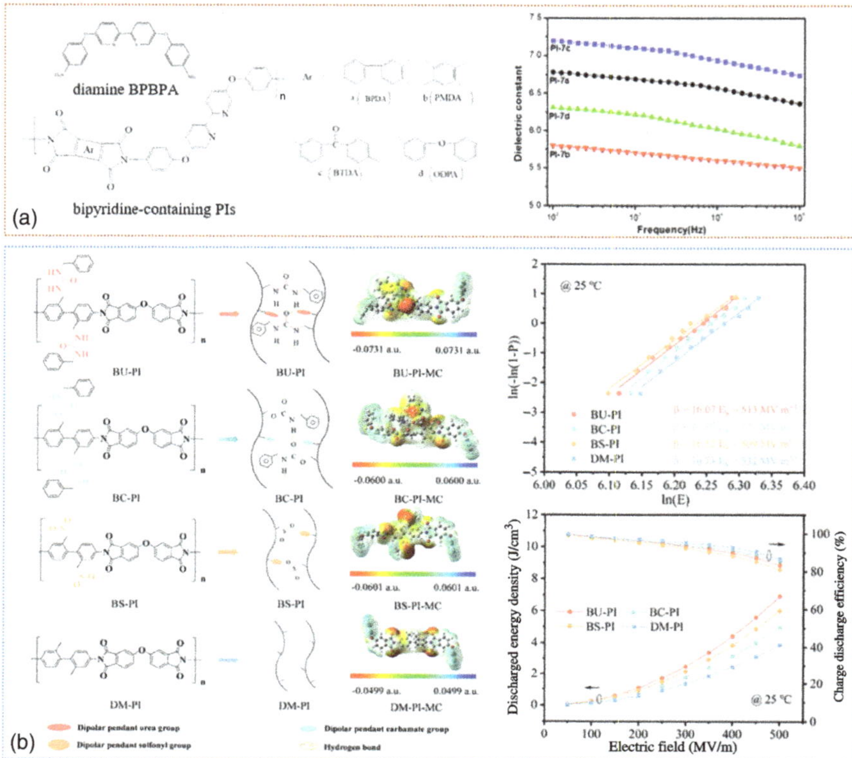

Figure 2.6 (a) Structures of PIs containing bipyidine and relevant dielectric constants. Source: Ref. [36]/with permission of Elsevier; (b) structures of PIs containing urea, carbamate, and sulfonyl groups, electrostatic potential distribution, Weibull breakdown strength, discharged energy density, and charge–discharge efficiency at 25 °C. Source: Ref. [37]/John Wiley & Sons.

stacking effect of these conjugated structures makes the E_g narrow, leading to a reduction in discharge energy density and an increase in leakage current. Unlike traditional PIs with the fully aromatic structures, introducing non-conjugated units into the main chain can disrupt the long-range conjugation of electrons, which can contribute greatly to the improvement of E_g and dielectric strength. The design strategies for introducing non-conjugated structures into polymer chain mainly include the incorporation of aliphatic and cycloaliphatic structures, both of which are semi-aromatic PIs.

2.2.1 Aliphatic Structure

Baldwin et al prepared two series of semi-aromatic PIs including homopolymers and copolymers by reacting dianhydride PMDA with short-chain diamines containing different alkyl groups or ether bonds (Jeffamines) with the purpose of producing PIs with high imide density [38]. The chemical structures of these semi-aromatic PI are displayed in Figure 2.7(a), and the relationship between structure and dielectric properties was investigated, especially analyzing the influence of aliphatic structure. The dielectric constants of such PIs can be increased from 3.96 to 6.57, much higher than that of BOPP. Researchers found that the dielectric constant is directly proportional to the percent functionality of imide groups. The higher density of imide groups corresponds to the greater molecular polarization and leads to the higher dielectric constant. By using short-chain aliphatic diamines, the density of imide groups containing in the molecular chain can be increased, thereby improving the dielectric constant of PI film. In addition, aliphatic structures such as the longer ether chains or the shorter alkyl groups greatly affect the solubility and thermal performance of PIs. The results show that PIs with long-chain ether linkages have better solubility but lower T_g, while exhibiting the reduced dielectric constants due to the larger free volume between flexible molecular chains. It is indicated that the dielectric loss of semi-aromatic PIs is mainly related to their structural rigidity. The rigidity of polymer chain can be increased by introducing short-chain aliphatic diamines, thereby reducing the loss factor that means less response to the switching electric field. Based on these series of PIs containing aliphatic structures, their experimental dielectric constants are also compared to calculation results through density functional theory (DFT), and it is proved that there is a close relationship between them.

Ma et al. reported a series of semi-aromatic PIs by polycondensation reactions with aromatic dianhydrides (PMDA, BTDA, ODPA, and 6FDA) and different aliphatic diamines including 1,3-DAP, DAH, Jeffamine D230, and HK511 [39]. These semi-aromatic PIs were synthesized by the guidance of high-throughput DFT calculations for rational design in terms of a high dielectric constant and E_g. The maximum T_g of such PIs is as high as 174 °C for PI (BTDA/1,3-DAP). The maximum dielectric constant is up to 7.8 with a dielectric loss of less than 0.01 for PI (BTDA/HK511), and its energy density is notably improved and can reach around 15 J/cm^3, which is 3 times that of BOPP. This is mainly ascribed to the increased dipole volume from the ether linkage and the longer conjugation length built on the carbonyl spacer inserted between the benzene rings in dianhydride moiety. When the dianhydride remains unchanged but the diamine is altered, PIs containing polyether chain segments have higher dielectric constants than those containing aliphatic chain segments. It is shown that the arrangement of benzene rings, ether bonds, and other functional groups in dianhydrides moiety has a significant impact on dielectric properties. For example, the prepared PI (BTDA/HK511) with the longest conjugation length and polyether segment exhibits the highest dielectric constant. It is also found that the incorporation of polyether chain segments may also increase the dielectric loss. Furthermore, PIs based on the 6FDA and long-chain ether diamines show the highest E_g values that are measured by UV absorption, which indicate lower electron mobility. As provided in Figure 2.7(b), the relationship between E_g values and dielectric properties was also analyzed by preliminary screening of PI structures through high-throughput DFT calculations. For PI (BTDA/DAH) and PI (BTDA/HK511), their Weibull breakdown strengths can reach 812 MV/m and 676 MV/m. This research provides insights into PIs containing polyether aliphatic structures as promising dielectrics for achieving higher energy density. However, it is also seen that their heat resistance

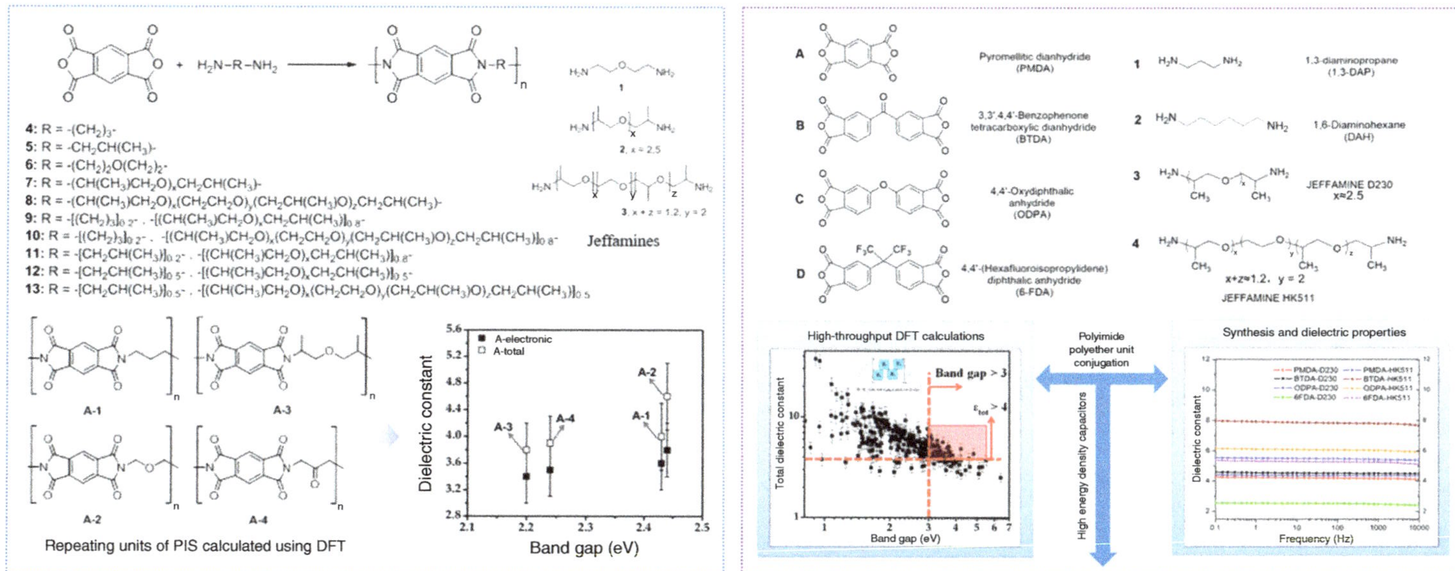

(a) (b)

Figure 2.7 (a) Structures of semi-aromatic PIs containing short-chain aliphatic diamines, the calculated band gap, and dielectric constant results. Source: Ref. [38]/John Wiley & Sons; (b) structures of semi-aromatic PIs based on different aromatic dianhydrides and aliphatic diamines, DFT-based initial screening results, and dielectric constants. Source: Ref. [39]/American Chemical Society.

is insufficient, and further copolymerization with rigid monomers is needed to improve the operating temperature.

Li et al. reported three kinds of all-organic polymers with high energy density and high efficiency by enhancing dipolar polarization at the molecular level, including polythiourea (PDTC/HK511), polyurea (TDI/EDR148), and PI (BTDA/HK25) [40]. This semi-aromatic PI is copolymerized by dianhydride BTDA, Jeffamine HK511, and 1,6-diaminohexane (DAH), and the molar ratio of two diamines is 0.25 : 0.75 in order to balance dielectric and thermal properties. The flexible segments of HK511 moiety increase the local freedom of motion of polymer chains and allow dipoles to rotate more easily under an electric field, thereby enhancing dipole polarization. Rigid segments including aromatic dianhydride and diamine DAH moieties ensure a sufficient heat resistance. This copolymerized structure design shows multiple strong relaxation peaks that are ascribed to sub-glass transition (sub-T_g), indicating an increase in the freedom of motion of the dipoles below the temperature of T_g. According to the results of unipolar *D-E* loop measurements, the dielectric constant estimated from the slope of charging loops is 4.5 for PI (BTDA/HK25). All three polymers show narrow *D-E* loops with the breakdown strengths of up to 780 MV/m. These high breakdown strengths are attributed to the synergistic effect of the high dielectric constants caused by the enhanced dipolar polarization and large E_g based on the selected rigid segments. The maximum energy density of PI (BTDA/HK25) is 12 J/cm^3 with efficiencies above 90% at breakdown field.

2.2.2 Cycloaliphatic Structure

Song et al synthesized two kinds of cycloaliphatic PIs by cycloaliphatic dianhydride with saturated four-membered ring (CBDA) and two diamines (TFMB and DCHM) in order to increase E_g that is greatly limited by the π-π stacking effect of the conjugated benzene ring structure [41]. Due to the non-coplanar and rigid alicyclic structures with saturated electrons, both fully cycloaliphatic PI (CBDA/DCHM) and semi-cycloaliphatic PI (CBDA/TFMB) present large band gap with optical E_g of 4.6 eV and 4.1 eV, respectively, and the structural influence is supplied in Figure 2.8(a), while maintaining excellent heat resistance with T_g above 275 °C. Compared to the fully aromatic PI (PMDA/ODA), these two PIs exhibit relatively lower dielectric constants in the wide frequency range. This can be attributed to the introduction of cycloaliphatic structure and strongly electronegative trifluoromethyl (-CF_3) groups, which disrupt the conjugated structure and lead to the increased inter-molecular chain spacing and reduced packing density. The breakdown strength of PI (CBDA/DCHM) and PI (CBDA/TFMB) is 620 MV/m and 556 MV/m at 150 °C, respectively. Moreover, both cycloaliphatic PI films exhibit excellent temperature stability in terms of their breakdown strength. By introducing cycloaliphatic structures into the main chain, energy storage density is enhanced and the dielectric loss can be simultaneously reduced. At high temperatures, the leakage current density of two PIs is significantly reduced, demonstrating that cycloaliphatic PIs with large E_g can effectively suppress electrical conduction losses at elevated temperatures. This indicates that cycloaliphatic PIs act like

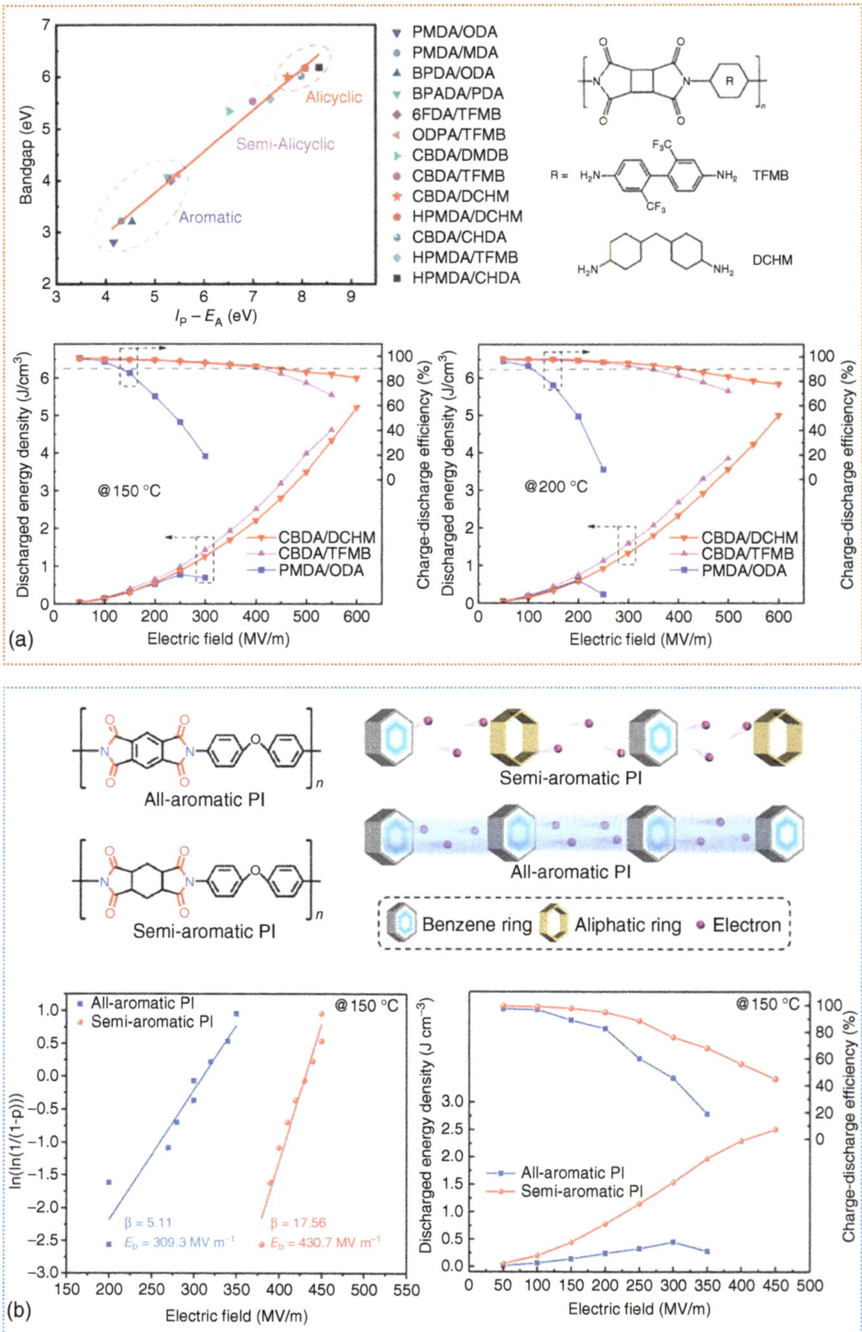

Figure 2.8 (a) Structures of cycloaliphatic PIs based on CBDA, calculated band gap (E_{LUMO}-E_{HOMO}) of PI model compounds *versus* difference (I_P-E_A) between ionization potential (I_P) of diamines and electron affinity (E_A) of dianhydrides, discharged energy density, and charge–discharge efficiency at 150 °C and 200 °C. Source: Ref. [41]/Royal Society of Chemistry; (b) structures of all-aromatic and semi-aromatic PIs, possible motion of electrons under electric field, Weibull breakdown strength, discharged energy density, and charge–discharge efficiency at 150 °C. Source: Ref. [42]/with permission of Elsevier.

deeper traps that can greatly inhibit the migration of charge carriers. Particularly, fully cycloaliphatic PI (CBDA/DCHM) achieves a high energy storage density of 5.01 J/cm^3 with efficiency of 78.1% at 600 MV/m and 200 °C. Notably, energy density can still reach 2.55 J/cm^3 with efficiency of 90% even at 200 °C, which is a significant improvement compared to the existing commercial PEI. In addition, fully cycloaliphatic PI (CBDA/DCHM) exhibits an excellent self-cleaning capability due to the lower ratio of carbon to hydrogen and oxygen, ensuring the long-term reliability in practical applications.

Yu et al. prepared a semi-aromatic PI with alicyclic structures by cycloaliphatic dianhydride with saturated six-membered ring (CHDA) and aromatic diamine (4,4′-ODA), and studied the mechanism of staggered stacking of benzene rings in this semi-aromatic PI to improve capacitive performance [42]. The introduction of cycloaliphatic units disrupts the conjugated structures, thereby reducing the delocalization effect of electrons on the molecular chain and enlarging the band gap. Based on the experimental simulation results, cycloaliphatic structures can increase the electron trap energy level with the decreasing number of thermally excited charge carriers, resulting in the reduction of leakage current density and improvement of breakdown strength. Furthermore, cycloaliphatic structures can change the stacking mode of PI molecular chain and avoid the overlap between benzene rings that can cause the conjugation effect between molecular chains, which is illustrated in Figure 2.8(b). Compared with fully aromatic PI (PMDA/4,4′-ODA), semi-aromatic PI demonstrates higher discharge energy density and efficiency at room temperature and high temperatures. The energy density of semi-aromatic PI (CHDA/4,4′-ODA) is 0.78 J/cm^3 with a high efficiency of 95% at the conditions of 150 °C and 200 MV/m, which is almost twice that of BOPP. This research demonstrates that by introducing a cycloaliphatic structure semi-aromatic PI dielectrics are ideal energy storage materials for the high-temperature condition.

Wang et al. reported a tailored combination strategy for the design of high-temperature polymer dielectrics through machine-learning methods and predicted 110 polyimide-derived polymer structures from diverse combinations of 21 structural units that possess a wide distribution of thermal and electrical properties [43]. Twelve representative PIs were screened and selected to be synthesized for direct experimental investigation, and the effect of each structural unit and various groups on the E_g, thermal stability and other properties of resultant polymer was determined quantitatively by analyzing the experimental and simulation results. Among these prepared PIs, PI-cyc-iso (A3-B7) is derived from cycloaliphatic dianhydride CHDA and aromatic diamine (ALPHA) containing diisopropylbenzene group that is shown in Figure 2.9, presenting a much higher E_g of 4.0 eV than fully aromatic PI (A3-B1) with a similar structure that is only replaced by aromatic benzene ring in the dianhydride moiety. The enlarged band gap is also superior to regular PI (A6-B1) derived from PMDA/4,4′-ODA or PEI (A10-B5) derived from BPADA/*m*-PDA. Although the heat resistance is significantly reduced compared with the fully aromatic series, PI-cyc-iso still maintains a high-temperature level with T_g of 274 °C. The results show that the replacement of benzene ring in dianhydride by saturated six-membered ring results in a significant increase in E_g

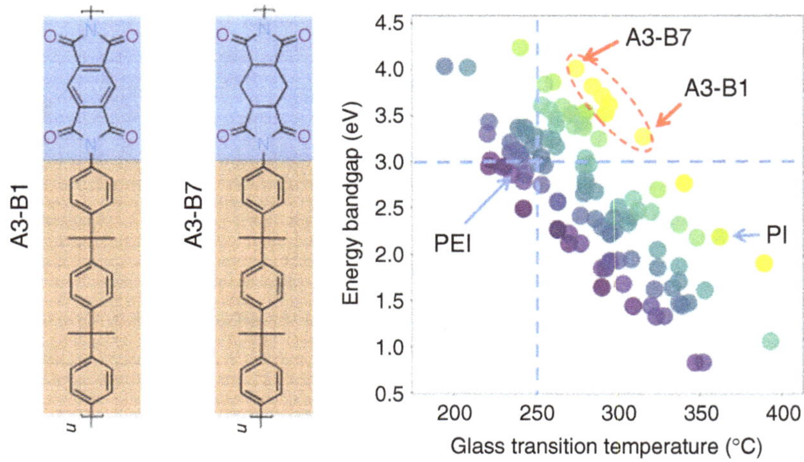

Figure 2.9 Structures of two selected PI (A3-B1) and PI (A3-B7) with the most balanced T_g and E_g, and the scatter plot of T_g *versus* E_g of all the 110 possible PI-derived polymers with two dashed lines marked corresponding to 3.0 eV and 250 °C. Source: [43]/Springer Nature/CC BY 4.0.

but a moderate decrease in T_g. In this study, researchers propose that the T_g value of polymer needs to be at least 10 °C higher than the target operating temperature in order to ensure the mechanical and structural stability of material at high temperatures. Thus, it is sufficient for applications of PI-cyc-iso as a high-temperature dielectric film.

2.3 Introducing Non-Coplanar Structures

In the common high-T_g polymers, the presence of conjugated benzene ring structures in the molecular chain leads to a remarkable π-π stacking effect. It makes the dielectric materials more prone to charge transport at high temperatures and high fields, resulting in increased leakage current and reduced energy density. Therefore, the core and key factor of the structural design of PI dielectrics with high T_g for high-temperature energy storage lies in how to decouple the π-π stacking effect caused by the conjugated structures. Introducing the bulky substituents with large-volume or spiro-ring units to disrupt the structure coplanarity is an effective strategy to disrupt the long-range conjugation and suppress intra- or inter-molecular charge transfer. The greatest advantage of this method is that the rigid aromatic backbone characteristic of PI is not destroyed, allowing for the maintenance of high heat resistance and thermal stability. This strategy is commonly used in the molecular structure design of transparent PI and has achieved good results.

2.3.1 Bulky Substituents

PIs containing trifluoromethyl groups (FPI) have attracted more and more attention in recent years due to their large substituent effect and good electron inhibition ability. Yang et al. used a kind of FPI derived from fluorinated dianhydride (6FDA) and diamine (TFMB) to prepare the high-performance polymer dielectrics, supplied by Kawamura Industries Co. Ltd. as the product of KPI-MX300F [25, 44]. These bulky fluorinated groups in the molecular structure endow FPI film with better stability and stronger charge carrier capture capability at high temperatures, achieving an energy storage density of about 5.5 and an efficiency of around 80% at 150 °C. Based on this FPI with chemical structure shown in Figure 2.10(a), a specific 3D small aromatic molecule (TE) is introduced as a physical crosslinking reagent to form physical crosslinking networks through electrostatic interactions between their oppositely charged phenyl groups. The composite film of FPI-3wt%TE was prepared by a solution casting process and exhibited high performance and excellent processibility for capacitive energy storage in harsh environments. It can maintain stable dielectric properties up to 250 °C, showing energy density of 8.05 J/cm^3 with efficiency of 90% at 150 °C, as well as energy density up to 5.12 J/cm^3 even at 200 °C. In another study, the researchers took phosphotungstic acid subnanosheets (PWNS) as fillers to prepare polymer-PWNS subnanocomposites [44]. Significantly, FPI-0.2 wt% PWNS presents the maximum energy density of 10.72 J/cm^3 at 150 °C and 8.84 J/cm^3 at 200 °C. More importantly, it has the ultrahigh energy density of 8.1 J/cm^3 with efficiency of 90% at 150 °C and 7.2 J/cm^3 at 200 °C.

Qin et al. also prepared the FPI film derived from 6FDA/TFMB and further explored the relationship between trifluoromethyl substituent characteristics and its superior capacitive performance at high temperatures [45]. Due to introduction of butyl trifluoromethyl groups, FPI film is colorless and transparent in contrast to the typical yellow PI (PMDA/4,4′-ODA), indicating that the trifluoromethyl groups significantly suppress the formation of charge transfer complexes (CTC). Simulation results show that the introduction of trifluoromethyl groups weakens the stacking of molecular chains and enlarges the inter-molecular spacing between polymer chains, which can help to increase the free volume fraction and reduce the migration of charge carriers at high temperatures. As expected, FPI has a high T_g of 329 °C and excellent thermal stability, which is attributed to the retained benzene rings and imide heterocycle structures in the molecular chain. Because of the strong electronegativity and large volume substituent effects of trifluoromethyl groups, the long-range conjugation in FPI is notably inhibited, thereby increasing the band gap to 3.6 eV that is presented in Figure 2.10(b). This structural change is beneficial to improve the electric field endurance at high temperatures, exhibiting noticeable Weibull breakdown strength values of 578 MV/m and 452 MV/m at 150 °C and 200 °C, respectively. FPI achieved energy density of 3.21 J/cm^3 with efficiency of 84% at 150 °C. As the temperature rises to 200 °C, its energy density reaches 2.42 J/cm^3 with discharging efficiency of 82%, which is higher than most existing heat-resistant polymer dielectrics. Meanwhile, it is demonstrated FPI shows excellent fatigue resistance and self-healing ability.

Wang et al. employed machine learning to screen polymer structures with a wide range of electrical and thermal characteristics, and provided quantitative description of the influence of each structural unit on the E_g and T_g of PIs [43]. The results show that phenyl rings with grafted side groups are beneficial for simultaneously

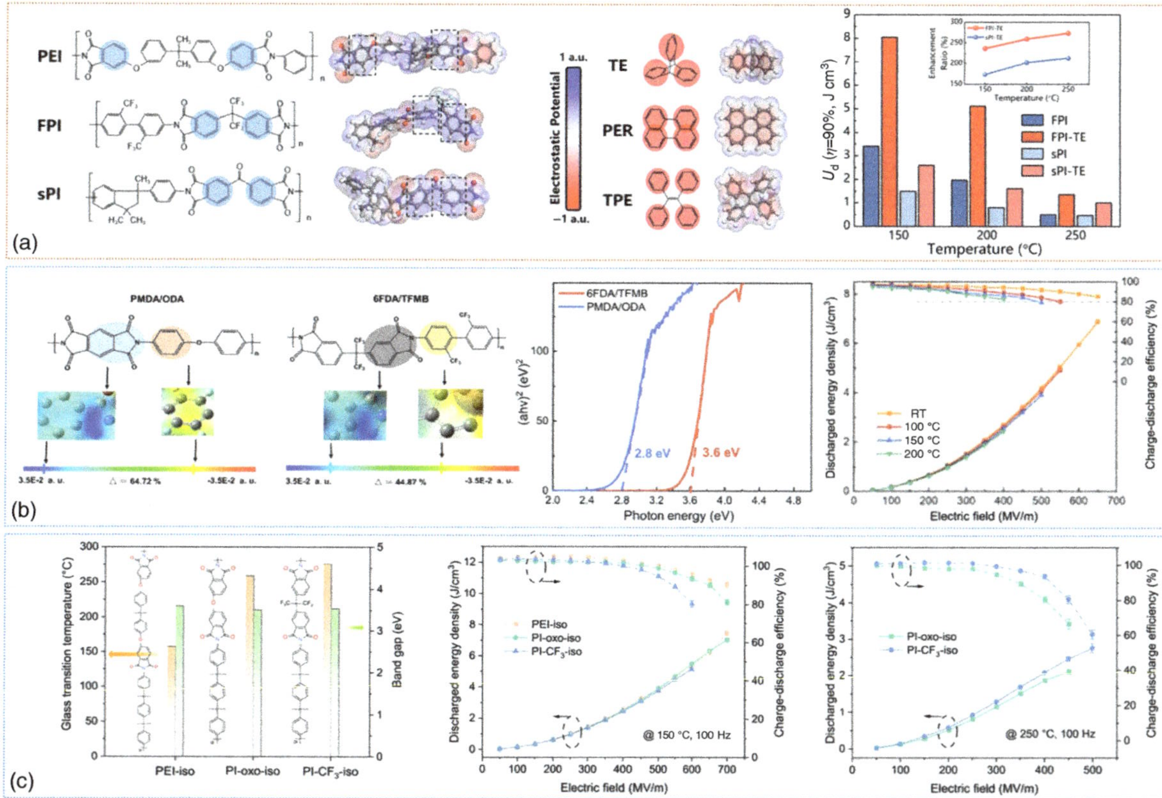

Figure 2.10 (a) Structures of aromatic PIs and small aromatic molecules, schematic diagram of electrostatic potential distribution with the oppositely charged phenyl groups marked in red, discharged energy density, and charge–discharge efficiency at different temperature. Source: [25]/John Wiley & Sons; (b) structures of PMDA/4,4′-ODA and 6FDA/TFMB, the electrostatic potential distribution in molecular chain, UV–Vis Tauc plots and calculated optical bandgaps, discharged energy density, and charging–discharging efficiency of 6FDA/TFMB at different temperatures. Source: Ref. [45]/John Wiley & Sons; (c) structures of three typical PIs, comparison of glass transition temperature and bandgap, discharged energy density, and charge–discharge efficiency at 150 °C and 250 °C. Source: Ref. [43]/Springer Nature/CC BY 4.0.

increasing E_g and T_g. It is further found that, when the polymer's E_g increases to a critical point (approximately 3.3 eV) or above, the high-temperature insulation performance will experience a diminishing marginal utility, and its main conduction mechanism shifts from Poole–Frenkel emission to hopping conduction. MD and DFT simulations results indicate that at extreme temperatures, the dihedral angle (θ) between adjacent conjugated planes in the molecular structure becomes the decisive factor affecting the capacitive performance. The trap depth is positively correlated with the θ value, which means that larger non-planarity makes electron tunneling more difficult, resulting in better insulating performance. This is significantly different from the previous notion that optimizing the E_g determines the capacitive performance at high temperatures. Based on the interrelationship of E_g, T_g, and θ, three different PIs suitable for high temperatures were designed by the tailored structural unit combinations. As shown in Figure 2.10(c), PI-oxo-iso (A3-B2) based on the flexible dianhydride (ODPA) and aromatic diamine with diisopropylbenzene group has the largest dihedral angle with θ about 81.3°, exhibiting the E_g of 3.5 eV and T_g about 262 °C. As for the maximum discharged energy density above 90% charge–discharge efficiency (U_{e90}), the PI-oxo-iso presents the highest result of 5.3 J/cm^3 at 200 °C. In addition, PI-CF_3-iso derived from fluorinated dianhydride of 6FDA and aromatic diamine with diisopropylbenzene group, was designed for application at 250 °C by replacing the ether linkage in PI-oxo-iso with a hexafluoroisopropyl methylene group. The experimental E_g and T_g results of PI-CF_3-iso are 3.53 eV and 275.5 °C, respectively; meanwhile, the outstanding energy storage performance is achieved with U_{e90} of 2.1 J/cm^3 at 250 °C, which is far higher than the polymers reported in the literature.

2.3.2 Spiro Structures

In addition to the incorporation of bulky substituents, introducing non-coplanar structures such as spiro units into the main chain can also effectively disrupt the long-range conjugation in aromatic PIs. The findings have shown that the band gap can be obviously increased by adopting spiro structures, resulting in the enhancement of dielectric strength and discharge energy density.

Ran et al. reported two kinds of spiral-structured dielectric PIs to avoid the π-π stacking effect caused by the conjugated benzene rings in the aromatic chain [46]. As inspired by the stable double-chain structure of DNA, the spiral and non-coplanar structural units were incorporated into the PI main chains in order to enhance the capacitive performance up to 200 °C. Two diamines containing the unique spiro units were used to be polymerized with an aromatic dianhydride containing ether bonds (ODPA), and the corresponding spiro-structured PI films (PI-1 and PI-2) were prepared with structures provided in Figure 2.11. It is found that the high content of spiral units during the polymerization leads to the increased brittleness of film. To ensure the processability of PI films, the experiment used PI-B (ODPA/4,4′-ODA) as the matrix and blended it with a small amount of spiral-structured PI-1 or PI-2 to prepare composite films. The optimal composition

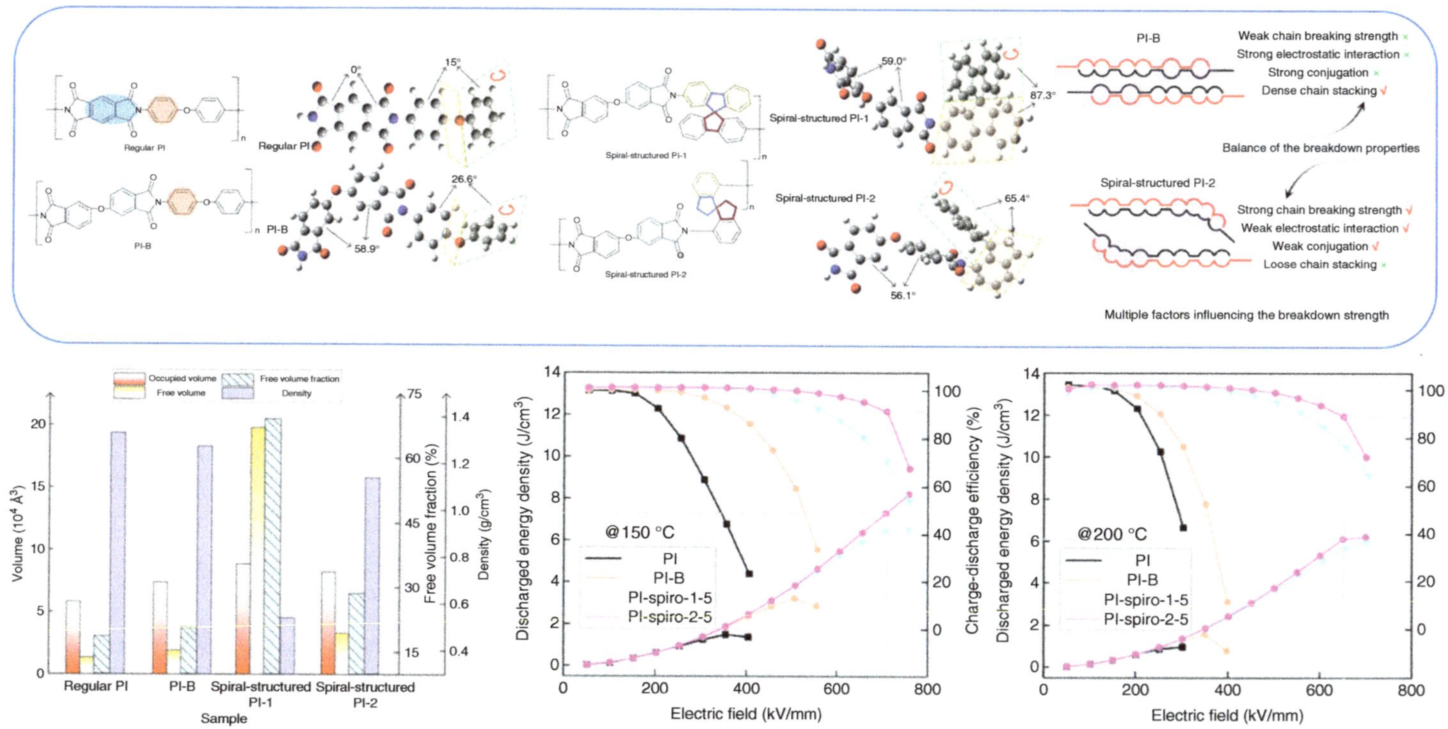

Figure 2.11 Structures of traditional and spiro-structured PIs, schematic diagram of different PI molecules at the lowest system energy, multiple influencing factors of breakdown strength, the relevant occupied and free volume data, discharged energy density, and charge–discharge efficiency at 150 °C and 200 °C. Source: Ref. [46]/John Wiley & Sons.

ratio was 5 mol% of the spiral-structured PI-1 or PI-2, referred to as PI-spiro-1-5 or PI-spiro-2-5, respectively.

As illustrated in Figure 2.11, the spiral structure is characterized by two single rings sharing a joint carbon atom, and it is prone to forming a stable and expansive three-dimensional topological structure because of its high rotational potential energy, showing a stronger non-coplanarity than the single alicyclic ring or flat benzene ring. Such spiro-structured design capitalizes on the inherent strength of two bound chains, where the dissociation of molecular chain is contingent upon the simultaneous destruction of both bound chains, thereby endowing it with excellent heat resistance. Experimental and simulations results confirm that the spiro structure can disrupt the conjugated planar segments and π-π stacking in PIs, and effectively decouple the electronic conjugation and achieve to regulate the carrier transport process by introducing the deeper trap. Meantime, the non-coplanar double-chain conformation of spiro structure greatly increases the free volume of molecular chains, which is helpful to weaken the inter-molecular electrostatic interactions and reduce the carrier migration between polymer chains. Thus, the spiro structure remarkably suppresses charge injection and transfer, and significantly improves the dielectric strength and energy storage density.

For spiral-structured PI-1 with excessively rigid fluorene units, the molecular chain curling or folding is severely obstructed, and its free volume fraction (FFV) has increased by more than 50% compared to regular PI (PMDA/4,4′-ODA). In addition to reducing molecular ordered arrangement and loosening chain stacking, the combination of spiral structures and non-planar units helps to increase the bandgap values of PI-spiro-1-5 and PI-spiro-2-5 to 3.42 eV and 3.50 eV, respectively. The polarization decreases due to the increased chain spacing and large free volume results that the dielectric loss factor of PI-spiro-2-5 is only 55.9% of that of the regular PI at 200 °C and 100 Hz. Moreover, the restriction of charge transfer also contributes to reducing conduction losses at lower frequencies. Owing to the stability of spiral double-chain structure, the Weibull breakdown strength of spiral-structured PIs at high temperatures is significantly increased. The breakdown strength of PI-spiro-2-5 is 667.5 MV/m at 200 °C, which is about 83.6% higher than that of the regular PI. At the high temperature and 300 MV/m electric field, the energy loss of spiral-structured PIs is significantly suppressed, and the charging and discharging efficiency is markedly improved which is presented in Figure 2.11. At 150 °C and 200 °C with an efficiency greater than 90%, the maximum energy density of PI-spiro-2-5 can reach 7.29 J/cm^3 and 6.13 J/cm^3, respectively, which is more than 10 times that of regular PI with 0.62 J/cm^3 at 150 °C and 0.61 J/cm^3 at 200 °C. It is worth noting that the energy storage performance at 200 °C is superior to the currently reported polymer dielectrics and is even better than BOPP at room temperature. This work provides a new approach for the design of polymer dielectrics exhibiting superior energy storage capabilities at elevated temperatures by integrating the spiral structure as the non-planar structural units into molecular chains.

2.4 Introducing Crosslinking Structures

Physical or chemical crosslinking is considered an effective method to construct deep traps, and extensive research has confirmed that it can significantly reduce the dielectric conduction loss of polymer dielectric films and improve the energy storage density [17].

Zhang et al. introduced the phenylethynyl groups (PEPA) that is thermally crosslinkable into molecular structure of PEI (BPADA/4,4′-ODA), and prepared a series of self-crosslinkable oligomers with different crosslinking functional group contents, and the chemical structures are presented in Figure 2.12 [47]. The crosslinked PEPA-PEI dielectric films with enhanced voltage resistance and low leakage current density are obtained by thermal curing crosslinking treatment under oxygen. The results show that, as the crosslinking degree increases, the dielectric loss of film decreases, and the β-relaxation peak moves toward a higher temperature. It is indicated that crosslinking effectively restricts the motion of molecular chains especially β-relaxation, thereby improving the dielectric properties of dielectric films. In addition, crosslinking in oxygen is more effective in restricting molecular chain motion than when treated in vacuum; thus, the β-relaxation is preferably suppressed. It is also found that excessively long molecular chains can lead to incomplete crosslinking reactions, while too short molecular chains can lead to large distances between molecular chains and inevitably brings defects. For the c-10%PEPA-PEI with proper chain length, a moderate number of interchain crosslinking reactions can be caused, resulting in low dielectric loss and higher discharge efficiency. Its dielectric loss values at 10^3 Hz are 0.0037 and 0.0043 at room temperature and 150 °C, respectively. This crosslinked film maintains a low loss factor below 0.010 from 10^2 to 10^6 Hz. Compared with the untreated film, appropriately crosslinked c-10%PEPA-PEI film exhibits better energy storage performance, possessing the improved energy density of 3.6 J/cm^3 with ultrahigh charge–discharge efficiency above 95.0% at 150 °C. It proves the feasibility of using chemical crosslinking approach to prepare high performance PI dielectric film for high-temperature applications.

Based on the same molecular structure of PEI (BPADA/4,4′-ODA) and crosslinking agent (PEPA), Zhu et al. studied the relationship of structures with various crosslinking degrees on the high-temperature energy storage performance by adjusting the heating time and temperature [48]. It is found that crosslinking networks and linear chain extension coexist during the crosslinking process. As the heating time and heating temperature increase, weight-average molecular weight (M_w) and the crosslinking degree increase firstly until a critical point is reached. After that, the gel content continues to increase while M_w decreases accordingly, leading to molecular chain scission. The performance of crosslinked films treated for a long time at lower temperatures is similar to that of films treated for a short time at higher temperatures. Moreover, the longer the linear chain segment, the better the energy storage performance. This is because long chain segments can improve the mechanical properties and breakdown strength, thereby enhancing energy storage performance. As the degree of crosslinking increases, the energy

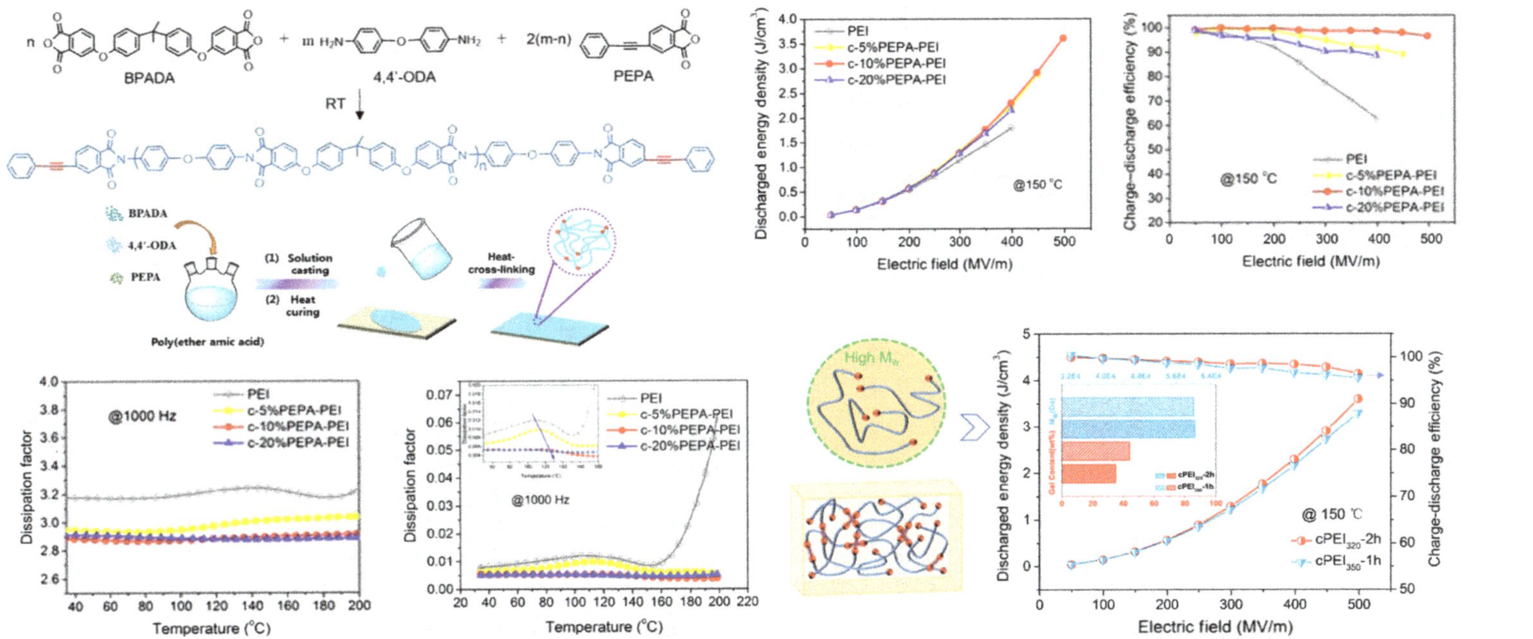

Figure 2.12 Structure of PEPA-terminated PEI (BPADA/4,4′-ODA) and crosslinking preparation process of cured films, dielectric constant, and dissipation factor at 10^3 Hz as a function of temperature, discharged energy density, and charge–discharge efficiency at 150 °C, the effect of M_w and heating times on energy storage performance at 150 °C [47, 48]/Chinese Chemical Society and Elsevier.

storage performance first strengthens and then weakens. The moderate crosslinking reaction can restrict the motion and relaxation of polymer chain segments, thereby suppressing dielectric loss and improving modulus and breakdown strength and energy storage performance. However, excessive crosslinking can lead to more short chains that is generated by chain scission and even the adverse effect of free radicals, resulting in an increased leakage current and the reduction of energy storage performance. As displayed in Figure 2.10(a), the appropriate degree of crosslinking is beneficial to form the optimal crosslinking network structure, thereby achieving the optimal energy density of about 3.6 J/cm^3 with charge–discharge efficiency up to 95% at 150 °C.

Wu et al. took the norbornene dicarboxylic anhydride (NA) as the end-capping agent to prepare the crosslinked XL-PI (ODPA/4,4′-ODA) films with different NA addition amount, and the structure is given in Figure 2.13(a) [49]. The crosslinking structures formed between molecular chains can restrict the polarization motion of polar groups and suppress the directional migration of charge carriers, reducing dielectric loss significantly. The dielectric loss of XL-PI films was significantly reduced to 0.003–0.006, much lower than that of pure PI films without crosslinking structure. The crosslinked network restricts the migration of free charges, delaying the development of electrical trees and the formation of internal breakdown paths. As a result, the breakdown strength of XL-PI films can reach up to 545 kV/mm. For XL-PI-3 film with 3 wt% addition of NA, the energy density is 9.4 J/cm^3 with an energy efficiency of 88% at room temperature and 525 kV/mm, which is 56% higher than that of PI with linear structure. In addition, XL-PI-3 film maintains an energy storage density of 6.5 J/cm^3 with efficiency of 82% at 150 °C, showing excellent high-temperature energy storage performance.

Xu et al. used 1,3,5-tris(4-aminophenyl) benzene (TAB) as a crosslinking agent to synthesize a series of crosslinked PI (PMDA/4,4′-ODA) films with different crosslinker contents (5–30 wt%), and investigated the enhancement of the energy storage performance at high temperatures by controlling the crosslinking density [50]. The results demonstrate that dielectric constant of crosslinking PIs firstly increases with the crosslinking degree, but will decrease when the crosslinking reaction is excessive due to the restricted directional polarization of dipoles. The crosslinking network also contributes to the inhibition of the free charges migration and breakdown paths formation which is shown in Figure 2.13(b). More importantly, the crosslinking structure can reduce dielectric loss and electrical conductivity, which helps to suppress polarization loss. DFT calculation results indicate that the introduction of crosslinking structure lowers energy level of the lowest unoccupied molecular orbital (LUMO), suggesting it is more likely to capture electrons and contribute to the formation of deeper charge traps. Experimental results prove that as the crosslinking agent content increases, both the energy storage density and efficiency of crosslinked PI increase firstly and then decrease. PI/30 with 30 wt% addition of TAB exhibits the maximum energy density of 5.87 J/cm^3 and efficiency of 81.5% at room temperature, much higher than the corresponding

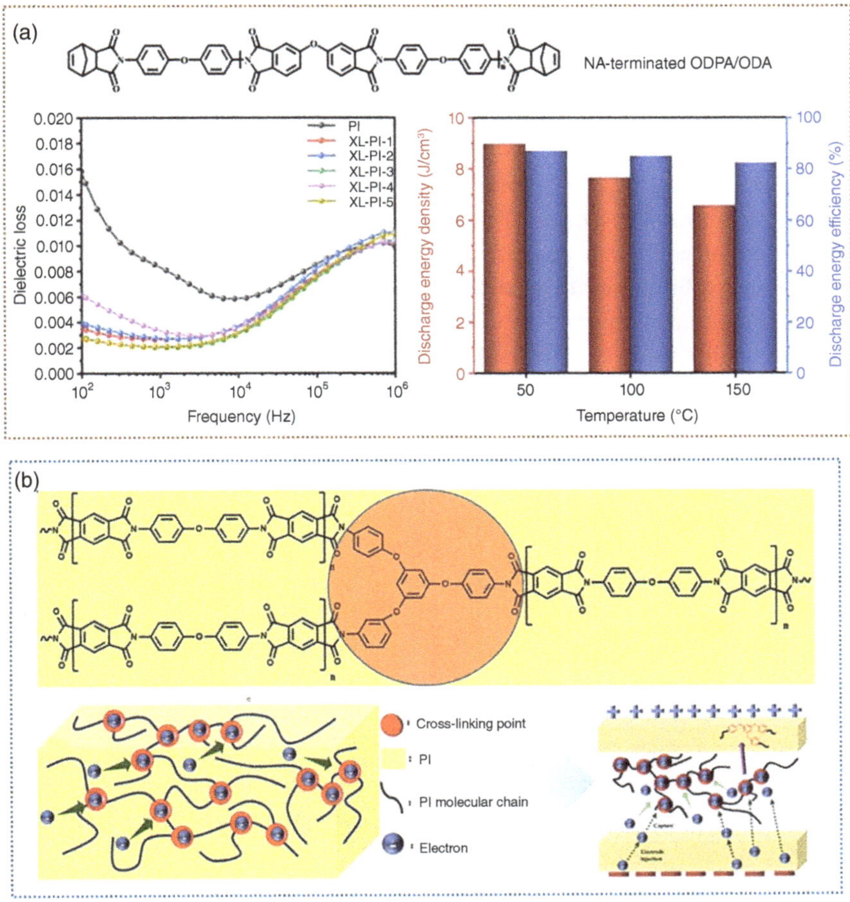

Figure 2.13 (a) Structures of NA-terminated PI (ODPA/4,4′-ODA), dielectric loss as a function of frequency, discharged energy density, and charge–discharge efficiency of XL-PI-3 film at 50 °C, 100 °C, and 150 °C [49]/Elsevier; (b) structures of crosslinking PI based on TAB, schematic diagram of the breakdown mechanism, and space charge-trapping mechanism of crosslinking PI. Source: Ref. [50]/John Wiley & Sons/CC BY 4.0.

values of 2.61 J/cm^3 and 50.2% for pure PI without crosslinking. Furthermore, PI/30 film can achieve energy density of 1.75 J/cm^3 with efficiency of 70% at 150 °C.

Duan et al. reported a crosslinked PEI (ODPA/4,4′-ODA) by using the 2,4,6-triamino-pyrimidine (TAP) as a crosslinker, and dielectric and energy storage properties of film were also improved effectively [51]. The results show that dielectric loss of crosslinked PEI gradually decreased with the increasing content of crosslinker. Compared with non-crosslinked PEI, the breakdown strength and maximum energy density of crosslinked PEI by adding 2 wt% TAP increased to 399.4 MV/m and 2.53 J/cm^3 at 150 °C, with the increment of 23.3% and 24.8%, respectively. All the above results indicate that the crosslinking method can effectively improve the energy storage performance of PI dielectrics at high temperatures.

2.5 Summary and Outlook

With the increasing demand for safe operation in harsh environments, polymer dielectric capacitors with high energy density and charge–discharge efficiency operating at high temperature and high electric field are urgently needed for potential electrical and pulsed power applications. PI dielectrics are considered to be the most promising heat-resistant materials that can operate stably under extreme working conditions and efficiently over the long term. The key properties of representative PI dielectrics reported in literature are summarized in Table 2.2, including glass transition temperature (T_g), dielectric constant (ε_r), dielectric loss (tanδ), energy band gap (E_g), breakdown voltage (E_b), maximum energy density (U_e) and efficiency (η). In the structure design of high-temperature PI dielectrics, the thermal resistance and capacitive performance tend to be mutually exclusive. Developing intrinsic high-performance PIs with high dielectric constant, low dielectric loss, high breakdown voltage, and excellent thermal stability is still a great challenge.

a) For the common PIs with high T_g, high thermal stability usually requires increasing the long-range conjugation of molecular chains. However, the π-π stacking of conjugated benzene rings can cause a significant reduction in the E_g, resulting in a decrease in discharge energy density and an increase in leakage current, and further lead to an exponential increase in conduction loss at elevated temperatures and high electric field.
b) Although the E_g is typically used to predict the intrinsic breakdown voltage of PIs, the main mechanism of charge transport is hopping conduction under high temperature and high electric field, and the tunneling rule of electron escape is not fully related to the E_g of material. Therefore, in the design of PI dielectrics for high-temperature capacitive energy storage, focusing solely on E_g optimization may not yield the best results.
c) Introducing polar functional groups can effectively increase the polarizability and improve the dielectric constant, but it is difficult to avoid the adverse effect of increasing dielectric loss that generally leads to the reduced charging–discharging efficiency. Incorporation of non-conjugated structures can greatly increase the bandgap and improve the breakdown strength, while the heat resistance of semi-aromatic PI film is inevitably sacrificed and it is necessary to introduce a moderate rigid structure to improve the operating temperature. In addition, PIs containing non-coplanar structures can expand the band gap and increase energy density, but the improvement by this method is limited and the dielectric loss is relatively high.

In the future, the precision structure design and regulation of the dielectric polarization and E_g will be required to achieve a balance of comprehensive properties, as the most important research direction for energy storage dielectrics. More and more researches show that some trends observed in the experimental findings are in good agreement with the theoretical calculation or machine-learning results [52],

Table 2.2 Summary of key data for representative polyimide dielectric materials reported in literature.

Polyimide	Dianhydride	Diamine	T_g (°C)	ε_r (1 KHz)	tanδ (1 KHz)	E_g (eV)	E_b (MV/m)	Maximum U_e (J/cm^3)	Efficiency η	References
Ultem 1000 (GE)	BPADA	*m*-PDA	~221	~3.3	~0.09	/	~468.72 (150 °C) ~424.13 (200 °C)	~3.2 (150 °C) ~2.3 (200 °C)	~73% (150 °C) ~65% (200 °C)	[25]
PEI	BPADA	*m*-PDA	220.2	~2.9	/	3.3	544.36 (150 °C) 464.29 (200 °C)	~3.8 (150 °C) ~3.2 (200 °C)	~87% (150 °C) ~68% (200 °C)	[26]
KPI-MX300F (Kawamura)	6FDA	TFDB	/	~2.8	~0.06	/	/	~5.5 (150 °C) ~4.2 (200 °C) ~1.9 (250 °C)	~80% (150 °C) ~80% (200 °C) ~48% (250 °C)	[25]
Matrimid 5218 (PolyK)	BTDA	PIDA	/	~3.1	~0.075	/	/	~3.0 (150 °C) ~2.1 (200 °C) ~1.3 (250 °C)	~30% (150 °C) ~33% (200 °C) ~42% (250 °C)	[25]
6FDA/ 3,4-APBN	6FDA	3,4-APBN	252	3.30	0.013	/	/	/	/	[27]
PMDA/ *p,p*-3BCN	PMDA	*p,p*-3BCN	341	3.7 (10 Hz)	/	/	/	~0.17 (190 °C)	/	[28]
BTDA/2CN	BTDA	2CN	325	4.80	0.0016	/	~235	1.03 (RT)	/	[30]
BPADA/BCA	BPADA	BCA	/	4.2	0.025	2.22	470 (RT) 433 (250 °C)	3.3 (RT) 3.0 (150 °C) 2.5 (250 °C)	96% (RT) 90% (150 °C) 65% (250 °C)	[31]
ODPA/ 4,4′-DDS	ODPA	4,4′-DDS	290	5.98	0.00373	/	536	7.60 (RT)	91.3% (RT)	[32]

Table 2.2 (Continued)

Polyimide	Dianhydride	Diamine	T_g (°C)	ε_r (1 KHz)	tanδ (1 KHz)	E_g (eV)	E_b (MV/m)	Maximum U_e (J/cm^3)	Efficiency η	References
BTDA/BIA/4,4′-DDS	BTDA	BIA/4,4′-DDS (85%/15%)	354	4.56	0.0060	/	352 (RT) ~250 (150 °C)	~3.27 (RT) 1.14 (150 °C)	~50% (RT) 74% (150 °C)	[33]
0.87PI-0.13PAA	PMDA	*m*-BAPS	/	/	/	~2.73	613	8.9 (RT) 4.6 (150 °C)	~45% (RT) ~15% (150 °C)	[34]
0.90PI-0.10PAA	ODPA	4,4′-DDS	/	4.17 (RT) 3.8 (150 °C)	/	/	535 (RT) 483 (150 °C)	5.14 (RT) 2.4 (150 °C)	90% (RT) 80% (150 °C)	[35]
BTDA/BPBPA	BTDA	BPBPA	296	7.2	0.038	/	295	2.77 (RT)	/	[36]
BU-PI	ODPA	BUBPA	238	6.14	0.0097	/	513	6.92(RT)	~83%(RT)	[37]
PMDA/1,3-DAP/D230	PMDA	1,3-DAP/D230 (20%/80%)	54	6.55	0.00626	/	/	/	/	[38]
BTDA/DAH	BTDA	DAH	150	3.57	0.0085	3.42	812	9.98	/	[39]
BTDA/HK511	BTDA	HK511	78	7.8	0.0056	3.48	676	15.77	/	[39]
BTDA/HK25	BTDA	HK25	/	4.5	/	/	/	12	>90%	[40]
CBDA/TFMB	CBDA	TFMB	347	2.7	<1%	4.1	645 (RT) 556 (150 °C) 497 (200 °C)	~6.0 (RT) ~4.61 (150 °C) ~3.86 (200 °C)	~90% (RT) ~70% (150 °C) ~70% (200 °C)	[41]
CBDA/DCHM	CBDA	DCHM	277	2.6	<1%	4.8	692 (RT) 620 (150 °C) 605 (200 °C)	~6.0 (RT) 5.21 (150 °C) 5.01 (200 °C)	~90% (RT) ~70% (150 °C) ~70% (200 °C)	[41]
CHDA/4,4′-ODA	CHDA	4,4′-ODA	321.82	~3.7	/	/	502 (RT) 431 (150 °C)	4.63 (RT) 2.51 (150 °C)	58% (RT) 47% (150 °C)	[42]

PEI-iso	BPADA	ALPHA	167.4	/	/	3.59	667 (150 °C)	~7.3 (150 °C)	~90% (150 °C)	[43]
PI-oxo-iso	ODPA	ALPHA	258.2	~3.1	/	~3.5	635 (150 °C) 606 (200 °C) 386 (250 °C)	~6.8 (150 °C) ~6.1 (200 °C) ~2.2 (250 °C)	~81% (150 °C) ~68% (200 °C) ~67% (250 °C)	[43]
PEI-CF_3-iso	6FDA	ALPHA	275.5	/	/	3.53	577 (150 °C) 458 (250 °C)	~5.0 (150 °C) ~3.5 (200 °C) ~2.5 (250 °C)	~81% (150 °C) ~68% (200 °C) ~60% (250 °C)	[43]
6FDA/TFMB	6FDA	TFMB	329	2.9	/	3.6	690 (RT) 578 (150 °C) 452 (200 °C)	6.88 (RT) 3.21 (150 °C) 2.42 (200 °C)	85% (RT) 84% (150 °C) 82% (200 °C)	[45]
PI-spiro-1-5	ODPA	Spiro-1/4,4′-ODA (5%/95%)	258	~3.1	/	3.38	674 (150 °C) 6445 (200 °C)	~6.1 (150 °C) ~5.9 (200 °C)	~57% (150 °C) ~62% (200 °C)	[46]
PI-spiro-1-5	ODPA	Spiro-2/4,4′-ODA (5%/95%)	256	~3.2	/	3.50	694 (150 °C) 668 (200 °C)	~8.3 (150 °C) ~6.2 (200 °C)	~66% (150 °C) ~71% (200 °C)	[46]
PEPA-PEI (BPADA/ 4,4′-ODA)	BPADA PEPA 10 mol%	4,4′-ODA	/	~2.9	0.0037	/	/	~3.7 (RT) ~3.6 (150 °C) ~2.1 (200 °C)	~96% (RT) ~95% (150 °C) 73% (200 °C)	[47]
PEPA-PEI-320-2 h (BPADA/ 4,4′-ODA)	BPADA PEPA 10 mol%	4,4′-ODA	/	2.9	0.003	/	533 (RT) 512 (150 °C)	3.81 (RT) 3.60 (150 °C)	97.4% (RT) 96.5% (150 °C)	[48]
NA-XL-PI-3	ODPA NA 3 wt%	4,4′-ODA	/	3.7	0.003		545 (RT)	9.4 (RT) 6.5 (150 °C)	88% (RT) 82% (150 °C)	[49]
TAB-PMDA/ 4,4′-ODA	PMDA	4,4′-ODA TAB 30 wt%	/	3.36	0.0095	/	566 (RT) 450 (150 °C)	5.87 (RT) 1.75 (150 °C)	81.5% (RT) 70% (150 °C)	[50]
TAP-ODPA/ 4,4′-ODA	ODPA	4,4′-ODA TAP 2 wt%	/	3.18	0.0015	/	399 (RT) 285 (150 °C)	3.02 (RT) 2.53 (150 °C)	88.1% (RT) 82.5% (150 °C)	[51]

indicating that PIs with excellent overall performance for the energy storage applications can be determined through reasonable design and single or multiple property requirements for specific applications can be predicted accurately. It is essential to identify the decisive structural factors of PI dielectrics that are directly related to the electrical performance at high temperatures. Local states such as trap depth that is associated with specific PI structures should be paid more attention. To expand the band gap and disrupt electron transitions in the conjugated structures, introducing non-conjugated cycloaliphatic units is considered to be an effective and fundamental approach, which can break the long-range conjugation in the fully aromatic molecular chain and maintain sufficient heat resistance due to its moderate rigidity of cyclic structure. Moreover, PIs with side chains containing dipolar groups is proved to be an effective method to balance thermal stability and dielectric properties. In summary, the advanced PI dielectric materials will be innovative by the modular structure design, including at least the following modular units: rigid units corresponding to the good heat resistance, flexible units corresponding to the large breakdown strength, polar units corresponding to the high dielectric constant, and deep-trap units corresponding to the low conduction loss. The contribution of each modular unit to energy storage performance at high temperature and electric field needs to be further clarified through a combination of theoretical simulation and experimental verification.

In addition, there is another important consideration, which is to control the aggregation structure of PI dielectric films through imidization from PAA to PI and film preparation processes. The heat resistance and breakdown strength should be further improved by regulating precisely the aggregation structure and reducing defects physical defects to improve the surface quality of film. The whole process of PI dielectric film should be considered more comprehensively including structure design, polymerization, imidization method, film preparation, stretching orientation, and post-treatment. The influence of various factors on energy storage characteristics also needs to be further clarified in order to achieve compatibility between high thermal stability, energy density, and discharge efficiency when operating in extreme environments. In addition, the mechanical properties of polyimide dielectric films should also be given sufficient attention, as these reflect their potential for large-scale production. It is worth noting that the methodology about molecular structure tailoring and the correlating mechanism of PI dielectric films is also applicable to the design of other polymer dielectrics.

References

1 Tan, D., Zhang, L., Chen, Q., and Irwin, P. (2014). High-temperature capacitor polymer films. *Journal of Electronic Materials* 43: 4569.

2 Li, Q., Chen, L., Gadinski, M.R. et al. (2015). Flexible high-temperature dielectric materials from polymer nanocomposites. *Nature* 523: 576.

3 Ho, J.S. and Greenbaum, S.G. (2018). Polymer capacitor dielectrics for high temperature applications. *ACS Applied Materials and Interfaces* 10: 29189.

4 Pan, H., Kursumovic, A., Lin, Y.H. et al. (2020). Dielectric films for high performance capacitive energy storage: multiscale engineering. *Nanoscale* 12: 19582.
5 Zhang, T., Sun, H., Yin, C. et al. (2023). Recent progress in polymer dielectric energy storage: from film fabrication and modification to capacitor performance and application. *Progress in Materials Science* 140: 101207.
6 Huan, T.D., Boggs, S., Teyssedre, G. et al. (2016). Advanced polymeric dielectrics for high energy density applications. *Progress in Materials Science* 83: 236.
7 Zha, J.W., Liu, X.J., Tian, Y.Y. et al. (2021). High-temperature polyimide dielectric materials for energy storage. In: *Polyimide for Electronic and Electrical Engineering Applications* (ed. S. Diaham), 1–20. IntechOpen.
8 Li, H., Zhou, Y., Liu, Y. et al. (2021). Dielectric polymers for high-temperature capacitive energy storage. *Chemical Society Reviews* 50: 6369.
9 Liu, X.J., Zheng, M.S., Chen, G. et al. (2022). High-temperature polyimide dielectric materials for energy storage: theory, design, preparation and properties. *Energy & Environmental Science* 15: 56.
10 Huang, B.D., Yu, J.C., Dong, J. et al. (2024). Improving charge storage of biaxially-oriented polypropylene under extreme electric fields by excimer uv irradiation. *Advanced Materials* 2311713.
11 Qiao, Y.L., Yin, X., Zhu, T. et al. (2018). Dielectric polymers with novel chemistry, compositions and architectures. *Progress in Polymer Science* 80: 153.
12 Fan, B., Zhou, M., Zhang, C. et al. (2019). Polymer-based materials for achieving high energy density film capacitors. *Progress in Polymer Science* 97: 101143.
13 Dong, J.F., Deng, X.L., Niu, Y.J. et al. (2020). Research progress of polymer based dielectrics for high-temperature capacitor energy storage. *Acta Physica Sinica* 69: 217701.
14 Wei, J. and Zhu, L. (2020). Intrinsic polymer dielectrics for high energy density and low loss electric energy storage. *Progress in Polymer Science* 106: 101254.
15 Feng, Q.K., Zhong, S.L., Pei, J.Y. et al. (2022). Recent progress and future prospects on all-organic polymer dielectrics for energy storage capacitors. *Chemical Reviews* 122: 3820.
16 Luo, H., Wang, F., Guo, R. et al. (2022). Progress on polymer dielectrics for electrostatic capacitors application. *Advancement of Science* 9: 2202438.
17 Meng, Z., Zhang, T., Zhang, C. et al. (2023). Advances in polymer dielectrics with high energy storage performance by designing electric charge trap structures. *Advanced Materials* 2310272.
18 Pei, J.Y., Yin, L.J., Zhong, S.L., and Dang, Z.M. (2023). Suppressing the loss of polymer-based dielectrics for high power energy storage. *Advanced Materials* 35: 2203623.
19 Wu, C., Deshmukh, A.A., Li, Z.Z. et al. (2020). Flexible temperature-invariant polymer dielectrics with large bandgap. *Advanced Materials* 32: 2000499.
20 Ho, J. and Schroeder, M. (2021). Polyimides as high temperature capacitor dielectrics. In: *Polyimide for Electronic and Electrical Engineering Applications* (ed. S. Diaham), 1–27. IntechOpen.

21 Zha, J.W., Tian, Y.Y., Zheng, M.S. et al. (2023). High-temperature energy storage polyimide dielectric materials: polymer multiple-structure design. *Materials Today Energy* 31: 101217.

22 Yang, M.Z., Yuan, F., Shi, W.X. et al. (2023). Sub-nanowires boost superior capacitive energy storage performance of polymer composites at high temperatures. *Advanced Functional Materials* 33: 2214100.

23 Yan, C.F., Wan, Y.T., Long, H.P. et al. (2024). Improved capacitive energy storage at high temperature via constructing physical cross-link and electron–hole pairs based on p-type semiconductive polymer filler. *Advanced Functional Materials* 34: 2312238.

24 Alamri, A., Wu, C., Mishra, A. et al. (2022). Improving the rotational freedom of polyetherimide: enhancement of the dielectric properties of a commodity high-temperature polymer using a structural defect. *Chemistry of Materials* 34: 6553.

25 Yang, M.Z., Zhou, L., Li, X. et al. (2023). Polyimides physically crosslinked by aromatic molecules exhibit ultrahigh energy density at 200°C. *Advanced Materials* 35: 2302392.

26 Zeng, T., Li, Q., Liu, D. et al. (2024). Improved capacitive energy storage performance in hybrid films with ultralow aminated molybdenum trioxide integration for high-temperature applications. *Materials Horizons* 11: 1539.

27 Wang, D.H., Riley, J.K., Fillery, S.P. et al. (2013). Synthesis and characterization of unsymmetrical benzonitrile-containing polyimides: viscosity-lowering effect and dielectric properties. *Journal of Polymer Science, Part A: Polymer Chemistry* 51: 4998.

28 Wang, D.H., Kurish, B.A., Treufeld, I. et al. (2015). Synthesis and characterization of high nitrile content polyimides as dielectric films for electrical energy storage. *Journal of Polymer Science, Part A: Polymer Chemistry* 53: 422.

29 Treufeld, I., Wang, D.H., Kurish, B.A. et al. (2014). Enhancing electrical energy storage using polar polyimides with nitrile groups directly attached to the main chain. *Journal of Materials Chemistry A* 2: 20683.

30 Zhu, T.W., Yu, Q.X., Zheng, W.W. et al. (2021). Intrinsic high-*k*-low-loss dielectric polyimides containing *ortho*-position aromatic nitrile moieties: reconsideration on Clausius–Mossotti equation. *Polymer Chemistry* 12: 2481.

31 Tian, Y., Zheng, M.S., Li, Y. et al. (2023). Intrinsic-designed polyimide dielectric materials with large energy storage density and discharge efficiency at harsh ultra-high temperatures. *Materials Horizons* 10: 5835.

32 Tong, H., Fu, J., Ahmad, A. et al. (2019). Sulfonyl-containing polyimide dielectrics with advanced heat resistance and dielectric properties for high-temperature capacitor applications. *Macromolecular Materials and Engineering* 304: 1800709.

33 Wu, D., Zhao, X., Li, X.T. et al. (2022). Polyimide film containing sulfone groups with high dielectric properties. *Polymer* 256: 125221.

34 Dai, Z., Bao, Z., Ding, S. et al. (2022). Scalable polyimide-poly(amic acid) copolymer based nanocomposites for high-temperature capacitive energy storage. *Advanced Materials* 34: 2101976.

35 Liu, X.J., Zheng, M.S., Wang, Q. et al. (2022). Enhancement of high temperature capacitive performance of sulfone-containing polyimide by suppressing carrier transport. *Applied Physics Letters* 121: 243902.

36 Peng, X.W., Xu, W.H., Chen, L.L. et al. (2016). Development of high dielectric polyimides containing bipyridine units for polymer film capacitor. *Reactive and Functional Polymers* 106: 93.

37 Tang, Y.D., Xu, W.H., Yao, H.Y. et al. (2024). Constructing novel high dielectric constant polyimides containing dipolar pendant groups with enhanced orientational polarization. *Macromolecular Rapid Communications* 45: 2300699.

38 Baldwin, A.F., Ma, R., Wang, C.C. et al. (2013). Structure–property relationship of polyimides based on pyromellitic dianhydride and short-chain aliphatic diamines for dielectric material applications. *Journal of Applied Polymer Science* 130: 1276.

39 Ma, R., Baldwin, A.F., Wang, C.C. et al. (2014). Rationally designed polyimides for high-energy density capacitor applications. *ACS Applied Materials Interfaces* 6: 10445.

40 Li, Z.Z., Treich, G.M., Tefferi, M. et al. (2019). High energy density and high efficiency all-organic polymers with enhanced dipolar polarization. *Journal of Materials Chemistry A* 7: 15026.

41 Song, J.H., Qin, H.M., Qin, S.Y. et al. (2023). Alicyclic polyimides with large band gaps exhibit superior high-temperature capacitive energy storage. *Materials Horizons* 10: 2139.

42 Yu, S., Zhou, J., Xu, A. et al. (2023). The scalable and high performance polyimide dielectrics containing alicyclic structures for high-temperature capacitive energy storage. *Chemical Engineering Journal* 469: 143803.

43 Wang, R., Zhu, Y.J., Fu, J. et al. (2023). Designing tailored combinations of structural units in polymer dielectrics for high-temperature capacitive energy storage. *Nature Communications* 14: 2406.

44 Yang, M.Z., Li, H.Y., Wang, J. et al. (2024). Roll-to-roll fabricated polymer composites filled with subnanosheets exhibiting high energy density and cyclic stability at 200°C. *Nature Energy* 9: 143.

45 Qin, S.Y., Song, J.H., Qin, H.M. et al. (2024). Fluorinated aromatic polyimide with large bandgap exhibiting superior capacitive performance at elevated temperatures. *Journal of Energy Storage* 87: 111458.

46 Ran, Z.Y., Wang, R., Fu, J. et al. (2023). Spiral-Structured dielectric polymers exhibiting ultrahigh energy density and charge–discharge efficiency at high temperatures. *Advanced Materials* 35: 2303849.

47 Zhang, Y., Liu, Z., Zhu, L.X. et al. (2020). Enhanced discharged efficiency and high energy density at elevated temperature in polymer dielectric via manipulating relaxation behavior. *CCS Chemistry* 2: 1169.

48 Zhu, L.X., Zheng, Z.L., Xu, W.H. et al. (2022). Optimizing high-temperature capacitive energy storage performance by constructing crosslinked structure in self-crosslinkable polyetherimides. *Materials Today Energy* 30: 101145.

49 Wu, Z., Peng, Y., Song, Y. et al. (2023). Polyimide dielectrics with cross-linked structure for high-temperature film capacitors. *Materials Today Energy* 32: 101243.

50 Xu, W., Zhou, C., Ji, W. et al. (2024). Anisotropic semicrystalline homopolymer dielectrics for high-temperature capacitive energy storage. *Angewandte Chemie, International Edition* 63: e202319766.

51 Duan, G.Y., Hu, F.Y., Zhang, R.N., and Hu, Z.M. (2023). Preparation of a novel cross-linked polyetherimide with enhanced breakdown strength and high-temperature energy storage performance. *High Voltage* 8: 630.

52 Chen, L.H., Kim, C., Batra, R. et al. (2020). Frequency-dependent dielectric constant prediction of polymers using machine learning. *npj Computational Materials* 6: 61.

3

Processing Technology of Polyimide Dielectrics: From Experimental to Industrial Scale

Xi Ren[1], Zhibin He[2], Shunqi Yuan[3], Zhenzhong Wang[1], Shujun Han[1], Yuexin Qi[1], and Jingang Liu[1]

[1] *China University of Geosciences, Engineering Research Center of Ministry of Education for Geological Carbon Storage and Low Carbon Utilization of Resources, School of Materials Science and Technology, No. 29, Xueyuan Road, Beijing 100083, China*
[2] *Peking University, School of Material Science and Engineering, Key Laboratory of Polymer Chemistry and Physics of Ministry of Education, No. 5, Yiheyuan Road, Beijing, 100871, China*
[3] *RAYITEK Hi-Tech Film Company, Co. Ltd., Songgang Street, Baoan District, Shenzhen 518105, China*

3.1 Introduction

Polyimide (PI) dielectrics (PID) represent a class of high-performance electrical and electronic insulating polymers characterized by the excellent combination properties, including good thermal shock resistance from −269 °C to 400 °C, high dielectric strength, high surface and volume resistivities, low dielectric constants (D_k), and dissipation factors (D_f) [1–3]. As shown in Figure 3.1, PIs have been used as H-class (long-term servicing temperature >180 °C) or higher levels of high-temperature-resistant dielectrics for more than half a century since the first commercialization by DuPont, USA, in 1960s [4]. Scientists and engineers in DuPont, including Walter M. Edwards, Ivan M. Robinson, and Cyrus E. Sroog, invented PIs at the beginning of 1960s [5]. Afterward, various PIs, including the Pyre-ML® enamels, Vespel® moldings, and Kapton® films (all registered trademarks of DuPont, USA) based on the pyromellitic dianhydride (PMDA) and 4,4′-oxydianiline (ODA), have been successively commercialized. DuPont's early fundamental and application research work on PI electrical insulating materials in 1960s presented at the electrical insulation conferences organized by the Institute of Electrical and Electronics Engineers (IEEE) established the leading position of the company in the field of PIDs [6–8]. At the same time, the General Electric, Westinghouse Electric Company, TRW Company, and other manufacturers also carried out systematic research on PIDs [9].

In the subsequent development of more than half a century, PIDs have made great progress in basic theory, formulation design, application formed, and manufacturing technology. For example, in terms of basic theory, in the past half century, research

Functional Polyimide Dielectrics: Structure, Properties, and Applications, First Edition.
Edited by Jun-Wei Zha.

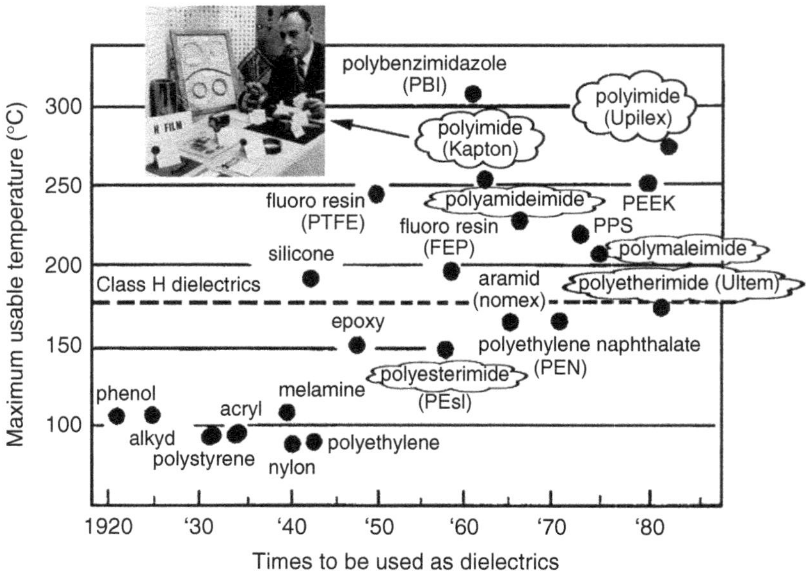

Figure 3.1 Development of polymer dielectrics and their maximum usable temperatures [4]. Source: Kapton® H film (1960)/DuPont, USA.

on the theory of PIDs has been greatly improved. The breakdown mechanism of PIDs, the space charge distribution and the influence on the electrical properties of the PIDs, the partial charge aging mechanism, the occurrence mechanism and preventive measures for corona aging, the inorganic–organic interfaces in PI nanodielectrics, and so on have all been established [10–15]. The basic theories well directed and supported the research and development (R&D) of advanced PIDs for various fields, such as high-temperature energy storage [16], high-temperature capacitors [17], organic and flexible electronic devices [18], and interlayer insulation for microelectronic devices [19]. On the other hand, in terms of the application forms, PIDs have also been developed from the initial impregnating varnish or enamels to films, tapes, insulating papers, molded parts, and even aerogels [20]. Figure 3.2 summarizes the common application forms of PIDs.

The reason why various PIDs can be successfully applied in modern industry depends largely on the development of advanced processing technology. The processing technology for different PIDs is often different. In order to achieve a good formula for the PIDs, appropriate processing technology is always indispensable. At the same time, even for the same PID, different processing technologies might result in different properties [21]. Therefore, the processing technology for PIDs has been continuously improved with the R&D of high-performance PIDs. Besides the conventional processing technologies for PIDs, such as casting for PI film dielectrics, compression or injection, and extrusion moldings for PI sheet dielectrics, new processing technologies, such as three-dimension (3D) additive manufacturing (AM) procedures, have been developed in recent years [22]. In this chapter, a brief introduction to the processing techniques for common PIDs,

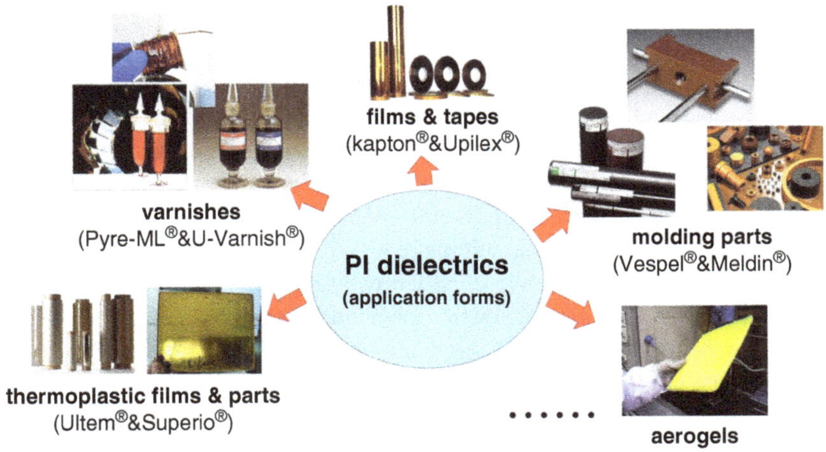

Figure 3.2 Classification of PI dielectrics according to the forms of applications.

including varnish dielectrics, film dielectrics, molding parts dielectrics, and recently developed aerogel dielectrics, will be reviewed and presented.

3.2 PID Synthetic Chemistry

The synthetic chemistry plays a very important role in determining suitable processing technologies for various PIDs [23, 24]. It has been well established in the literature for the synthetic chemistry of PIs after several decades of basic research in this field. The three most commonly used synthetic routes for PIs are summarized in Figure 3.3.

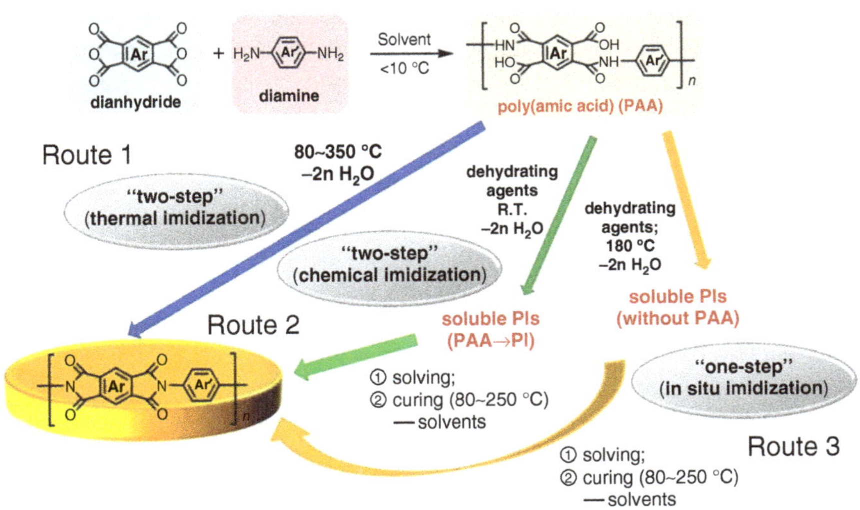

Figure 3.3 Common synthesis chemistry for PI dielectrics.

The first synthetic route is called "two-step thermal imidization" procedure, which is the most commonly used synthetic process and could theoretically be used for the synthesis and any PIs. In this process, the precursor of PI, which is often polyamide acid (PAA), is obtained first by the polycondensation reaction of the dianhydride and diamine monomers at low temperatures. In practical applications, the soluble PAA solution could often be directly used on the devices, which was then thermally cyclized at elevated temperatures up to 300 ~ 350 °C to afford the final PIDs. For example, Tong et al. reported the sulfonyl-containing PIDs with advanced heat resistance and dielectric properties for high-temperature capacitor applications [17]. The PAA solution was first prepared by the polycondensation of the dianhydrides of 3,3′,4,4′-diphenylsulfonyl tetracarboxylic dianhydride (DSDA), 4,4′-oxydiphthalic anhydride (ODPA), and various aromatic diamines. Then, the PAA precursors were thermal imidized at temperatures up to 250 °C to convert into the PIDs. The sulfonyl-containing PIDs with different flexible linkages exhibit simultaneously a high dielectric constant (4.50 ~ 5.98), low dissipation factor (0.00298 ~ 0.00426), and outstanding breakdown strength (most above 500 MV/m), as well as superior heat resistance with glass transition temperatures (T_g) in the range of 244 ~ 304 °C. Although the "two-step thermal imidization" route is generally applicable, it is most suitable for the monomers with high polymerization reactivity when considering the molecular weights of the afforded PAAs and the mechanical properties of the final PIDs. Higher molecular weights are usually critical for PAAs because they could guarantee the flexibility and toughness of the imidized PIDs. The molecular weights of the PAAs are mainly influenced by the polymerization reactivities of the monomers if the purity and molar equivalent of the monomers were fixed. During the polymerization reaction, the electron-rich amino groups will attack the electron-deficient anhydride carbonyl carbon atoms, followed by the opening of the anhydride rings to afford the PAAs [25, 26]. It has been well established in the molecular orbital (MO) theory that the lowest unoccupied molecular orbital (LUMO) energy levels (ε_{LUMO}) of one dianhydride and the highest occupied molecular orbital (HOMO) energy levels (ε_{HOMO}) of one diamine were closely related with their polymerization reactivity in the nucleophilic substitution reaction [27]. A lower ε_{LUMO} value for the dianhydride and a higher ε_{HOMO} value for the diamine could roughly indicate higher reactivity of the monomers. According to the criteria, the polymerization reactivities of common monomers for PAA preparation have been reported in the literature, as shown in Figure 3.4 [28, 29]. For the wholly aromatic dianhydrides, such as pyromellitic dianhydride (PMDA), 3,3′,4,4′-biphenyl tetracarboxylic dianhydride (BPDA), 3,3′,4,4′-benzophenone tetracarboxylic dianhydride (BTDA), and ODPA, the relatively high polymerization reactivities could usually provide PAAs with extremely high molecular weights. Thus, most of the commercially available PIDs are based on the dianhydrides mentioned above. However, for many of the other monomers with relatively low polymerization reactivities, such as the alicyclic dianhydrides and aliphatic or fluorinated diamines, it is often difficult to obtain PAAs with high molecular weights and the PIDs with good properties by the "two-step thermal imidization" procedure. On the other hand, it should be noted that sufficient studies have shown that in the imidization procedure

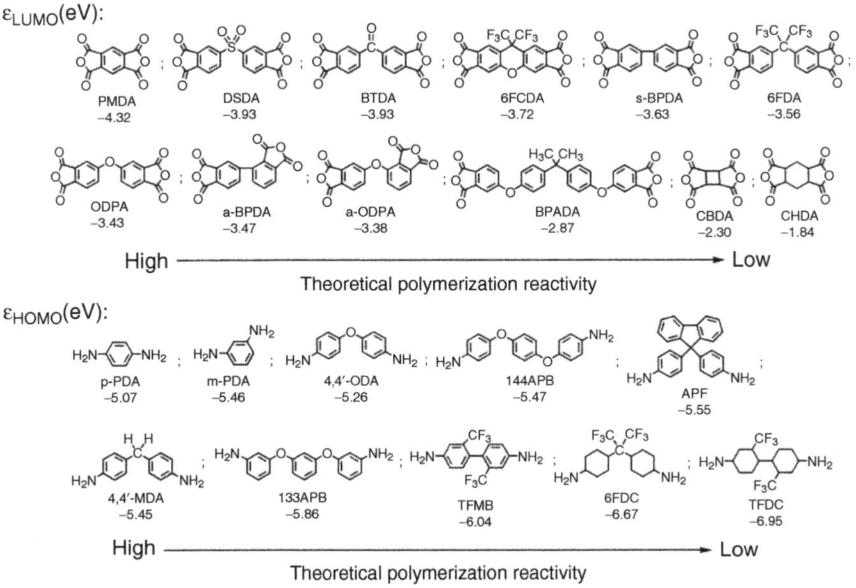

Figure 3.4 Theoretical polymerization reactivity predicted by the molecular orbit theory.

from PAA into PI, in order to meet the thermodynamic and kinetic conditions of the cyclization reaction, the thermal imidization temperature is usually higher than 300 °C [30]. This harsh reaction condition will have a significant impact on the color and appearance of the final PIDs. At the same time, it also causes the high cost and high energy consumption of industrial manufacturing equipment. For this reason, other low-temperature imidization routes have gradually been developed.

Route 2 is similar to Route 1, with the difference being the imidization pathway of the PAA precursors. Unlike Route 1, Route 2 uses chemical dehydrating agents, such as acetic anhydride, that allows the PAA to complete the imidization process in solution at room temperature. This characteristic makes Route 2 a "two-step chemical imidization" process. The PI solution obtained after chemical imidization can be precipitated in a poor solvent to afford the solid PI resins. The pre-imidized PI resins are further dissolved in polar aprotic solvents, such as N-methyl-2-pyrrolidone (NMP), N,N-dimethylformamide (DMF), N,N-dimethylacetamide (DMAc), or other suitable organic solvents, to form the PI solutions (not PAA solutions). The PI solutions can be physically cured to afford the final PIDs at temperatures which are only necessary to remove the organic solvents. This temperature is usually below 250 °C, which is much lower than that of the imidization temperature of PAA (300 ~ 350 °C). Therefore, this route is particularly suitable for applications where the curing temperature is more sensitive and the optical properties of the final PIDs are required. For example, Zou et al. reported the PI-based gate dielectrics for high-performance organic thin-film transistors (OTFT) [31]. The organo-soluble PIs were prepared from the 4,4′-(hexafluoroisopropylidene)diphthalic anhydride (6FDA) with moderate reactivity and the aromatic diamines via the chemical imidization procedure with acetic anhydride as the dehydrating agent and pyridine as the catalyst. The PI solutions were prepared from the 6FDA–PI resins in cyclopentanone (boiling point: 130 ~ 131 °C). The 6FDA–PI solutions underwent thermal curing at 180 °C to afford the wholly imidized PIDs. The low-temperature curing feature of such PIs greatly met the requirements of the temperature-sensitive OTFT devices. It is not difficult to deduce from the characteristics of the "two-step chemical imidization" process that the successful application of such procedure is highly dependent on the solubility of the imidized PIs in the reaction medium. This requires that the monomers used in the procedure have special structural characteristics, such as low-conjugated molecular skeleton, flexible molecular chains, and high molar volumes.

Apparently, when preparing PIDs with Routes 1 and 2, the PAA precursors need to be prepared first, and then thermal or chemical processes are used to convert the PAA precursors into the final PIs. This requires that the monomers used possess moderate-to-high polymerization reactivity. For the monomers with very low reactivity, such as the alicyclic dianhydride of hydrogenated pyromellitic dianhydride (HPMDA) (or CHDA: 1,2,4,5-cyclohexane tetracarboxylic dianhydride), it is often difficult to supply sufficient thermodynamic and kinetic conditions for the polycondensation reactions when using Routes 1 and 2. In this case, a third route of Route 3 emerged. As shown in Figure 3.3, the biggest difference between this route and the other two routes is that this route does not require the synthesis of PAA precursors alone, but directly converts PAA precursors generated by the reaction of dianhydride and diamine into PI under the action of dehydrating agents (toluene, xylene, chlorobenzene, etc.) in the polymerization process. The water by-product generated during the conversion from PAA to PI can be distilled out the polymerization system directly or by forming azeotrope with toluene or other dehydrating agents, so that the polymerization reaction can proceed and finally get PI solutions in the reaction medium (NMP, m-cresol, etc.). The subsequent treatment process is similar to that of the chemical imide process (Route 2). Due to the high polymerization temperature (≥180 °C) used and the polymerization reaction being completed in one step, the process is often referred to as "one-step high-temperature solution polycondensation" process. Baldwin reported the synthesis of next-generation dielectric materials through rational exploration of chemical space [32]. A series of PIDs with high dielectric constants up to 7.8 were prepared from the aromatic dianhydrides and aliphatic diamines via the one-step high-temperature solution polycondensation procedure by using NMP as the solvent. No water azeotropic agents were used in the experiments. Chung et al. reported the soluble PIs containing benzimidazole rings for inter-level dielectrics for microelectronic devices [33]. The PIDs were prepared from BTDA, the rigid-rod 6,4′-diamino-2-phenylbenzimidazole (BIA), and the fluoro-containing 2,2′-bis(trifluoromethyl)-4,4′-diaminobiphenyl (TFMB) with low polymerization reactivity (Figure 3.4) via the one-step high-temperature solution polycondensation procedure with NMP as the solvent and xylene or chlorobenzene (in the case of higher BIA contents) as the water azeotropic agents. The PI homopolymer and copolymers with high molecular weights (inherent viscosities: 0.86 ~ 1.74 dL/g) were obtained. The PIs showed good solution processability in polar solvents of NMP, DMAc, and DMF. The afforded PIDs exhibited the low dielectric constants in the ranges of 2.74 ~ 2.94. In addition, the PIDs containing a higher content of benzimidazole rings exhibited higher adhesion to copper.

Years of development in the field of PI synthesis chemistry have provided theoretical basis and feasible measures for the R&D of processing technologies for various PIDs. The developing status and future trends of the manufacturing technology for the common PIDs will be briefly introduced below.

3.3 Processing Technologies for PIDs

3.3.1 PI Precursor Varnishes

Enamel-insulated copper and aluminum magnet wires are used principally in the electrical and electronics industries for coils, inductors, transformers, generators, armatures, solenoids, and other windings. Conductive wires are usually dipped in a varnish bath of the electrically insulating resin and then thermally cured at elevated temperatures. The dipping and curing procedures might be repeated many times to obtain the desired thickness of the insulating layers. For instance, in most of the electrical and electronic insulating applications, the PIDs have been used with the form of the precursor varnishes, such as PAA [34], poly(amic acid ester) (PAE) [35], or poly(amic acid salt) (PAS) [36] due to the insoluble and infusible nature of the conventional PIDs [37]. For the common PID applications, PAA varnishes or enamels are usually used, such as the commercially available Pyre-ML series of PAAs developed by DuPont, USA; the PIQ® and PI2600 series developed by HD MicroSystems, Japan–US joint venture; the Semicofine® series developed by Toray, Japan; the Durimide® 100 series by Fujifilm, Japan; the CT® series by Kyocera, Japan; and the ZKPI® series by POME, China [38, 39]. For the functional insulating applications, such as the interlayer dielectrics for microelectronic or display devices, the photodefinable PAE varnishes are often used, such as the Photoneece® series of photosensitive polyimides (PSPIs) (PW-1200, PW-1500, etc.) developed by Toray, Japan; the HD®-4100 series by HD MicroSystems, Japan–US joint venture [40, 41].

Figure 3.5 illustrates the typical processing technology for PAA varnishes. The processing is usually performed in a clean environment in order to eliminate the negative effects of the dusts and impurities on the electrical properties of the produced PIDs. Meanwhile, in order to afford a high molecular weight and a stable viscosity for the afforded PAA varnish, the processing is often carried out in an environment with constant temperature around 25 °C and a relative humidity (RH) below 50%. Polymerization solvents, such as NMP and DMAc, should be ultra-dry class with the water content below 200 ppm so as to afford the PAAs with high molecular weights. Generally, the diamine monomer was first dissolved into the polymerization solvent in a reactor, and then the dianhydride was added with several batches to avoid the serious exothermic reaction in the reactor. If necessary, cold water can be passed into the jacket of the reactor to cool the heat generated by polymerization, and the polymerization temperature is controlled within 20 °C. The high temperature in the polymerization reaction might cause the depolymerization of PAA molecular chains, resulting in the decrease of molecular weight and the widening of the

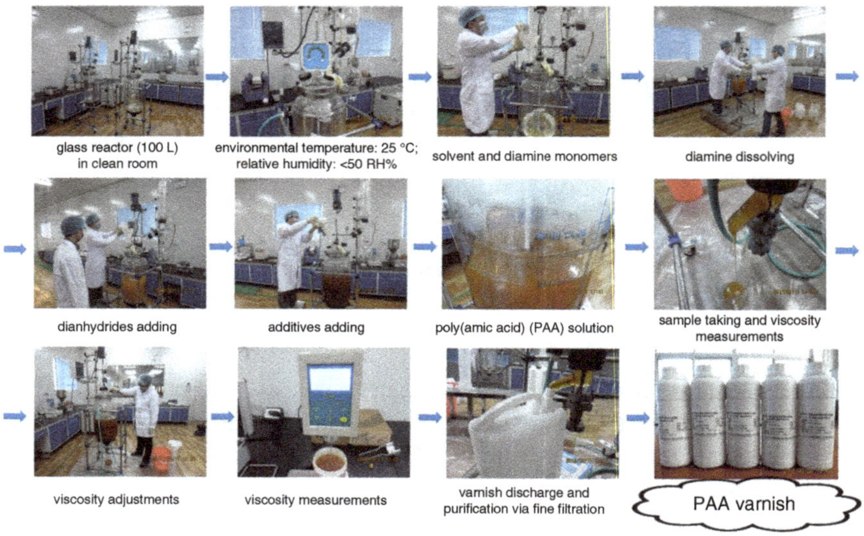

Figure 3.5 Processing illustration for the PAA-type of dielectric varnish at a pilot scale.

molecular weight distribution of the polymers [42]. In addition, some additives, such as the surface-active agents, coupling agents, and imidization accelerators, can be added into the system after the dianhydride monomer was added. Generally, with the addition of the dianhydride monomer, the polymerization reaction starts, PAA is gradually formed, and the viscosity of the reaction solution begins to increase. At this time, the viscosity of PAA can be adjusted by the amount of dianhydride monomer added until polymerization reaches the required viscosity. The viscosity value is often based on the practical requirements. With all of the monomers and additives added into the polymerization system, additional solvents are added to reach the required solid content of the PAA varnishes. When the viscosity and solid content of the PAA varnish reach the required values, and the molecular weight no longer fluctuates, the polymerization reaction ends. The afforded PAA solution is filtered to remove the insoluble impurities, and the PAA dielectric varnish can be used for the subsequent insulating applications.

In the early1980s, Wilson in Texas Instruments Incorp., USA, reviewed the use of PIDs as insulators in multilevel interconnections [43]. The chemistry of commercially available PI resins of semiconductor purity and of known chemical composition was also discussed. Various commercially available PAA varnishes with NMP and/or aromatic hydrocarbon solvents, including the DuPont PI2545, PI2555, the Monsanto Skybond 703, and the Rhodia Nolimid 605, were introduced. In recent years, with the rapid development of aviation, high-speed rail, new energy vehicles, and other industries, the requirements for highly reliable electrical insulating coatings are increasing. The development of PAA dielectric varnishes has also shown a rapid growth trend. Various PIDs with specific functionalities based on PAA varnishes, such as the corona resistance, the hydrolysis resistance, the oil resistance, the low dielectric constant, the high partial discharge inception voltage (PDIV), and other products, have been developed [44–46]. Besides the PAA precursors, several soluble PIs have also been summarized as the dielectric varnishes in the review mentioned above, including the Ciba-Geigy XU 218, Rhodia Kermid 600, and Upjohn 2080 [43]. Recently, Wu and coworkers reported preimidized semi-alicyclic PI varnishes with low curing temperatures and desirable processing viscosities at high solid contents [47]. The soluble PI resins were first prepared from hydrogenated 3,3′,4,4′-biphenyltetracarboxylic dianhydride (HBPDA) and various aromatic diamines via Route 3 shown in Figure 3.3. The prepared PI resins were all soluble in polar aprotic solvents, such as NMP and DMAc. Various PI varnishes with the solid contents as high as 35 ~ 45 wt% were successfully prepared by dissolving the PI resins in NMP. The PI varnishes were successfully used as the thick-film electrical insulating layers for light-emitting diodes (LED) chip protection. The HBPDA–PI insulating layers exhibited good thermal stabilities with the glass transition temperatures (T_g) in the range of 171.5–296.3 °C and 5% weight loss temperatures higher than 470 °C. In addition, they showed good optical transparency with transmittance higher than 72% at 450 nm at a thickness of 25 μm and low yellow indices and haze values.

3.3.2 PI Film Dielectrics

3.3.2.1 PI Films

PI films were originally developed for use in the field of electrical insulation. The PI films can be used as the insulating coatings of the wires or as the inter-turn insulators of the motors. The applications of the PI films as dielectrics in electrical and electronic fields have been well reviewed in the literature [48–50]. Since the commercialization of PI films nearly half a century ago, processing technologies have also been rapidly developed. Generally, according to the different starting materials for making the PI films, two processing technologies have mainly been developed for the PI films.

For the standard PI films, the main processing technology is to use the soluble PAA precursor solution, as shown in Figure 3.6a. First, the PAA solution is prepared by the polycondensation procedure of dianhydride and diamine monomers in the solvent of DMAc according to the Route 1 shown in Figure 3.3. Various additives, such as the steel belt–releasing agent and the antistatic agent, are also added into the polymerization system. Then, after the defoaming treatment, the PAA solution is cast through the die on the continuous running stainless steel belt. The partial DMAc solvent is evaporated and collected to afford the semi-solid PAA film. Then, the self-supported PAA film is peeled off from the steel belt and introduced into the high-temperature imide furnace through the guide roller. Then, the PAA film is stretched from both the machine direction (MD) and the transverse direction (TD) at high temperatures up to 350 ~ 450 °C. Meanwhile, the PAA film is dehydrated and imidized to afford the final PI films. It can be seen that the PI film processing process is a parallel process of film making, the change of polymer condensed state, and the chemical imidization. After the high-temperature biaxial stretching, the mechanical properties, electrical properties, and high-temperature dimensional stability of the PI films are obviously improved and stabilized [52]. In addition to the PI film prepared by the above-mentioned thermal imidization process, chemical dehydrating agents, such as acetic anhydride, can also be added to the PAA solution, and then the partially gelated PAA film is cast onto the steel belt, and then the biaxial stretching is carried out in the gel state, and finally the final PI film is obtained by the following high-temperature imidization treatment. In principle, due to the higher ratio of stretching can be achieved with the PAA films in the gel state, the degree of molecular chain orientation in the PI films might be higher. In this way, the final obtained PI films often show the better dimensional stability, especially in high-temperature and high-humidity environment. Most of the commercially available PI films are all prepared from the PAA precursors, such as the poly(pyromellitic dianhydride-4,4′-oxydianiline) ($PI_{PMDA-ODA}$) films (Kapton® by DuPont, USA; Aptical® by Kaneka, Japan; etc.), the poly(biphenyltetracarboxylic dianhydride-*p*-phenylenediamine) ($PI_{BPDA-PDA}$) film (Upilex®-S, Ube, Japan), and the poly(biphenyltetracarboxylic dianhydride-4,4′-oxydianiline) ($PI_{BPDA-ODA}$) film (Upilex®-R, Ube, Japan). Furutani et al. reported a series of ester-containing PI films designed for the applications as superconducting magnet insulation adhesive tapes for the Large Hadron Collider (LHC) in European Organization for Nuclear Research (CERN) [53]. The PI films were prepared via the PAA intermediates from 2,2-bis(4-hydroxyphenyl)propane-dibenzoate-3,3′,4,4′-tetracarboxylic acid anhydride (ESDA) and various diamines. The PI films, especially the one derived from ESDA and the diamine of 4,4′-bis(3-aminophenoxy)diphenylsulfone (mBAPS) (PIXEO® BP-S), exhibited high radiation resistance and superior cryogenic properties and contributed to the discovery of the Higgs Boson, which was awarded the Nobel Prize in Physics in 2013.

The advantage of the process using the PAA intermediates is that it has good versatility, and in principle all types of PI films can be prepared by this process. In addition, the process also has good designability, people can adjust the casting process, imide (thermal imidization, chemical imidization, etc.) process, biaxial stretching process, post-treatment process (plasma, corona, etc.) to control the characteristics of PI film, so as to meet different application needs. The disadvantage of this process is that the imidization temperature of PAA is higher (>350 °C); thus, a high energy consumption is accompanied. More importantly, it is not suitable for the preparation of colorless and transparent or temperature-sensitive PI films.

In view of the defects of the standard PI film processing technology via the PAA precursors, the alternative manufacturing technology based on the organo-soluble PI resins was developed in recent years. The typical processing route for preparing PI films with soluble PI resin is shown in Figure 3.6b [51]. Different from the traditional PI film preparation via the PAA intermediates, the new process directly dissolves the soluble PI resin in DMAc or other solvents to afford the PI solution with suitable viscosities and solid contents. Then, the PI films are prepared according to the similar procedure with the PAA-derived one. The main difference is that the high-temperature treatment of the PI-based films is a physical drying process instead of the chemical dehydration course for the PAA-based one. Thus, the thermal treatment temperature is often below 250 °C for the PI-based films, which is usually high enough to evaporate the solvents and the residual water

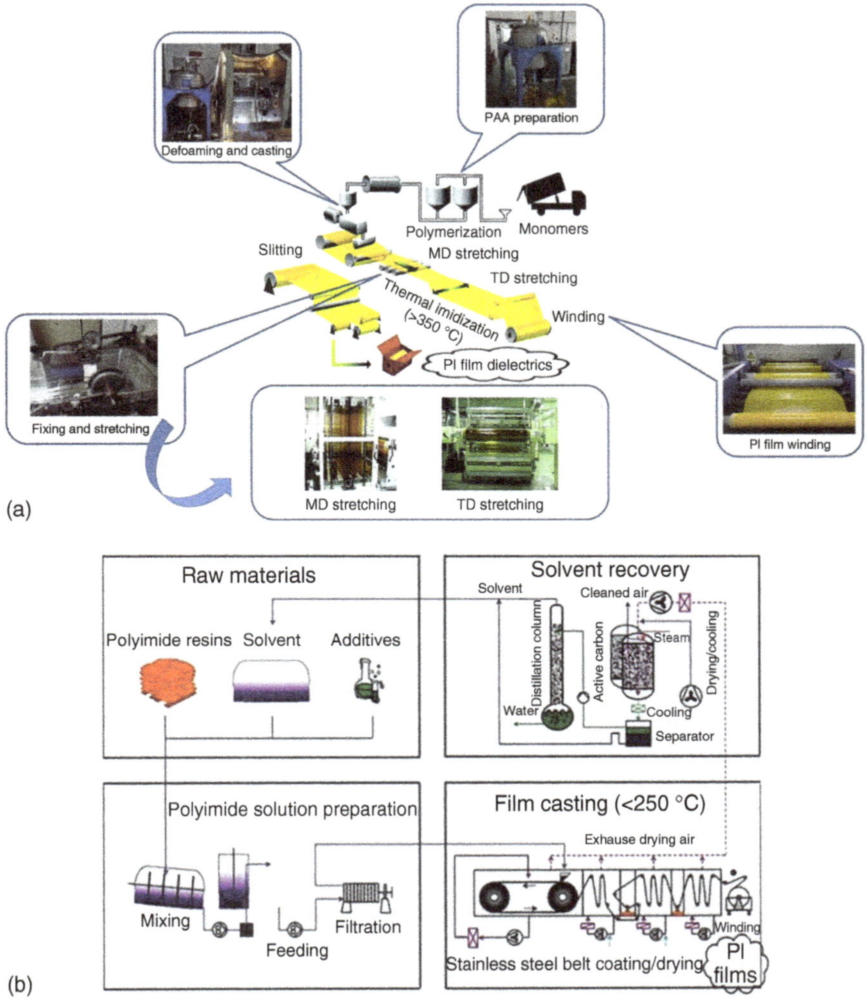

Figure 3.6 Processing technology for PID films with industrial scale. (a) PAA route; (b) soluble PI route. Source: Ref. [51]/Springer Nature.

by-products. The obviously reduced processing temperatures for the PI-based films are quite beneficial for the preparation of temperature-sensitive PI films, such as the colorless and transparent PI (CPI) films. The soluble PI resins used in this process can generally be prepared by the two-step chemical imidization (Route 2 in Figure 3.3) or one-step high-temperature polycondensation procedure (Route 3 in Figure 3.3). Since the soluble PI resin has been imidized nearly completely, the baking and curing temperatures of the derived films ($\leq$250 °C) are far lower than those of the conventional PAA-based route ($\geq$350 °C). However, the new route is only applicable for the PI systems with good solubility in organic solvents in the form of resins. Meanwhile, in order to prepare PI films with excellent mechanical properties, the soluble PI resins with high molecular weights are often required.

However, the increase of the molecular weights of the resins will usually reduce the solubility of the resins to a certain extent, and then further affect the quality of the afforded films. The high-temperature dimensional stability and flame retardancy of the soluble PI-based films are often deteriorated in the course of solubility improvements for the PI resins. In practical applications, it is always necessary to find the best balance among the required properties of the soluble PI-based films, such as the high thermal stability (high-T_g), high tensile properties, low linear coefficients of thermal expansion (low-CTE), high optical transparency, high flame retardancy, low dielectric constants (low-D_k), and high dielectric strength. Wu et al. reported the continuous manufacturing of nonflammable semi-alicyclic PI films modified with phenoxy-phosphazene (PPZ) oligomer flame retardant via the procedure shown in Figure 3.6b [54]. The soluble PI resin of PI (CHDA-ODA) was first prepared from 1,2,4,5-cyclohexanetetracarboxylic dianhydride (CHDA) and ODA according to the Route 3 shown in Figure 3.3. Then, the PI resin was dissolved into DMAc, together with the PPZ flame retardant. Colorless and transparent PI composite films with a width of 1000 mm were successfully prepared from the PI/PPZ solution. The composite films exhibited good thermal stability with glass transition temperature (T_g) and 5% weight loss temperature ($T_{5\%}$) of 325.0 °C and 404.3 °C, respectively, good flame retardancy with the limiting oxygen indices (LOI) of 41.2%, and flammability rating of UL94 VTM-0. Meanwhile, the PI films exhibited good electrical insulating and dielectric properties with the breakdown strength of 324 V/μm, the volume resistivity of 1.76×10^{15} Ω cm, the D_K and D_f values 3.59 and 0.044 at the frequency of 10 GHz, respectively. Good comprehensive properties make the PI films good candidates as high-performance dielectrics for electrical and (opto)electronic applications.

3.3.2.2 PI Nanodielectric Films

With the gradual popularization of high-speed transportation and new-energy vehicles, the reliability requirements for motor systems are rising rapidly. The reliability of a motor system largely depends on the motor insulation system and the associated materials [55]. Variable frequency drive (VFD) motors with energy saving, high efficiency, and easy maintenance have been widely used in high-performance motor systems. However, the unique working characteristics of the VFD motors make the insulating polymers survive in the harsh environments, such as high temperature, corona, electrical surge, and chemical (automatic transmission oil, etc.) erosion [56]. In terms of electrical characteristics, dielectrics for VFD motors insulation should have specific characteristics, such as high dielectric strength, corona resistance, high partial discharge inception voltage (PDIV), and high electrical surge resistance. Generally, it is quite difficult for simple polymers to meet all the property requirements mentioned above, even for the PID materials with excellent combined properties. Therefore, composite dielectric materials, especially the nanodielectrics, came into being [57]. Studies have shown that blending SiO_2, Al_2O_3, TiO_2, and other nano-fillers with polymer matrix can significantly reduce the deterioration of the dielectrics caused by the partial discharge, corona, and other factors [58].

Nanodielectric is known as the one of the most important classes of high-performance electrical insulators and thought to be the insulating material of the future [59]. The concept of "nanodielectric" was proposed at the beginning of 2000s, which is a combination of "nanotechnology" and "dielectric." Nanodielectric is usually defined as "a nanodielectric would consist of a multicomponent dielectric possessing nanostructures, the presence of which results in the change of one or several of its dielectric properties." Or it is defined as "a dielectric exhibiting behavior in one of its properties that is different than expected due to an alteration on a nanoscale" [60, 61]. Either definition indicates the characteristics of the nanodielectrics which are different from traditional dielectrics. In general, introduction of specific nano-fillers has an obvious effects on the electrical properties of the dielectrics, which are mainly reflected by the following results [62]: (i) effectively increase the free volume of nano-dielectric, resulting in a decrease in dielectric constant; (ii) effectively improve the alternative circuit (AC) and direct circuit (DC) breakdown strength of the dielectrics; (iii) significantly improve the long-term aging performance and corona resistance of the dielectrics; and (iv) effectively prohibit the occurrence of the partial discharge inside the dielectrics. In the past decades, the research and development of nanodielectrics have been thoroughly reviewed in the literature [63–67]. Hence, the processing technology of the PI nanodielectrics is mainly summarized in the current work.

The preparation of nanodielectrics usually takes into account many factors, including the selection of polymer matrix and nanofillers according to the different application requirements, the dispersion process between nano-filler and polymer matrix and the manufacturing procedures of the nanodielectrics. The polymer matrix, nano-filler, and the interfacial compatibility between the matrix and the filler all have important effects on the properties of the prepared nanodielectrics. For the PI nanodielectrics, both the thermoplastic and the thermosetting resins can be used. For the nanometer fillers, according to the chemical structures of the fillers, they usually include non-metal oxide (SiO_2, etc.), metal oxides (Al_2O_3, TiO_2, MgO, ZrO_2 etc.), and hydroxides ($Al(OH)_3$, $Mg(OH)_2$, etc.), nano-mineral fillers (mica, montmorillonite, etc.), and other species [68]. Nano-fillers with various shapes, such as spherical nanoparticles, whisker, rod particles, and layer particles, are all used. In practical applications, in order to achieve uniform dispersion and distribution of the nano-fillers in the PI matrix, the surface of the nano-fillers is usually treated with coupling agents, which can act as a bridge for efficiently combining the organic polymer and the inorganic fillers [69]. The coupling agent can reduce the surface free energy of the nanofillers and the interfacial tension between the polymer matrix, so that the dispersion of the nanofillers is easier. The interface compatibility between the polymer matrix and the nanofillers can be effectively improved. Meanwhile, the physical, electrical, and thermal properties of the nanodielectrics can also be improved.

During the past decades, people have accumulated rich experience in preparing the PI nanodielectrics. For example, PI nanodielectrics with excellent insulating characteristics have been developed with nano-fillers with high breakdown strength, such as the nano-clay [70]. PI nanodielectrics with high dielectric constants for capacitors have been developed from the PI matrix and barium titanate nanofillers [71]. In the preparation of the PI nanodielectrics, the dispersion of nanofillers in the PI matrix has always been one of the important factors affecting the properties of the final products. Nanofillers are easy-to-form agglomerations in the PI matrix because of the high surface energy and strong van der Waals force, electrostatic force, and hydrogen bond between fillers. The agglomerated nanofillers will lose the original function to a large extent. At present, the effective methods to prohibit the agglomeration of nanofillers include the pathways of ultrasonic dispersion, high-speed mechanical blending, exfoliation/intercalation method, and sol–gel method [72]. Ultrasonic and high-speed mechanical blending procedures have good universality for the disaggregation of nanofillers and are most widely used in the preparation of PI nanodielectrics, although the effects on promoting the uniform dispersion of nanofillers are limited in practical applications. Exfoliation/intercalation method is mainly used for lamellar nanoclay fillers (montmorillonite, kaolinite, palyclite, palygorskite, sepiolite, vermiculite, etc.). In this method, the ion-exchange characteristics of the nanoclay and the extensibility of interlayer distance are utilized to replace the original metal ions between the clay layers with organic groups so as to expand the interlayer distance and even completely peel them off to form nanoscale sheets [73]. Sol–gel method is one of the effective methods to prepare nanoparticles in situ. The method is to hydrolyze and condense the precursors such as alkoxy metal or metal salt into sol under certain conditions, and then volatilize the solvent or heat the sol into an oxide gel with a network structure. After further aging and drying of the oxide gel, nano-sized oxide particles can be obtained [74]. Different from the traditional process of directly using the nanofillers, the sol–gel method is to generate nanoparticles in situ in the nanodielectrics. Thus, it can effectively prevent the agglomeration of nanoparticles, and the formed nanoparticles can be evenly distributed in the nanodielectrics. In practical application, the above methods can often be used together.

In recent years, a special nanofiller of colloidal silica ($cSiO_2$) or silica sol has been paid increasing attention for the R&D of PI nanocomposites [75]. As a substance widely used in major fields such as synthetic chemistry, separation science, aerospace, electronics, and biomedicine, colloids are popular in modern industry for their uniform size, shape, and composition [76]. According to the definition of the International Union of Pure and Applied Chemistry (IUPAC), colloids are stable particles in the size range of 1–1000 nm, suspended/dispersed in continuous phases of different compositions or states [77]. Colloids include sols, gels, and emulsions. A sol is a collection of stable solid colloidal particles suspended in a liquid, while a gel is defined as a continuous solid skeleton enclosing a continuous liquid phase. Colloidal silica is one of the most important colloids which can be applied to the composite modification [78]. In general, $cSiO_2$ is a water or organic solvents suspension of colloidal SiO_2 nanoparticles. The particles are amorphous, but the surface of the particles is composed of silanol groups, which are hydroxyl groups attached to silicon atoms [79]. In commercial $cSiO_2$, the diameter of SiO_2 particles is usually between 5 and 100 nm. Huang et al. reported the PI–silica nanocomposite films based on the $PI_{PMDA-ODA}$ matrix and the commercially available colloidal silica [80]. The colloidal silica nanoparticles were well dispersed into the PI matrix with a loading content up to 30 wt%, and the derived PI–silica nanocomposite films showed obviously improved thermal stability and decreased CTE values with the increasing silica contents. Tsai and coworkers reported the PI–silica hybrid films based on the colloidal silica and PI matrix derived from [4,4′-(4,4′-isopropylidenediphenoxy)bis(phthalic anhydride)] (IDPA) and 2,2′-bis(4-[4-aminophenoxy]phenyl)propane (BAPP) [81]. The loading contents of the colloidal silica could reach 50 wt% without adding any coupling agents and surfactants. The authors attributed the good dispersion of the $cSiO_2$ nanoparticles to the hydrogen bonding between the Si–OH groups in the silica and the C=O in imide rings of the PI matrix. This structural feature on the one hand effectively increases the compatibility between the silica nanofiller and the PI matrix, on the other hand also increases the stability of the silica dispersion and prevents the agglomeration. Therefore, the loading amounts of $cSiO_2$ in the PI matrix can be as high as 50 wt% without causing significant deterioration of the physical and chemical properties of the PI matrix films. This far exceeds the amounts of the conventional silica nanoparticles added in the PI nanocomposite films. Meanwhile, the dielectric properties (breakdown strength, corona resistance, etc.) and high-temperature dimensional stability of the derived PI nanocomposite films can usually be obviously improved. This is of great significance to the R&D of PI nanodielectric films.

Figure 3.7 illustrates the typical processing technology of PI nanodielectric films based on colloidal SiO_2 nanoparticles at an experimental scale. First, the organo-soluble PI resins were designed and prepared. Then, the resins were dissolved into DMAc or other good solvents at a solid content of 15 ~ 25 wt% to afford the matrix solution. Then, the $cSiO_2$ filler dispersed into DMAc was added into the PI solution and treated with ultrasonic and mechanical blending. The homogeneous and transparent PI/$cSiO_2$ composite solution was obtained and then cast onto the clean borosilicate glass and thermally dried in a clean oven with the heating procedure of 80 °C/3 h, 150 °C/1 h, 180 °C/1 h, 200 °C/1 h, and 250 °C/1 h. After the thermal treatment, the glass substrate was cooled to room temperature and immersed into the deionized water. Free-standing PI/$cSiO_2$ composite films automatically peeled off the substrate, which were further dried in vacuum at 120 °C for 24 h. Films with various thicknesses in the range of 10 ~ 100 μm could be obtained.

The industrial processing technology of the PI nanodielectric films is similar to the traditional PI film process shown in Figure 3.6, except that the surface treatment and dispersion steps of the nanofillers are always needed. If the colloidal nanoparticles are used as the fillers, the pretreatment processes might be significantly simplified [82].

3.3.3 Thermoplastic PI Dielectrics

As mentioned in Section 3.1, in electrical and electronic industry, the conductive wires are usually dipped in a varnish bath of the electrically insulating resin and then thermally cured at elevated temperatures to form the insulating layers. Due to the use of impregnated varnish with a certain solid content, the process is

Figure 3.7 Processing technology for PI nanodielectric films based on cSiO_2 nanoparticles at an experimental scale.

complicated, the production efficiency is low, and the content of volatile organic compounds (VOC) is high in practical application. Therefore, in recent years, a new solvent-free composite process using the high-temperature resistant thermoplastic resin as the insulating layers for the wires has been gradually developed [83]. In the process, thermoplastic resins, such as polyether ether ketone (PEEK), polyphenylenesulfide (PPS), liquid crystal polyester (LCP), and thermoplastic polyimide (TPI), are extruded over the conductor wires to form the insulation layers. Basically, this process is a thermo-mechanical fabrication. The conductive wires are pulled through the die equipped in the extruder so as to be covered with the molten polymers at constant pressure and temperature [84]. Compared with the enameled and taped insulation, long wire length could be continuously achieved. Meanwhile, the process allows to reduce the void spaces around the conductors and prevents the vacuoles within the wires. The new process is recently dramatically promoted by the wide applications of rectangular wires for the requirements of hybrid electric vehicles (HEVs) and electric vehicles (EVs) market [85].

Among various thermoplastic resins, TPI is one of the most important species which is well known in aerospace, aeronautical engineering, and other high-tech

Table 3.1 Chemical structures of typical TPI dielectrics.

Number	Typical chemical structures	Trademark or abbreviations	References
#1	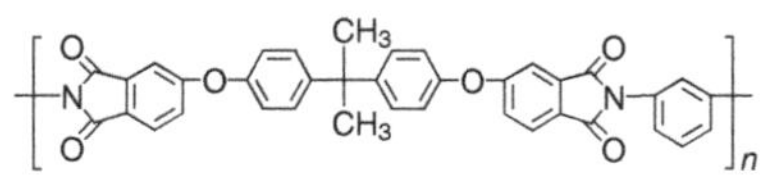	Ultem1010 ($PI_{BPADA-mPDA}$)	[88]
#2	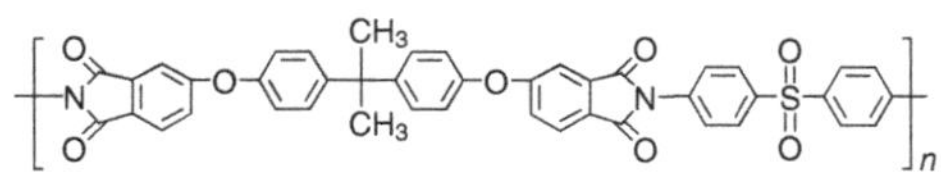	Extem ($PI_{BPADA-DDS}$)	[88]
#3	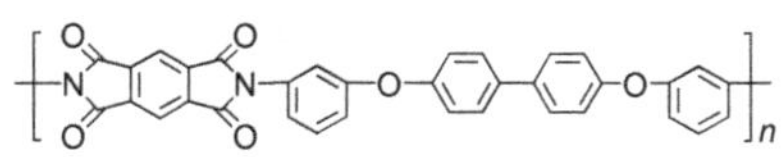	Aurum PL450C ($PI_{PMDA-BAPB}$)	[88]
#4	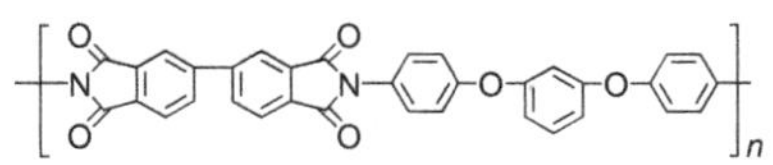	SuperAurum ($PI_{BPDA-134APB}$)	[88]
#5	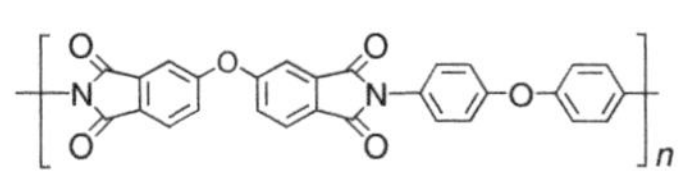	Ratem YS-20 ($PI_{ODPA-ODA}$)	[88]
#6	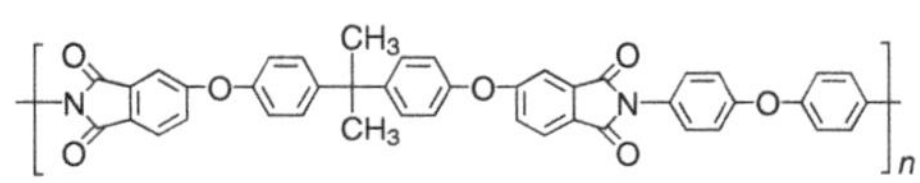	Ratem YS-30 ($PI_{BPADA-ODA}$)	[88]

fields [86]. Various TPIs have achieved the commercialization in the past decades, such as the Ultem® and Extem® series of TPIs developed by Sabic Corporation, Saudi Arabia; the Aurum® and SuperAurum® series of TPIs developed by Mitsui Chem. Co. Ltd., Japan; the Ratem® series of TPIs developed by Shanghai Synthetic Resin Institute, China; and the Therplim® series of TPIs developed by Mitsubishi Gas Chem. Co. Ltd., Japan [87]. The typical chemical structures of the commercially available TPIs are shown in Table 3.1 [88]. TPIs have become one of the most important dielectrics to insulate instrumentation and magnet wires working under demanding environmental and operating conditions since they usually exhibit good mechanical properties both at high and low temperatures and good melt processability [89, 90].

In the recent years, various new TPIs have also been reported in the literature. Yu and coworkers reported the TPIs with ether and ketone moieties in the main chains [91]. The afforded TPIs showed excellent thermoplasticity and T_g in the range of $184 \sim 225\,°C$ and modulus up to 3.0 GPa. Zhou et al. reported the melt-processable TPIs derived from diaminodiphenylsulfone and BTDA dianhydride [92]. The derived TPIs showed excellent heat resistance with the T_g values in the range of $266 \sim 278\,°C$. Basically, the R&D for high-performance TPIs is mainly focused on the methodologies on how to achieve the balance among the desired properties, including low melting viscosities (good melting processability), high T_g values (good thermal stability), and good dielectric and high mechanical properties. Thus, in practical molecular designs for TPIs, multi-copolymerization procedures have to be adopted. For example, Liu and coworkers reported a series of TPIs based on the quaternary copolymerization chemistry from isomeric dianhydrides, 2,3,3′,4′-biphenyltetracarboxylic dianhydride (a-BPDA) and 3,3′,4,4′-biphenyltetracarboxylic dianhydride (s-BPDA), and bicomponent flexible diamines, ODA, and 1,3-bis(4′-aminophenoxy)benzene (TPE-R) [93]. The derived TPIs exhibited good comprehensive properties with T_g up to 283 °C, good hot-pressing processability, and good adhesion to copper matrix. In the traditional molecular design for TPIs, certain degree of the crystallization is often required in order to achieve the good melt-processing properties [94]. Thus, many commercially available TPIs are semi-crystalline polymers. Besides the semi-crystalline TPIs, the species with amorphous molecular structures have also been developed [95]. The developed amorphous TPIs usually exhibited enhanced optical properties. For example, Kamitani et al. reported the soluble and processable TPI sheets containing polyhedral oligomeric silsesquioxane (POSS) units in the main chain [96]. The TPI derived from fluoro-containing dianhydride, 6FDA, and the POSS-containing diamine showed good optical transparency in the visible light region (wavelength: 380 ~ 780 nm) at the thickness of 216 μm. Wu and coworkers reported the TPIs with good optical transparency based on the copolymerization of 4,4′-(4,4′-isopropyldiphenoxy)bis(phthalic anhydride) (BPADA), 6FDA, 1,2,3,4-cyclobutanetetracarboxylic dianhydride (CBDA) and the fluoro-containing diamine TFMB [97]. The developed TPI sheets showed optical transmittance over 80% at wavelength of 650 nm at thickness from 110 to 150 μm.

Very recently, Qi and coworkers reported the TPI engineering plastics and molded sheets derived from the stereoisomeric hydrogenated pyromellitic dianhydrides, including 1S,2R,4S,5R-hydrogenated pyromellitic dianhydride (ccHPMDA) and 1R,2S,4S,5R-hydrogenated pyromellitic dianhydride (ctHPMDA) and a fluorine-containing diamine, 2,2-bis[4-(4-aminophenoxy)phenyl]hexafluoropropane (BDAF) [98]. The preparation pathway for the TPI sheets is shown in Figure 3.8. First, the silky PI-1 (ccHPMDA-BDAF) and PI-2 (ctHPMDA-BDAF) resins (Figure 3.8b) with average molecular mass (M_n) around 10^5 g/mol were prepared according to the Route 3 shown in Figure 3.3. Then, the PI resins were made into the ultrafine powders via the precipitation method. The fully dried PI powders exhibited good melt flowability in the temperature range of 300 ~ 350 °C according to the rheological plots shown in Figure 3.8d. The PI powders were added into a rectangular stainless steel mold. The mold was then placed into the molding machine. The PI sheets with thickness of ~3 mm were successfully prepared after the hot press procedure (Figure 3.8a). The PI sheets exhibited better optical transparency than the common TPI sheets. For example, PI-2 sheet showed the optical transmittance value at the wavelength of 760 nm (T_{760}) of 69.6%, which was 50.3% and 65.7% higher than those of the commercially available Ratem YS-20 (#5, Table 3.1) sheet (T_{760} = 19.3%) and the Aurum PL450C (#3, Table 3.1) sheet (T_{760} = 3.9%), respectively. Meanwhile, the PI-1 and PI-2 sheets showed heat deflection temperatures (HDT) of 261.0 °C and 265.0 °C, respectively, which were obviously higher than those of the commercial TPIs. Finally, the PI sheets exhibited good mechanical and dielectric properties, including high flexural strength (130.0 ~ 132.0 MPa), flexural modulus (2.87 ~ 2.96 GPa), compression strength (124.4 ~ 149.3 MPa), compression modulus (1.49 ~ 1.59 GPa), good impact strength (129.0 ~ 139.6 KJ/m^2), and low dielectric constants (2.8 ~ 3.0@10GHz).

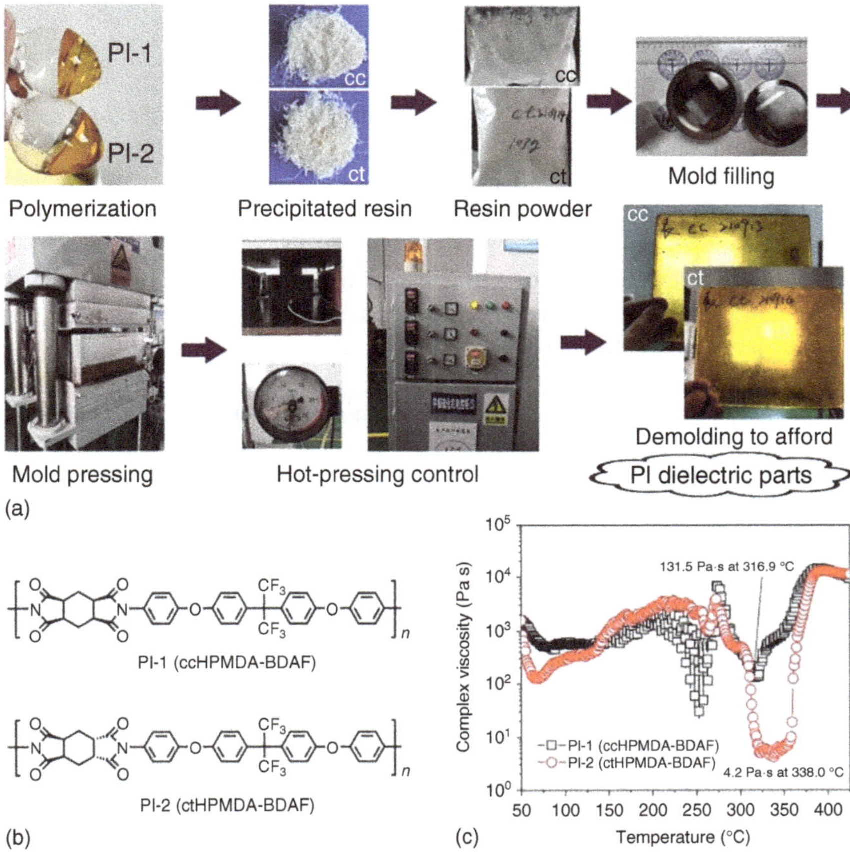

Figure 3.8 Processing technology for PI molding parts dielectrics based on thermoplastic PIs at a pilot scale: (a) process flow; (b) chemical structures of the PI dielectrics; (c) rheological plots of the polymers. Source: Ref. [98]/Springer Nature/CC BY 4.0.

TPIs have been investigated as the insulating layers for conductors for several decades [99]. Although common PI tape and coatings have had a long successful application history in wire insulating, the extrudable TPIs wire coating might result in lower motor weight and a high manufacturing efficiency [100, 101]. In future, the application of TPI dielectric materials in the field of wire insulation will gradually increase with the demand of electric vehicle motor insulation and other fields. How to improve the reliability of TPI dielectric materials under environmental and mechanical stress is the main problem to be solved in the development of high-performance TPI materials [102].

3.3.4 PI Aerogel Dielectrics

Gas insulation has been widely used in modern electrical engineering due to the great potentiality of size reduction, reliability, and economic efficiency improvements for the power facilities [103, 104]. Various gas media, including sulfur hexafluoride (SF_6), dry air, nitrogen, carbon dioxide, and fluorinated gases, have been investigated as the gas dielectrics for high-voltage systems [105]. However, using gas alone for electrical insulation generally leads to the problems of low dielectric strength and easy breakdown of the gas insulators under high pressure. Therefore, in recent years, porous electrical insulating materials consisting of filled insulating gases and inorganic or organic polymer frameworks have been paid increasing attention. Polymeric porous materials, such as polymer foams of polydimethylsiloxane, polyurethane, and epoxy, have been investigated as high-voltage engineering dielectrics in the literature [106]. Polymer foams insulators are usually characterized by the pore diameters at the level of micrometers. It has been established that void diameters lower than 50 μm or at the level of nanometers and high gas contents are highly desirable for practical applications [107]. Thus, an alternative porous dielectric of polymer aerogels has gained increasing attention in the electrical insulating fields.

Aerogels are a class of solid-state materials with a degree of porosity up to 99% in the composition [108]. This structural feature endows the aerogels extremely low density (close to air), low thermal conductivity, low dielectric constant (close to 1.0), high specific surface area, high acoustic attenuation, and many other distinctive properties. It is different from the inorganic aerogels, the polymer aerogels often exhibit good flexibility and thus have been widely used in aeronautics, microelectronic, and other modern high-tech fields [109]. Among the various polymer aerogels, PI aerogels possess both the property specialties of PI and aerogel and therefore exhibit excellent comprehensive properties in practical applications [110–112]. Comparing the PI aerogels with linear molecular structures, the crosslinking PI aerogels exhibited higher dimensional stability and low shrinkage. Crosslinking PI aerogels have also been the main species for the dielectric applications up to now.

Figure 3.9 illustrates the processing technology for the crosslinked PI aerogels. First, the excessive dianhydride monomer was reacted with diamine monomer in polar aprotic solvents to give an anhydride-capped PAA solution. Then, the stoichiometric multi-amine crosslinker was added to the PAA solution to achieve an equivalent molar ratio of the total amino and anhydride groups. When the dehydrating agent (acetic anhydride/pyridine, or acetic anhydride/triethylamine system) was added, imidization and crosslinking reactions of the PAA occurred within several to dozens of minutes to afford the PI gel. Then, the wet PI gel was immersed into ethanol to extract off the residual reaction solvent. At last, the residual reaction solvent and ethanol trapped in the PI gel was dried with supercritical carbon dioxide ($scCO_2$) fluid to give the final PI aerogel. The preparation of PI aerogels with varying temperature resistance, flexibility, and porosity can be achieved through the careful selection of polymerization monomers, crosslinking agents, as well as supercritical drying pressure and time.

The electrical conductivity of PI aerogels is quite low because the tenuous solid structure provides poor conduction paths and few charge carriers. The volume resistivity of PI aerogels is expected to be high for the same reason. Similarly, the dielectric strength of PI aerogels is also expected to be high due to the high volume resistivity. These properties guarantee that PI aerogels might find various applications as dielectric materials for electrical engineering. Zhang et al. investigated the dielectric behavior of PI aerogels [113]. The PI aerogel was first prepared from the rigid dianhydride of sBPDA, the diamine of 5-amino-2-(4-aminophenyl)benzoxazole

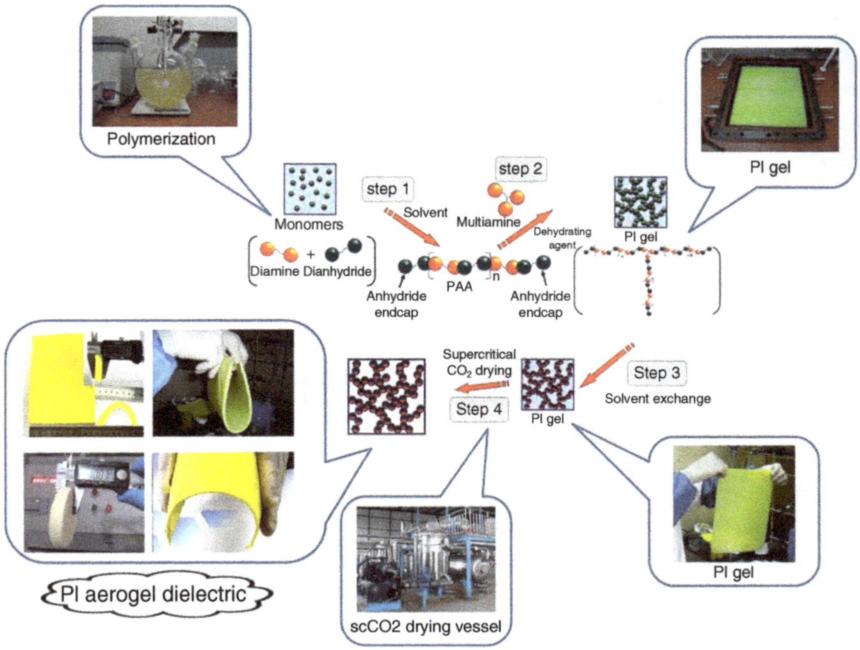

Figure 3.9 Processing technology for crosslinking PI aerogel dielectrics at a pilot scale. Source: Ref. [111]/Nova Science Publishing Inc.

(APBO), and the crosslinker of octa(aminophenyl)silsesquioxane (OAPS) according to the procedure shown in Figure 3.9. The afforded PI aerogel showed excellent thermal stability with a T_g of 358.3 °C, a 5% weight loss temperature over 500 °C, and a residual weight ratio of 66.7% at 750 °C in nitrogen. The developed PI aerogel exhibited surface and volume resistivities of $8.74 \times 10^{13}\ \Omega$ and $5.45 \times 10^{14}\ \Omega$ cm, respectively, indicating good electrical insulation properties. The dielectric breakdown strength of the aerogel is 132 kV/cm at a thickness of 2.5 mm. The dielectric breakdown behavior for the PI aerogel was further studied. Figure 3.10 compares the surface macro- and micro-topographies of PI aerogel after dielectric breakdown under high voltage. Obvious pinholes were observed at the dielectric breakdown area after measurement. The dielectric breakdown for PI aerogel was thought to be ascribed to the simultaneous actions of high voltage to the trapped air and the dense PI solid. High temperature induced by the air discharge caused the surface ablation of the aerogel. At the same time, the instantaneous high temperature caused the expansion of the trapped air in the aerogel, resulting in the swelling at the surface of the aerogel.

3.4 Summary

The laboratory, pilot, and industrial processing technologies of various PI dielectrics, including the PAA varnishes, PI films, TPI engineering plastics, and PI aerogels, are briefly introduced. As mentioned earlier, the processing technology usually has a

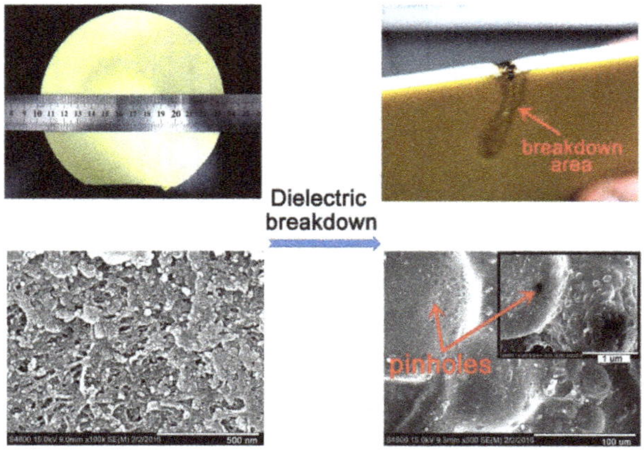

Figure 3.10 Dielectric breakdown of PI aerogel dielectrics under high voltage. Source: Ref. [111]/Nova Science Publishing Inc.

crucial impact on the properties of PI dielectrics. For the same PI dielectric, different manufacturing processes often bring about different properties. The manufacturing process level of PI dielectrics is also developing rapidly with the continuous expansion of the applications in high-tech fields. New manufacturing technologies, such as 3D printing, are also emerging. At the same time, the rapid popularization and application of new technologies such as big data, artificial intelligence, and computer simulation will also bring innovation to the manufacturing technology of PI dielectrics. Of course, the R&D of molecular structure design, synthesis, and composite technologies of new PI dielectrics will also have an impact on the manufacturing technology. In summary, the future research on PI dielectrics processing technology is full of opportunities and challenges.

Acknowledgments

Financial support from Shenzhen Science and Technology Program (No. JSGG20210629144539012) is gratefully acknowledged.

Declarations

Conflict of Interest Statement

The authors declare that there are no conflicts of interest existing in the submission of the chapter; also we have no financial and personal relationships with other people or organizations that can inappropriately influence our work, and there is no professional or other personal interest of any nature or kind in any product, service, and/or company that could be construed as influencing the position presented in, or the review of the chapter entitled.

References

1 Liu, X.J., Zheng, M.S., Chen, G. et al. (2022). High-temperature polyimide dielectric materials for energy storage: theory, design, preparation and properties. *Energy & Environmental Science* 15: 56–81.

2 Ji, D., Li, T., Hu, W., and Fuchs, H. (2019). Recent progress in aromatic polyimide dielectrics for organic electronic devices and circuits. *Advanced Materials* 31 (15): 1806070.

3 Zha, J.W., Tian, Y., Zheng, M.S. et al. (2023). High-temperature energy storage polyimide dielectric materials: polymer multiple-structure design. *Materials Today Energy* 31: 101217.

4 Mitsui, H. (1996). Progress in Japan in electrical insulation at high temperatures. *IEEE Electr Insual Magazine.* 12 (3): 16–27.

5 Sroog, C.E., Endrey, A.L., Abramo, S.V. et al. (1965). Aromatic polypyromellitimides from aromatic polyamic acids. *Journal of Polymer Science Part A: General Papers* 3: 1373–1390.

6 Schweitzer, F.E., Chalmers, J.R., Einstman, R.V. et al. (1962). An evaluation of insulation systems based on polymer ML. *1962 EI Electrical Insulation Conference Materials and Application*, pp. 30–33, Washington, DC, USA (19–22 February 1962). IEEE.

7 Tatum, W.E., Amborski, L.E., Gerow, C.W. et al. (1963). H film– Du Ponts new polyimide film. *EI Electrical Insulation Conference Materials and Application*, pp. 1–4, Chicargo, IL, USA (16–19 September 1963). IEEE.

8 Gumerman, C., Turner, V.L. (1969). Properties of high temperature magnet wire insulated with PI films and system usability. *9th Electrical Insulation Conference*, pp. 178–179, Boston, MA, USA (8–11 September 1969). IEEE.

9 Burns, R.L. (1969). Advances in high temperature polymers. *9th Electrical Insulation Conference*, pp. 263–265, Boston, MA, USA (8–11 September 1969). IEEE.

10 Diaham, S., Zelmat, S., Locatelli, M.L. et al. (2010). Dielectric breakdown of polyimide film: area, thickness and temperature dependence. *IEEE Transactions on Dielectrics and Electrical Insulation* 17 (1): 18–27.

11 Xiao, M. and Du, B.X. (2016). Review of high thermal conductivity polymer dielectrics for electrical insulation. *High Voltage* 1 (1): 34–42.

12 Sili, E., Cambronne, J.P., Naude, N., and Khazaka, R. (2013). Polyimide lifetime under partial discharge aging: effects of temperature, pressure and humidity. *IEEE Transactions on Dielectrics and Electrical Insulation* 20 (2): 435–442.

13 Lebey, T., Rumi, A., and Cavallini. (2022). Challenges for electrical insulation systems in high voltage aviation applications. *IEEE Electrical Insulation Magazine* 38 (6): 5–11.

14 Ren, H., Tanaka, Y., Miyake, H. et al. (2021). Effect of sinusoidal waveform and frequency on space charge characteristics of polyimide. *High Voltage* 6 (5): 760–769.

15 Li, S., Yu, S., and Feng, Y. (2016). Progress in and prospects for electrical insulating materials. *High Voltage* 1 (3): 122–129.

16 Wan, B., Dong, X., Yang, X. et al. (2023). Rising of dynamic polyimide materials: a versatile dielectric for electrical and electronic applications. *Advanced Materials* 35 (39): 2301185.

17 Tong, H., Fu, J., Ahmad, A. et al. (2019). Sulfonyl-containing polyimide dielectrics with advanced heat resistance and dielectric properties for high-temperature capacitor applications. *Macromolecular Materials and Engineering* 304 (4): 1800709.

18 Park, S., Chang, H.Y., Rahimi, S. et al. (2018). Transparent nanoscale polyimide gate dielectric for highly flexible electronics. *Advanced Electronic Materials* 4 (2): 1700043.

19 Kuntman, A. and Kuntman, H. (2000). A study on dielectric properties of a new polyimide film suitable for interlayer dielectric material in microelectronics applications. *Microelectronics Journal* 31 (8): 629–634.

20 Zhang, X.M., Liu, J.G., and Yang, S.Y. (2016). A review on recent progress of R&D for high–temperature resistant polymer dielectrics and their applications in electrical and electronic insulation. *Reviews on Advanced Materials Science* 46: 22–38.

21 Appelhans, L.N., Keicher, D.M., Lavin, J.M. (2016). Comparison of dielectric properties of additively manufactured vs. solvent cast polyimide dielectrics. *2016 IEEE Conference on Electrical Insulation and Dielectric Phenomena (CEIDP)*, pp. 62–65, Toronto, ON, Canada (16–19 October 2016). IEEE.

22 Zhang, F., Tuck, C., Hague, R. et al. (2016). Inkjet printing of polyimide insulators for the 3D printing of dielectric materials for microelectronic applications. *Journal of Applied Polymer Science* 133 (18): 43361.

23 Deligoz, H., Ozgumus, S., Yalconyuva, T. et al. (2005). A novel cross–linked polyimide film: synthesis and dielectric properties. *Polymer* 46: 3720–3729.

24 Zhu, L., Zheng, Z., Xu, W. et al. (2022). Optimizing high–temperature capacitive energy storage performance by constructing crosslinked structure in self–crosslinkable polyetherimides. *Materials Today Energy* 30: 101145.

25 Ghosh, A., Sen, S.K., Banerjee, S., and Voit, B. (2012). Solubility improvements in aromatic polyimides by macromolecular engineering. *RSC Advances* 2: 5900–5926.

26 Munoz, D.M., de la Campa, J.G., de Abajo, J., and Lozano, A.E. (2007). Experimental and theoretical study of an improved activated polycondensation method for aromatic polyimides. *Macromolecules* 40 (23): 8225–8232.

27 Ando, S., Matsuura, T., and Sasaki, S. (1997). Coloration of aromatic polyimides and electronic properties of their source materials. *Polymer Journal* 29 (1): 69–76.

28 Ando, S. (2010). Optical properties of polyimides. In: *The Newest Polyimides: Fundamentals and Applications* (ed. Japan Polyimide & Aromatic Polymers Study Group), 102–128. Tokyo: NTS Incorp. (in Japanese).

29 Kafafi, S.A., Lafemina, J.P., and Nauss, J.L. (1990). Electronic structure and conformation of polymers from cluster molecular orbital and molecular mechanics calculations: polyimide. *Journal of the American Chemical Society* 112: 8742–8746.

30 Chen, W.J., Chen, W., Zhang, B. et al. (2017). Thermal imidization process of polyimide film: interplay between solvent evaporation and imidization. *Polymer* 109: 205–215.

31 Zou, J., Wang, H., Zhang, X. et al. (2019). Polyimide-based gate dielectrics for high-performance organic thin film transistors. *Journal of Materials Chemistry C* 7: 7454–7459.

32 Baldwin, A. (2014). Synthesis of next generation dielectric materials through rational exploration of chemical space. Ph D. thesis. University of Connecticut.

33 Chung, I.S., Park, C.E., Ree, M., and Kim, S.Y. (2001). Soluble polyimides containing benzimidazole rings for interlevel dielectrics. *Chemistry of Materials* 13: 2801–2806.

34 Kokubu, R., Hara, N., Yamamoto, A. et al. (2024). Study on the diffusion coefficients of molecules in precursor varnishes and solid polyimides for understanding the polyimide curing process. *Industrial and Engineering Chemistry Research* 63 (11): 4934–4941.

35 Kim, Y., Goh, W.H., Chang, T. et al. (2004). Optical and dielectric anisotropy in polyimide nanocomposite films prepared from soluble poly (amic diethyl ester) precursors. *Advanced Engineering Materials* 6 (1–2): 39–43.

36 Yin, L., He, Y., Guo, W. et al. (2023). Low dielectric constant ultrathin polyimide films alternated by poly (amic acid) salt/poly(allylamine hydrochloride)/imogolite through layer by layer deposition. *Advanced Composites and Hybrid Materials* 6: 212.

37 Hicyilmaz, A.S. and Bedeloglu, A.C. (2021). Applications of polyimide coatings: a review. *SN Applied Sciences* 3: 363.

38 Niu, S., Zhu, L., Cai, Q., and Zhang, Y. (2022). Polymer composites for high-temperature applications. In: *Polymer Composites for Electrical Engineering* (ed. X. Huang and T. Tanaka), 63–98. New Jersy: John Wiley& Sons Ltd.

39 Diaham, S. (2021). Polyimide in electronics: applications and processability overview. In: *Polyimide for Electronic and Electrical Engineering Applications* (ed. S. Diaham), 1–18. London: IntechOpen.

40 Gao, R., Ma, R., Li, J. et al. (2022). Selection and characterization of photosensitive polyimide for fan-out wafer-level packaging. *IEEE Transactions on Components, Packaging and Manufacturing Technology* 12 (2): 368–374.

41 Karim, Z. (2021). Low-temperature polyimide processing for next-gen backend applications. *Chip Scale Review* 7–8: 1–5.

42 Zheng, Y., Zhai, Y., Li, G. et al. (2011). Synthesis and properties of a high-molecular-weight poly (amic acid) and polyimide based on 2,2-bis (3-amino-4-hydroxyphenyl) hexafluoropropane. *Journal of Applied Polymer Science* 121 (2): 702–706.

43 Wilson, A.M. (1981). Polyimide insulators for multilevel interconnections. *Thin Solid Films* 83: 145–163.

44 Chen, H., Li, L., Zhao, W. et al. (2022). Corona resistance mechanism of nano-modified polyimide. *Polymers* 14: 5469.

45 Berger, C. (2024). High-performance polyimides – solutions for battery and hydrogen powered vehicles. *MTZ Worldwide* 85: 38–41.

46 Diaham, S., Lambkin, P., O'Sullivan, L., and Chen, B. (2024). Intrinsic AC dielectric breakdown strength of polyimide films at the extreme: new breakthrough insights on thickness dependence. *Advanced Materials Interfaces* 11 (5): 2300822.

47 Wu, X., Liu, J., Jiang, G. et al. (2019). Highly transparent preimidized semi-alicyclic polyimide varnishes with low curing temperatures and desirable processing viscosities at high solid contents: preparation and applications for LED chip passivation. *Journal of Materials Science: Materials in Electronics* 30: 549–560.

48 Khazaka, R., Locatelli, M.L., Diaham, S., and Bidan, P. (2013). Endurance of thin insulation polyimide films for high-temperature power module applications. *IEEE Transactions on Components, Packaging and Manufacturing Technology* 3 (5): 811–817.

49 Tan, D.Q. (2020). Review of polymer-based nanodielectric exploration and film scale-up for advanced capacitors. *Advanced Functional Materials* 30 (18): 1808567.

50 Okamoto, T. and Uehara, H. (2023). Partial discharge current measurements with small discharge gaps over polyimide film. *IEEE Transactions on Dielectrics and Electrical Insulation* 30 (1): 158–164.

51 Siemann, U. (2005). Solvent cast technology– a versatile tool for thin film production. *Progress in Colloid and Polymer Science* 130: 1–14.

52 Wang, L., Yu, X., Wang, D. et al. (2013). *Materials Chemistry and Physics* 139: 968–974.

53 Fururani, H., Tsuji, H., and Sogabe, K. (2017). Synthesis and properties of polyester imides based on 2,2-bis (4-hydroxyphenyl) propane-dibenzoate-3,3′,4,4′-tetracarboxylic acid dianhydride. *Polymer Journal* 49: 587–591.

54 Wu, X., Wu, L., Qi, L. et al. (2021). Preparation, characterization, and continuous manufacturing of nonflammable colorless and transparent semi-alicyclic polyimide film modified with phenoxy-phosphazene oligomer flame retardant. *Express Polymer Letters* 15 (4): 329–342.

55 Sato, A., Iiduka, S., and Kimura, K. (2020). Magent wires for driving motors in electric vehicles. *SEI Technical Review* 90: 17–21.

56 Zhu, W., Gaetano, D.D., Chen, X. et al. (2022). A review of modeling and mitigation techniques for bearing currents in electrical machines with variable-frequency drives. *IEEE Access* 10: 125279–125297.

57 Tanaka, T. (2005). Dielectric nanocomposites with insulating properties. *IEEE Transactions on Dielectrics and Electrical Insulation* 12 (5): 914–928.

58 Tanaka, T., Montanari, G.C., and Mulhaupt, R. (2004). Polymer nanocomposites as dielectrics and electrical insulation-perspectives for processing technologies, material characterization and future applications. *IEEE Transactions on Dielectrics and Electrical Insulation* 11 (5): 763–784.

59 Nelson, J.K. (2007). Overview of nanodielectrics: Insulating materials of the future. *Electrical Insulation Conference and Electrical Manufacturing Expo*, pp. 229–235, Nashville, TN, USA (22–24 October 2007). IEEE.

60 Fréchette, M.F. (2013). What are nanodielectrics? *IEEE Electrical Insulation Magazine* 29 (6): 8–11.

61 Reed, C.W. (2013). Polymer nanodielectrics–Basic concepts. *IEEE Electrical Insulation Magazine* 29 (6): 12–15.

62 Tanaka, T. and Imai, T. (2013). Advances in nanodielectric materials over the past 50 years. *IEEE Electrical Insulation Magazine* 29 (1): 10–23.

63 Cao, Y., Irwin, P.C., and Younsi, K. (2004). The future of nanodielectrics in the electrical power industry. *IEEE Transactions on Dielectrics and Electrical Insulation* 11 (5): 797–807.

64 Calebrese, C., Hui, L., Schadler, L.S., and Nelson, J.K. (2011). A review on the importance of nanocomposite processing to enhance electrical insulation. *IEEE Trans. Dielec. Elec. Insual.* 18 (4): 938–945.

65 Cherney, E.A. (2013). Nanodielectrics applications–Today and tomorrow. *IEEE Electrical Insulation Magazine* 29 (6): 59–65.

66 Lau, K.Y., Vaughan, A.S., and Chen, G. (2015). Nanodielectrics: opportunities and challenges. *IEEE Electrical Insulation Magazine* 31 (4): 45–54.

67 Li, S., Yin, G., Chen, G. et al. (2010). Short–term breakdown and low–term failure in nanodielectrics: a review. *IEEE Transactions on Dielectrics and Electrical Insulation* 17 (5): 1523–1535.

68 Zhong, S.L., Dang, Z.M., Zhou, W.Y., and Cai, H.W. (2018). Past and future on nanodielectrics. *IET Nanodielectrics* 1 (1): 41–47.

69 Liu, J. and Zhang, D. (2020). Effect of directly added coupling agent on the dielectric properties of ceramic–polymer nanocomposites. *Journal of Materials Science: Materials in Electronics* 31: 9803–9811.

70 Wang, Y., Li, Z., Wu, C., and Cao, Y. (2020). High–temperature dielectric polymer nanocomposites with interposed montmorillonite nanosheets. *Chemical Engineering Journal* 401: 126093.

71 Yuan, P., Xue, R., Wang, Y. et al. (2024). Horizontally–oriented barium titanate@polydomine/polyimide nanocomposite films for high–temperature energy storage. *Journal of Colloid and Interface Science* 662: 1052–1062.

72 Anandraj, J., Joshi, G., and M. (2018). Fabrication, performance and applications of integrated nanodielectric properties of materials– a review. *Composite Interfaces* 25 (5–7): 455–489.

73 Zhang, K., Ma, Z., Deng, H., and Fu, Q. (2022). Improving high-temperature energy storage performance of PI dielectric capacitor films through boron nitride interlayer. *Advanced Composites and Hybrid Materials* 5: 238–249.

74 Hones, H.M., Cook, J.T., McCaffrey, M.J. et al. (2021). Effects of material preparation in polyimide/SiO_2 nanocomposites for low–temperature dielectric applications. *IEEE Transactions on Nanotechnology* 20: 695–702.

75 Abdalla, S., Almarzouki, F., Obais, A., and Gamal, S. (2016). Effect of addition of colloidal silica to films of polyimide, polyvinylpyridine, polystyrene, and polymethylmethacrylate nanocomposites. *Materials* 9 (2): 104.

76 Thickett, S.C. and Teo, G.H. (2019). Recent advances in colloidal nanocomposite design via heterogeneous polymerization techniques. *Polymer Chemistry* 10: 2906–2924.

77 Beltran-Osuna, A.A. and Perilla, J.E. (2016). Colloidal and spherical mesoporous silica particles: synthesis and new technologies for delivery applications. *Journal of Sol-Gel Science and Technology* 77: 480–496.

78 Hyde, E.D.E.R., Seyfaee, A., Neville, F., and Moreno0Atanasio R. (2016). Colloidal silica particle synthesis and future industrial manufacturing pathways: a review. *Industrial and Engineering Chemistry Research* 55 (33): 8891–8913.

79 Samuel, J.G.C. (2013). Colloids science-a toolbox for processing nanodielectrics. *IEEE Electrical Insulation Magazine* 29 (6): 16–20.

80 Huang, J.W., Wen, Y.L., Kang, C.C., and Yeh, M.Y. (2007). Preparation of polyimide–silica nanocomposites from nanoscale colloidal silica. *Polymer Journal* 39: 654–658.

81 Tsai, M.H., Huang, Y.C., Tseng, I.H. et al. (2011). Thermal and mechanical properties of polyimide/nano-silica hybrid films. *Thin Solid Films* 519: 5238–5242.

82 Zhang, S., Pelligra, C.I., Feng, X., and Osuji, C.O. (2018). Directed assembly of hybrid nanomaterials and nanocomposites. *Advanced Materials* 30 (18): 1705794.

83 Khowja, M.R., Vakil, G., Ahmad, S.S. et al. (2024). Life characterization of PEEK and nanofilled enamel insulated wires under thermal ageing. *IEEE Access* 12: 39470–39483.

84 Kenfaui, D., Esteban, J., Baena, P. et al. (2024). Insulting materials for electrical components used for power transmission on board the aircraft. *IEEE Transactions on Transportation Electrification* 10 (3): 6141–6161.

85 Ito, K., Shibata, T., and Kawasaki, T. (2016). Development of high voltage wire for new structure motor in full hybrid vehicle. *SAE International Journal of Alternative Powertrains* 5 (2): 272–277.

86 Das, A., Chatham, C.A., Fallon, J.J. et al. (2020). *Additive Manufacturing* 34: 101218.

87 Nazarychev, V.M. and Lyulin, S.V. (2023). The effect of mechanical elongation on the thermal conductivity of amorphous and semicrystalline thermoplastic polyimides: atomistic simulations. *Polymers* 15 (13): 2926.

88 Feng, X. and Liu, J. (2018). Thermoplastic polyimide (TPI). In: *High Performance Polymers and Their Nanocomposites* (ed. P.M. Visakh and A.O. Semkin), 149–219. New Jersey: John Wiley & Sons.

89 Pfreiffenberger, N., Milandou, F., Niemeyer, M. et al. (2018). High temperature dielectric polyetherimide film development. *IEEE Transactions on Dielectrics and Electrical Insulation* 25 (1): 120–126.

90 Lopez, G. (2021). High-performance polymers for aeronautic wires insulation: current uses and future prospects. *Recent progress in Materials* 3 (1): 2101005.

91 Yu, X., Zhao, X., Liu, C. et al. (2010). Synthesis and properties of thermoplastic polyimides with ether and ketone moieties. *Journal of Polymer Science, Part A: Polymer Chemistry* 48: 2878–2884.

92 Zhou, H., Zhang, L., Mao, Y. et al. (2021). Synthesis and characterization of melt-processable polyimides derived from diaminodiphenylsulfone and benzophenone tetracarboxylic dianhydride with excellent heat resistance and thermoplasticity. *Journal of Applied Polymer Science* 138 (41): 51219.

93 Liu, X., Cao, X.W., He, G.J. et al. (2019). Preparation and properties of a novel quaternary copolymerizaed thermoplastics polyimide. *Express Polymer Letters* 13 (6): 524–532.

94 Zhang, H., Wang, W., Chen, G. et al. (2017). Melt-processable semicrystalline polyimides based on 1,4-bis(3,4-dicarboxyphenoxy)- benzene dianhydride (HQDPA): synthesis, crystallization, and melting behavior. *Polymers* 9 (9): 420.

95 Brostow, W., D'Souza, N.A., Gopalanarayanan, B., and Jacobs, E. (2000). Thermal expansivity and thermal conductivity of amorphous thermoplastic polyimide and polymer liquid crystal blends. *Polymer Engineering and Science* 40 (2): 490–498.

96 Kamitani, T., Imoto, H., and Naka, K. (2022). Soluble and processable thermoplastic hybrid polyimides containing POSS in main chains. *Polymer Chemistry* 13: 5145.

97 Wu, G., Wang, Z., Zi, Y. et al. (2023). Design and preparation of thermoplastic polyimides with high transmittance based on 4,4′-(4,4′-isopropyldiphenoxy) bis (phthalic anhydride) (BPADA). *European Polymer Journal* 184: 111798.

98 Qi, Y., He, Z., Han, S. et al. (2024). Preparation and properties of semi-alicyclic thermoplastic polyimide sheets from stereoisomeric hydrogenated pyromellitic dianhydrides and fluorinated diamine. *Journal of Polymer Research* 31: 82.

99 Takei, S., Tanabe, Y., Ohki, Y. (1998). Electrical properties of thermoplastic polyimide. *IEEE International on Conference Conduction and Breakdown in Solid Dielectrics*, pp. 349–352, Vasteras, Sweden (22–25 June 1998). IEEE.

100 Kemari, Y., Belijar, G., Valdez-Nava, Z. et al. (2024). Review on high-temperature polymers for cable insulation: state-of-the-art and future developments. In: *High Temperature Polymer Dielectrics: Fundamentals and Applications in Power Equipment* (ed. J.W. Zha and Z.M. Dang), 103–148. Weinheim: Wiley-VCH.

101 Lima, G.P.D.S., Ail-Amar, S., Velu, G., Frezel, P. (2023). Characterization of classical and extruded enameled wires. *14th International Electrical Insulation Conference (INSUCON)*, pp. 119–124, Birmingham, UK (18–20 April 2023). IEEE.

102 Piccin, R., Rigaud, J.S., Santillana, I.A. et al. (2024). Environmental stress cracking of thermoplastic polyimide insulated wires. *IEEE Transactions on Applied Superconductivity* 34 (5): 7700405.

103 Mazzanti, G., Stomeo, G., and Mancini, S. (2016). State of the art insulation of gas insulated substations: main issues, achievements, and trends. *IEEE Electrical Insulation Magazine* 32 (5): 18–31.

104 Xiao, S., Shi, S.Y., Li, Y. et al. (2021). Review of decomposition characteristics of eco-friendly gas insulating medium for high-voltage gas-insulated equipment. *Journal of Physics D: Applied Physics* 54: 373002.

105 Pan, B., Wang, G., Shi, H. et al. (2020). Green gas for grid as an eco-friendly alternative insulation gas to SF_6: a review. *Applied Sciences* 10: 2526.

106 Tay, R.Y.T., Li, H., Lin, J. et al. (2020). Lightweight, superelastic boron nitride/polydimethylsiloxane foam as air dielectric substitute for multifunctional capacitive sensor applications. *Advanced Functional Materials* 30 (10): 1909604.

107 Fromm, U., Li, M., Schutte, T. (1998). Electric field control by polymer foam. IEEE International Conference on Conduction and Breakdown in Solid Dielectrics, pp. 460–463, Vasteras, Sweden (22–25 June 1998). IEEE.

108 Akhter, F., Soomro, S.A., and Inglezakis, V.J. (2021). Silica aerogels: a review of synthesis, applications and fabrication of hybrid composites. *Journal of Porous Materials* 28: 1387–1400.

109 Idumah, C.I., Ezika, A.C., and Okpechi, V.U. (2021). Emerging trends in polymer aerogel nanoarchitectures, surfaces, interfaces and applications. *Surfaces and Interfaces* 25: 101258.

110 Ghaffari-Mosanenzadeh, S., Tafreshi, O.A., Karamikamkar, S. et al. (2022). Recent advances in tailoring and improving the properties of polyimide aerogels and their application. *Advances in Colloid and Interface Science* 304: 102646.

111 Liu, J.G., Zhang, X.M., Chen, F.X. et al. (2017). A review on recent progress of research and applications for polyimide aerogels. In: *Polyimides: Synthesis, Applications and Research* (ed. C. Murphy), 105–130. New York: Nova Publishers.

112 Hu, R., Chen, Y., Zhang, C. et al. (2024). Porous monoliths from polyimide: synthesis, modifications and applications. *Progress in Materials Science* 144: 101284.

113 Zhang, X.M., Liu, J.G., and Yang, S.Y. (2016). Synthesis and characterization of flexible and high-temperature resistant polyimide aerogel with ultra-low dielectric constant. *Express Polymer Letters* 10 (10): 789–798.

4

Polyimide Dielectrics for High-Temperature Energy Storage

Wenjie Huang, Wenshuai Zhao, and Jun-Wei Zha

University of Science and Technology Beijing, School of Chemistry and Biological Engineering, Department of Chemistry, 30 Xueyuan Road, Haidian District, Beijing 100083, P. R. China

4.1 Introduction

Polymer dielectrics with high insulation, high breakdown strength, and low dielectric loss are ideal materials for electrostatic capacitors. Compared to supercapacitors and lithium-ion batteries, electrostatic capacitors have a relatively low storage energy density, but the inherent advantage of the highest power density (>10^6 W/kg) makes them stand out among various energy storage devices [1, 2]. Semi-crystalline biaxially oriented polypropylene (BOPP) dielectrics serve as the most widely used dielectric material in electrostatic capacitors. However, BOPP with a low melting point ($T_m \sim 165$ °C) limits its maximum operating temperature to less than 105 °C. In addition, other capacitor polymer materials such as polyethylene terephthalate (PET), polycarbonate (PC), and polystyrene (PS) are also limited to a maximum operating temperature of 125 °C [3]. Currently, commercial capacitor film materials are unable to meet the growing demand for high-temperature (>150 °C) applications, and the development of new polymer dielectrics with high energy density and excellent temperature resistance is imminent.

In order to meet the urgent requirements for high-temperature dielectric capacitors, a variety of high glass transition temperature (T_g) or high T_m polymer dielectrics have been extensively investigated, including polyimide (PI) [4], polyetherimide (PEI) [5], fluorene polyester (FPE) [6], and polytetrafluoroethylene (PTFE) [7]. PI dielectrics are considered optimal potential materials for high-temperature dielectrics due to their excellent insulating properties, good mechanical properties, and outstanding thermal stability [8–11]. The combination of aromatic and imide rings endows PI dielectrics an extremely high T_g (> 200 °C). Ma et al. [12] employed high-throughput density functional theory (DFT) to screen 267 polymers with unique and rational structures for potential application in electrostatic capacitors. PI was selected for further investigation based on the relative

Functional Polyimide Dielectrics: Structure, Properties, and Applications, First Edition.
Edited by Jun-Wei Zha.

equilibrium dielectric permittivity (ε_r) and bandgap (E_g). In addition, PIs offer excellent structure design advantages, where reactive monomers can be selected to introduce groups with special features (e.g., high polarization, wide E_g) as required [8, 13–15]. Furthermore, the development of PI-based multiphase composites shows promise [16, 17]. For instance, doping wide E_g and high thermal conductivity nanofillers such as boron nitride nanosheet (BNNS) [18] and alumina (Al_2O_3) [19], introducing organic polymers [20, 21] and small-molecule fillers [22, 23], and constructing multilayered structures [24, 25] can improve the high-temperature energy storage properties of PIs. The improved properties of PIs are promising to allow applications in oil exploration, power grids, electric vehicles, aircraft, wind turbines, electronic circuits, and other advanced power electronic applications, as shown in Figure 4.1.

This chapter first introduces the key characteristics and mechanisms of PI dielectrics for high-temperature energy storage applications. Subsequently, the recent progress in the design and development of PI and PI-based composite dielectrics for high-temperature energy storage is systematically reviewed and divided into the following categories: (i) the design of intrinsic polyimide dielectrics; (ii) the development of all-organic polyimide composite dielectrics; and (iii) the construction of PI-based nanocomposite dielectrics. In addition, attention was given to self-healing high-temperature PI dielectrics. Finally, the challenges and future research interests of PI dielectrics for high-temperature energy storage are summarized.

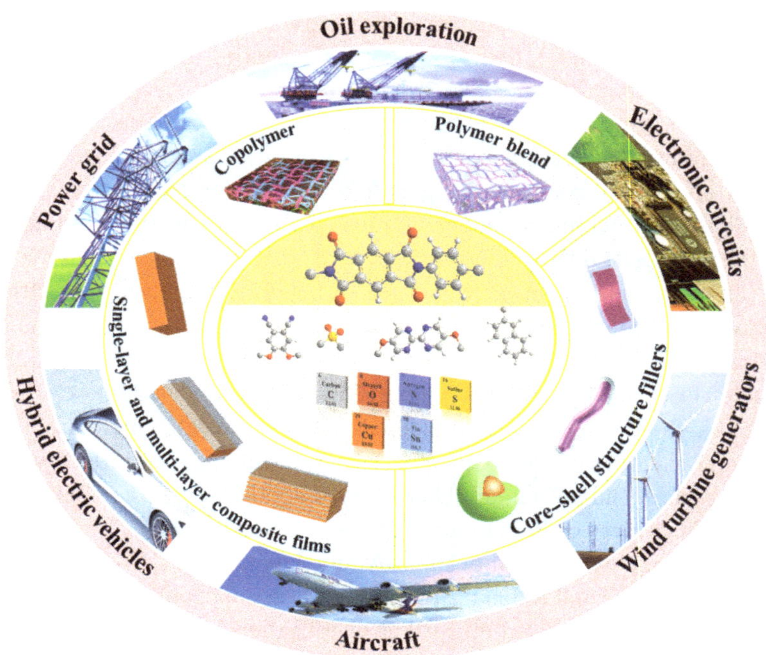

Figure 4.1 Structure and application of PI-based dielectrics for high-temperature capacitive energy storage [26].

4.2 Key Characteristics and Mechanisms of Polyimide Dielectrics for Energy Storage

4.2.1 Dielectric Energy Storage Mechanism

As shown in Figure 4.2a, under the action of an applied electric field, polarization occurs inside the dielectric, and energy is stored in the dielectric in the form of the electric field [27, 28]. The energy storage capacity of a dielectric capacitor is related to the capacitance (*C*) and the polarization value (*P*) of the material, as well as the applied electric field strength (*E*). The *C* can be expressed as the incremental change of charge (*Q*) with respect to voltage (*V*), as follows:

$$C = \frac{\mathrm{d}Q}{\mathrm{d}V} = \varepsilon_0 \varepsilon_r \frac{A}{d} \tag{4.1}$$

where ε_0 is the vacuum dielectric permittivity, *A* is the area, and *d* is the thickness of the dielectric material. Thus, the energy storage density of the dielectric material can be expressed in terms of energy stored per unit volume (*U*) as follows:

$$U = \frac{W}{Ad} = \frac{1}{Ad} \int_0^Q V\mathrm{d}Q = \int_0^D E\mathrm{d}D$$
$$\frac{Q}{A} = D = \varepsilon_0 \varepsilon_r E, E = \frac{V}{d} \tag{4.2}$$

Charge and discharge of the capacitor can be achieved by increasing or decreasing the applied electric field. For a linear dielectric, the equation can be simplified as:

$$U_{\mathrm{d}} = \frac{1}{2} \varepsilon_0 \varepsilon_r E^2 \tag{4.3}$$

In summary, high ε_{r} and high breakdown field (E_{b}) are necessary to increase the energy density of dielectric materials. For nonlinear dielectrics such as ferroelectrics, relaxation ferroelectrics, or antiferroelectrics, the charge energy density (*U*) and the discharged energy density (U_{d}) can be calculated by integrating the potential shift–electric field strength curves (*D-E* curves, Figure 4.2b) with the following equations:

$$U_{\mathrm{d}} = \int_{D_r}^{D_{\max}} E\mathrm{d}D \tag{4.4}$$

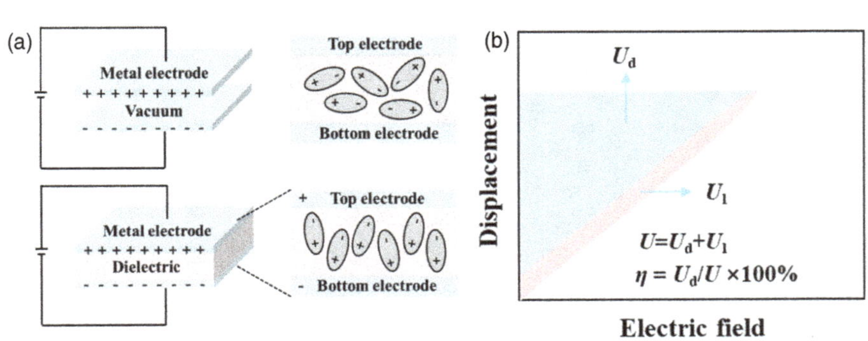

Figure 4.2 (a) Schematic diagram of electric polarization of dielectric capacitor [28]. (b) Electric displacement–electric field hysteresis loop.

where D_{max} is the maximum electrical displacement and D_r is the residual electrical displacement [29, 30]. The energy storage properties of dielectrics are mainly evaluated by the discharged energy density and the energy storage efficiency. The energy storage efficiency (η) is the ratio of U to U ($\eta = U_d/U$). U_l is the energy loss during the charge–discharge cycle, which mainly includes the polarization loss generated by overcoming the intermolecular force during the dipolar polarization and the conduction loss generated by the conduction current inside the dielectric materials.

4.2.2 Electrical Polarization and Dielectric Permittivity

The dielectric permittivity is usually defined as the ratio of the relative permittivity to the vacuum permittivity of a dielectric materials, and consists of a real permittivity and an imaginary permittivity, as follows [31]:

$$\varepsilon_r = \varepsilon_r' - i\varepsilon_r'' \tag{4.5}$$

The real permittivity represents the capacitance of the material, whereas the imaginary permittivity is related to the loss due to polarization that does not respond to frequency changes. In addition, the ε_r is related to the polarization behavior. Polarization is defined as the total dipole moments per unit volume of the dielectric material. P is related to ε_r as follows:

$$P = (\varepsilon_r - 1)\varepsilon_0 E \tag{4.6}$$

According to the relaxation and resonance mechanisms, there are four main types of polarization in dielectric materials including electronic (P_e), ionic (P_i), dipolar (P_d), and interfacial (P_{int}) polarization [26, 32] (Figure 4.3). Polarization is closely related to the motion of the molecular chains and the frequency of the applied electric field. Each polarization is associated with a dielectric loss at a specific frequency.

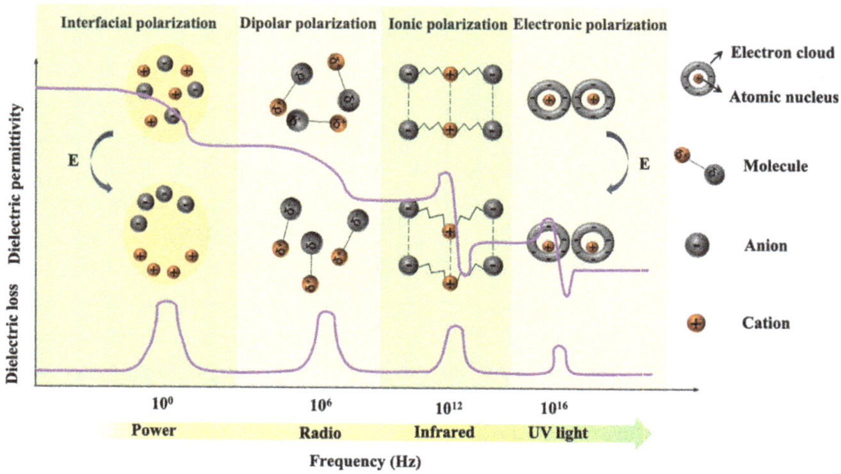

Figure 4.3 Polarization types and their frequency dependencies of dielectric materials [26].

In the absence of an electric field, the positive and negative charge centers of atoms coincide, indicating an electrically neutral state. Under the action of an electric field, the displacement of the electron cloud leads to a change in the dipole moment, a process known as electronic polarization. Electronic polarization is widespread in all dielectric materials, but the resulting dipole moment is very small [33].

Without an external electric field, solid dielectrics do not exhibit net polarization due to the electrostatic attraction of cations and anions [31]. The anions move toward the positive plate and the cations toward the negative plate under an external electric field. The polarization produced in this case is called ionic polarization. Ionic polarization produces a large dipole moment, which allows the material to achieve a high polarization value. This type of polarization is found mainly in ionic crystals.

Dipolar polarization is produced by molecules or chains of molecules, which usually have an inherent dipole moment [28]. When no electric field is applied, the dipole moment in a molecule or chain of molecules containing an inherent dipole is oriented randomly and the total dipole polarization is zero. In the presence of an applied electric field, the dipoles in the molecular chain are rotated and oriented in the direction of the electric field, thus exhibiting a dipolar polarization. Increasing dipole density by introducing polar groups into polymer molecular chains is a common method to increase dipolar polarization.

Interfacial polarization occurs mainly at the phase interface of heterogeneous dielectric materials [34, 35]. In the presence of an applied electric field, space charge accumulates at the interface, leading to interfacial polarization. Multiple factors affect the interfacial polarization, including the interface between the crystalline–amorphous state, the interface between the film and the electrode, and internal defects in the material. Increasing the phase interface within the polymer by adding fillers can enhance the interfacial polarization.

According to the Clausius–Mossotti equation [36], the relationship between ε_r and the molecular polarization (α) is given by

$$\frac{\varepsilon_r - 1}{\varepsilon_r + 2} = \frac{1}{3\varepsilon_0}\sum_j N_j\alpha_j \tag{4.7}$$

where N is the number of molecules per unit volume of dielectric. Therefore, the value of ε_r can be adjusted by changing the α or N of the dielectric.

4.2.3 Dielectric Breakdown Strength

The lowest critical voltage that causes breakdown is called the breakdown voltage or breakdown strength. To determine the dielectric breakdown strength, a two-parameter Weibull distribution function is widely used for analysis [27, 37]. The cumulative function of the Weibull distribution is expressed as follows:

$$P(E_i) = 1 - \exp\left[-\left(\frac{E_i}{E_b}\right)^{\beta}\right] \tag{4.8}$$

where $P(E_i)$ refers to the cumulative probability of electrical failure, E_i is the experimental breakdown strength, E_b represents the Weibull characteristic breakdown strength with a cumulative failure probability of 63.2%, and the shape parameter β related to the scatter and reliability of the data.

Although the complex mechanisms of dielectric breakdown behavior are not fully understood, some traditional breakdown theories have been used to explicate the breakdown process in dielectric materials [38]. Depending on the time after a voltage is applied to the material, the type of polymer breakdown can be categorized into short-term and long-term breakdown [31, 38], as shown in Figure 4.4a. Short-term breakdown includes electrical, thermal, and electromechanical breakdowns, while long-term breakdown includes partial discharge, electrical tree, and water tree.

Electrical breakdown includes electronic breakdown and avalanche breakdown [26] (Figure 4.4b). Electronic breakdown is related to the E_g of the dielectric. Electrons obtain enough energy from the electric field to transition from the valence band to the conduction band. As electrons accumulate in the conduction band, it results in electronic breakdown. Electronic breakdown increases rapidly as the electric field increases, causing the dielectric to lose its insulating properties. Electrons gain energy as they move through the electric field and collide with atoms. When the energy of the collision exceeds the ionization energy of the atom, the bound charge on the atom is released. The process of collisional ionization is repeated to form an avalanche of electrons, which ultimately leads to the avalanche breakdown of the dielectric [39].

Thermal breakdown often occurs at low electric fields, depending on the dielectric's conductivity, shape and size, operating temperature, and thermal conductivity. A dielectric under an electric field generates heat due to energy loss.

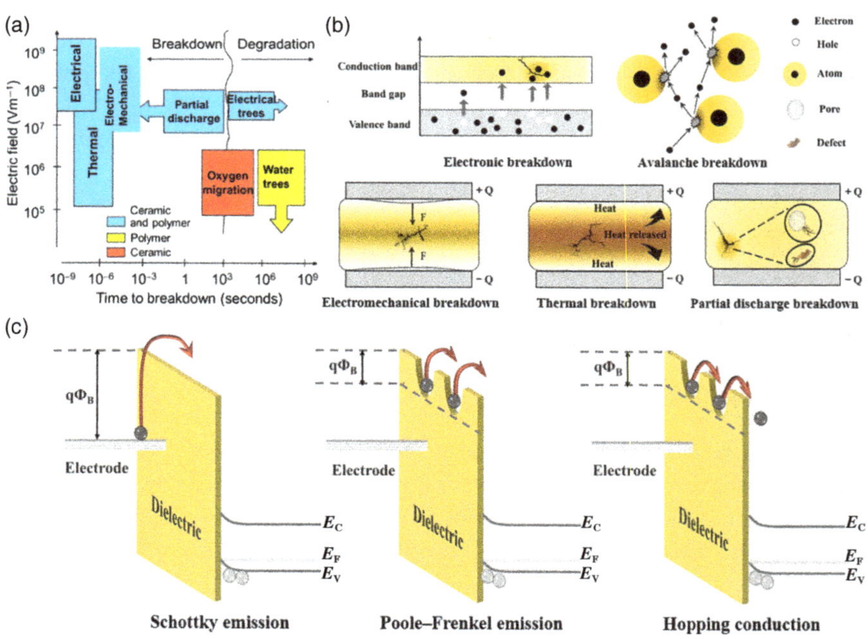

Figure 4.4 Schematic diagram of (a) breakdown types [31], (b) electric breakdown mechanisms, and (c) conduction loss mechanism of dielectric materials [26] ($q\Phi_B$ is the barrier height, E_c, E_v, and E_F are conduction band, valence band, and Fermi level, respectively).

When the electric field increases to the point where the rate of heat generation is greater than the rate of heat dissipation, the internal temperature of the dielectric rises even higher, resulting in a loss of insulation and irreversible damage [26] (Figure 4.4b). The relationship between the heat generation rate and heat dissipation rate is described by the heat and current continuity equation, which is given as follows [40]:

$$C_v \frac{\mathrm{d}T}{\mathrm{d}t} - \nabla \cdot (\lambda \cdot \nabla T) = \sigma E^2 \tag{4.9}$$

where C_v is the specific heat per unit volume, T is the temperature of the dielectric, and K is the thermal conductivity. It can be found that the breakdown field strength of dielectric materials decreases exponentially with the increase in temperature. Therefore, thermal breakdown can be mitigated by suppressing the conductivity loss of the dielectric material at elevated temperatures and improving the thermal transfer efficiency of the material [41]. PI dielectrics are commonly used at high temperatures but have shown fatal problems in application. For example, PI exhibits good dielectric properties at room temperature, but at high temperatures the conduction current increases rapidly, resulting in a dramatic increase in dielectric loss. When the temperature continues to rise and reaches a critical current density, this leads to a breakdown of the PI dielectric material. Therefore, thermal breakdown is the key mechanism of PI breakdown.

Under an applied electric field, the attraction between the positive and negative charges on both sides of the dielectric produces a static compressive force on the dielectric. When this electrostatic compressive force exceeds the compressive capacity of the dielectric, resulting in electromechanical breakdown [26, 39] (Figure 4.4b). Mechanical breakdown (E_EM) is related to Young's modulus (Y) as follows:

$$E_\mathrm{EM} = 0.606\sqrt{\frac{Y}{\varepsilon_0 \varepsilon_r}} \tag{4.10}$$

From the above equation, it can be seen that Young's modulus is positively correlated with electromechanical breakdown and that Young's modulus of a polymer material usually decreases significantly near the T_g or T_m. Therefore, electromechanical breakdown e tends to occur more at high temperatures.

Structure defects such as internal pores in the material lead to partial discharge and partial discharge breakdown in PI dielectric materials [26] (Figure 4.4b). Short-term partial discharge has minor effect on the insulating properties of the dielectric material. However, long-term partial discharge can result in electrical treeing and degradation of the polymer material, causing a decrease in the overall dielectric insulation properties.

4.2.4 Dielectric Loss

The dielectric produces partial energy loss during the polarization and depolarization cycles [42]. Electronic polarization and ionic polarization occur in the ultrahigh-frequency ranges of the UV light and infrared, respectively, and the loss is negligible [32]. Loss due to dipolar and interfacial polarization is the main type of polarization loss in polymer dielectrics. Dipolar polarization loss occurs at lower frequencies (radio level), and peak loss from interfacial polarization occurs in the lowest frequency range (power level). At low frequency, the polarization of the dielectric responds quickly to the change in the alternating electric field, and polarization loss is low. At high frequency, the polarization of the dielectric responds slowly to the change in the external electric field and hysteresis occurs, resulting in a significant increase in polarization loss. The dipolar polarization loss is caused by the friction that occurs when the dipole is oriented in the electric field. The dipolar polarization loss of dielectric materials can be reduced by providing enough space for the rotation of the groups to reduce the friction [43, 44].

Polymer dielectrics generate leakage currents that lead to conduction loss under an applied electric field. With the combined increase in temperature and electric field, the energy loss increases exponentially, which not only significantly reduces charge and discharge efficiency, but also leads to thermal breakdown of the dielectric material. Therefore, the suppression of high-temperature dielectric conduction loss has become a key strategy to improve high-temperature storage properties. Depending on the source of the leakage current, dielectric conduction can be divided into two categories [45, 46]. Conduction mechanisms that depend on the electrical properties at the electrode–dielectric interface are known as electrode-limited conduction mechanisms and include Schottky emission, Fowler–Nordheim tunneling, direct tunneling, and thermionic field emission. Conduction mechanisms that depend on the properties of the dielectric itself are called bulk-limited conduction mechanisms and include Poole–Frenkel emission, hopping conduction, ohmic conduction, space charge-limited conduction, and ionic conduction.

At high temperatures and high electric fields, the dielectric loss is mainly caused by conduction loss. The conduction mechanism transforms from linear ohmic conduction to a nonlinear conduction mechanism dominated by Schottky emission, Poole–Frenkel emission, and hopping conductance, as shown in Figure 4.4c.

Schottky emission, as the most common conduction mechanism, belongs to the electrode-limited conduction mechanism. As the ambient temperature increases, carriers in the electrode gain energy to overcome the energy barrier at the electrode/dielectric interface and are injected into the dielectric to participate in the formation of the leakage current. Poole–Frenkel emission and hopping conduction are bulk-limited conduction mechanisms, which are determined by the electrical properties of the polymer dielectric. Poole–Frenkel emission means that the charge trapped by the trap in the dielectric acquires sufficient energy to overcome the barrier with the increase of the applied electric field, becomes free charges, and participates in the conduction process. Carriers captured by a trap in the dielectric "hop" into another trap through a tunneling effect, creating a leakage current, which can lead to hopping conduction. Poole–Frenkel emission corresponds to the thermionic effect and hopping conduction corresponds to the tunneling effect. For Poole–Frenkel emission, carriers can overcome the trap energy barrier by the hot electron mechanism. However, for hopping conduction, when the energy obtained by the carriers is lower than the maximum energy of the potential between the two traps, it is still possible to transfer through the tunneling mechanism.

When the temperature is higher than a certain threshold, the increase in temperature leads to an exponential increase in the leakage current within the polymer dielectric. In some cases, the damage to the dielectric material is more severe at elevated temperatures than at electric field. Increasing the Schottky barrier at the interface between the electrode and the dielectric, introducing deep trap energy levels inside the dielectric, and decreasing the density of defects inside the dielectric can effectively suppress the conduction loss in the dielectric material, thus effectively improving the breakdown and energy storage properties of the dielectric material at high temperatures and high electric fields.

4.2.5 Breakdown Self-Healing Mechanism

The electrodes of a metalized film capacitor (MFC) consist of a very thin layer of metal (usually aluminum or zinc-aluminum alloy) deposited onto a dielectric film. The extremely thin deposited electrode (~nm) is the basis for the successful self-healing of the MFC. The metalized film self-healing process is shown in Figure 4.5. The dielectric film of the MFC contains lattice defects or impurities that exhibit lower breakdown field strengths than the surrounding areas. When the applied voltage increases, the electrical weakness of the dielectric film undergoes breakdown to form a discharge channel, the charge through the discharge channel forms a large current, generating Joule heat, so that the surrounding metal layer is heated by the evaporation and diffusion outward, the metal vapor is easy to ionize

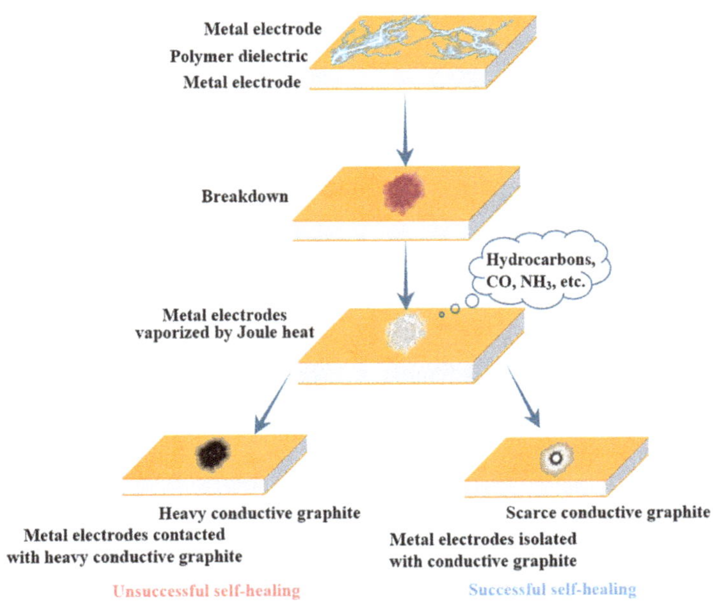

Figure 4.5 Schematic diagram of self-healing process of metalized dielectric film.

the formation of a plasma, along with the expansion of the area of evaporation, the plasma discharge electric arc extinguished, the MFC to restore the insulation, a process known as the "self-healing" [47]. After successful self-healing, the MFC can continue to work. Since the self-healing process will cause the MFC to lose a portion of the evaporation electrode, the capacitance of the MFC will gradually decrease with the increase in the number of times of operation [48]. Factors such as applied voltage, capacitance, electrode thickness, interlayer pressure, and thermoforming process play an important role in the "self-healing" of the MFC, while the chemical properties of the polymer film dielectrics are also critical [49]. The ratio of hydrocarbon content in the molecular structure of the dielectric material has an important influence on the reliability of self-healing [50]. A low oxygen content dielectric material may produce graphite molecules deposited around the breakdown point during the combustion of the arc, resulting in incomplete self-healing and short-circuiting, and high oxygen content dielectric material can effectively avoid the above situation [51]. Polypropylene is a non-polar material and a low hydrocarbon ratio will not produce almost no graphite deposition and can achieve the successful self-healing [52].

4.3 Intrinsic Polyimide Dielectrics

PIs are typically prepared by condensation polymerization of electron-donating diamine and electron-withdrawing dianhydride, and this particular structure endows PIs with excellent structural tunability, and designing new PI dielectric materials from the intrinsic structure is the fundamental approach for improving its high-temperature energy storage properties.

4.3.1 Dipolar Glass Polyimide Dielectrics

The dramatic increase in conduction loss due to the action of high temperatures and high electric fields is partly attributed to the large-scale movement of the polymer backbone. To overcome this problem, dipole glass polymers have been proposed, as the main chain motion is largely frozen to limit the dielectric loss [53]. At the same time, the polar groups of dipolar glass polymers are rotated locally to increase the orientational polarization and ε_r values. However, high T_g dielectric polymers with aromatic and heterocyclic backbone structures have $\varepsilon_r < 4$ at 1 kHz. In recent decades, dielectric polymers composed of different polar groups have been developed, based on non-polar and polar polymer backbones, which can be classified into main-chain and side-chain dipole-grafted polymers based on the location of the dipole moieties [54]. Typical polar units include hydroxyl (—OH), carbonyl (—C[=O]—), nitro (—NO_2), cyano (—CN), sulfone (—SO_2—), and thiourea (—NHC[=S]NH—) groups with increasing dipole moments from ~1.7D, ~2.3D, ~3.6D, ~3.8D, ~4.3D to ~4.9D.

Zhu et al. [55] designed and synthesized a novel diamine monomer with two —CN groups located ortho to the benzene ring. The —CN exhibits the advantage of enhanced dipole polarization, and the small spatial volume hardly affects the stacking density of the molecular chains. The resulting polyimide has a high dielectric permittivity of 4.8, a low dielectric loss of 1.57×10^{-3}, and a maximum discharged energy density of 1.02 J/cm^3. Subsequently, Zhang et al. [56]. prepared sulfone-containing polyimide with high dielectric permittivity (STP-PI) and similar structure polyimide without sulfone group (TP-PI) by monomer design and condensation polymerization reaction (Figure 4.6a). Compared with the model compound of TP-PI (TP-MC), which shows a uniform distribution of surface electrostatic potential, the positive and negative charges on the side chains of the model compound of STP-PI (STP-MC) are separated into dipoles, and the polarization of the dipoles is enhanced, resulting in a higher ε_r. As shown in Figure 4.6b, STP-PI presents a ε_r of 6.0 at room temperature and 100 Hz, which is much higher than the 3.6 of TP-PI (Figure 4.6c). As a result, the maximum discharged energy density of SBP-PI at room temperature reaches 8.16 J/cm^3. In addition, the rigid backbone endows STP-PI excellent thermal stability, and STP-PI obtains a high discharged energy density of 3.68 J/cm^3 with an efficiency of 84% at 200 °C (Figure 4.6d).

4.3.2 Wide Bandgap Polyimide Dielectrics

Conventional PI dielectrics feature high T_g value and excellent thermal stability attributed to the abundance of conjugated aromatic structures in the molecular

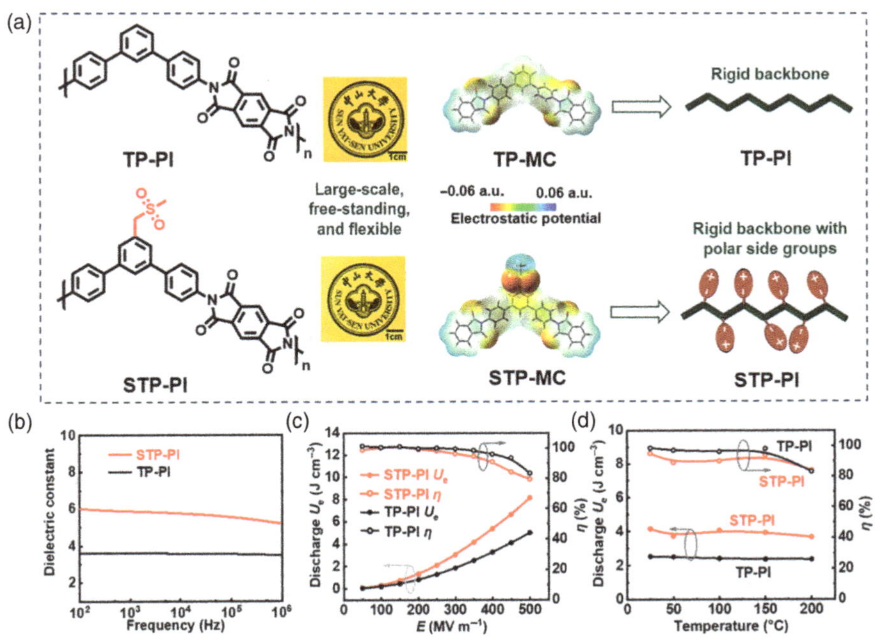

Figure 4.6 (a) Left: Chemical structures of TP-PI and STP-PI and digital photographs of the prepared films. Right: electrostatic potential distributions of TP-MC and STP-MC calculated based on DFT, and the structures of STP-PI and TP-PI [56]. (b) Dielectric permittivity of TP-PI and STP-PI films at room temperature. (c) Energy storage properties at room temperature. (d) Discharged energy density and charge–discharge efficiency at different temperatures.

backbone [20]. However, the charge transfer effect between the conjugated backbones leads to a decrease in E_g and the energy storage properties, especially at high temperatures and high electric fields. Recent studies have shown that the E_g affects the conductivity and intrinsic E_b of polymer dielectrics [57–60].

The flexible structural groups [61], bulky groups [62], and alicyclic structure fragments [63] are incorporated into the PI molecular backbone can effectively disrupt the strong conjugation between aromatic rings and regulate the charge transfer effect, resulting in PI dielectrics with a wide E_g. However, the introduction of these groups and structures may result in a decrease in the T_g of PI dielectrics. Wang et al. [61] designed and prepared 12 PI-derived polymers by machine learning screening. The nomenclature of each PI-derived polymer depends on the modification of the molecular structure of classical PI or PEI, e.g., PI-methyl refers to the structure formed by replacing the ether bond in classical PI with a methylene bond. To evaluate the effect of structural unit fragments on the E_g and T_g values of PI-derived polymers, the 12 PI-derived polymers were divided into four groups for analysis. In the first group (Group 1), which includes PI, PI-m, PEI, and PEI-p, the main comparison is between the meta- and para-substituted benzene rings in the polymers. The second group (Group 2), which includes PI, PI-methyl, PI-ether, and PI-iso, is designed to show the differences between ether, methylene, and isopropyl bonds in the E_g and T_g values of the polymers. The third group (Group 3), including PI, PI-oxo, PI-oxo-m, and PI-oxo-iso, focuses on the comparison of flexible and rigid dianhydride units. The fourth group (Group 4), to understand the effect of benzene ring modification on the properties of polymers, includes grafting of side groups onto the benzene ring (PEI-o-M) or replacement of the benzene ring with a saturated alicyclic ring (PI-cyc-iso). In the analyses of Group 1, although the T_g values of the polymers containing p-benzene structural fragments were significantly larger than those of the polymers containing m-benzene structural fragments (e.g., 360.5 °C for PI, 349.3 °C for PI-m, 232.2 °C for PEI-p, 221.3 °C for PEI), their E_g values were very similar (e.g., 2.77 eV for PI, 2.75 eV for PI-m, 3.33 eV for PEI-p, and 3.29 eV for PEI). The results of group 2 show that the E_g and T_g of isopropyl structural fragments in diamine monomers are slightly higher than those of ether structural units, e.g., the E_g and T_g of PI -iso are higher than those of PI-ether by 0.16 eV and 22.2 °C, respectively. In contrast, the introduction of the ether structural fragment into the dianhydride monomer resulted in a dramatic change in E_g (increase of 0.53 eV) and T_g (decrease of 94.1 °C), as can be seen from the Group 3 comparison. Group 4 comparisons revealed that the side groups on the benzene ring favored the simultaneous improvement of T_g and E_g (increase of 17.2 °C and 0.04 eV, respectively), and the substitution of the benzene ring by the cyclohexane structure caused a significant increase in E_g (increase of 1.16 eV) and a significant decrease in T_g (decrease of 49.7 °C). Based on the results of the above experiments, the exact effect of the addition (or replacement) of each structural unit on the E_g and T_g values of the PI-derived polymers can be extracted and quantitatively compared with those of plain PI or PEI (Figure 4.7a). Subsequently, Dong et al. [64]. prepared a semi-aliphatic PI copolymer (SO-PI-14.3) by incorporating alicyclic structures into sulfonated polyimide (SO-PI) through molecular structure

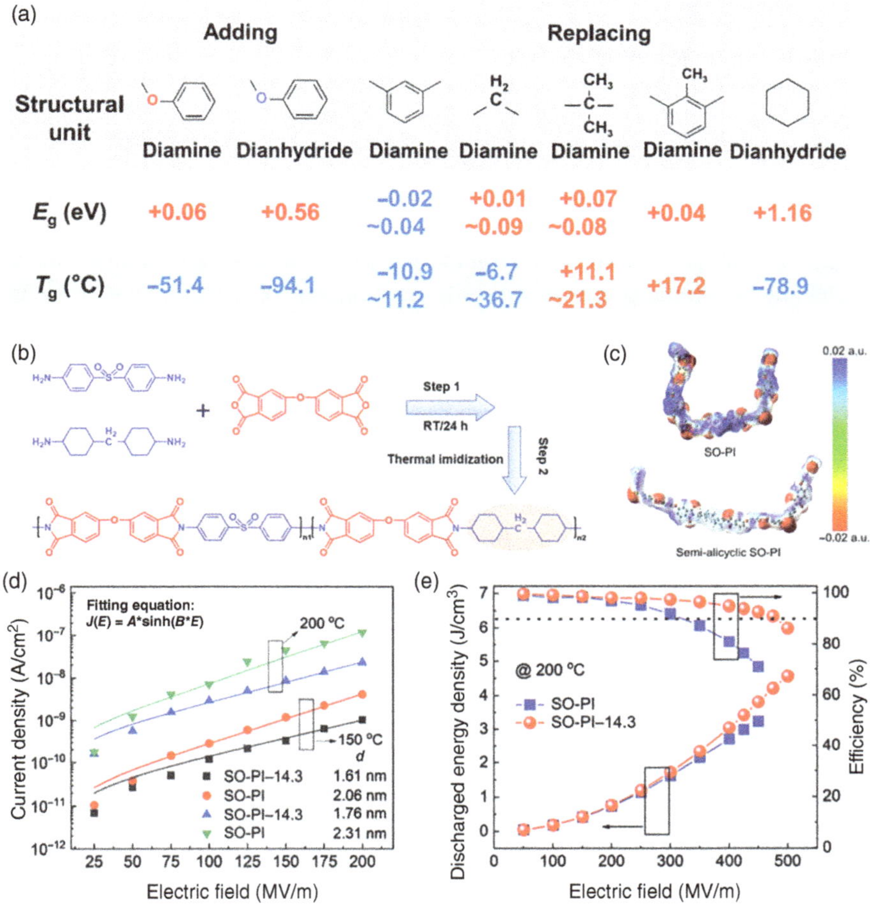

Figure 4.7 (a) The relationship between structure with E_g and T_g [61]. (b) Schematic diagram for the synthesis route of SO-PI–14.3. (c) Electrostatic potential distribution. (d) Current density of SO-PI and SO-PI-14.3 at 150 °C and 200 °C. (e) Energy storage properties for SO-PI and SO-PI-14.3 at 200 °C [64].

design (Figure 4.7b). The alicyclic structure restricts the electron delocalization of the SO-PI-14.3 molecular skeleton and effectively inhibits CT interactions, reducing the current density and achieving a wide E_g of 3.4 eV (Figure 4.7c,d). Ultimately, SO-PI-14.3 attains a U_d of 4.3 J/cm^3 and a η of more than 90% at 200 °C and 485 MV/m (Figure 4.7e).

4.4 All-Organic Polyimide Composite Dielectrics

All-organic dielectrics have attracted much attention because of their facile preparation and suitability for large-scale production. In this section, all-organic polyimide composite dielectrics will be introduced in terms of polymer blending, doping with organic molecular semiconductor fillers, and designing multilayer structures.

4.4.1 Blended Polyimide-Based Composite Dielectrics

Electrostatic interactions enable the blend chains to be compactly stacked to reduce the weak points, resulting in higher breakdown strength. Zhang et al. [65]. improved the chain stacking morphology by utilizing the strong electrostatic interactions between the PI and PEI molecular chains to reduce the voids and free volume (Figure 4.8a). As a result, the breakdown properties of PEI/PI blended polymers were significantly improved, with the 1 : 1 ratio of PI/PEI blends achieving a breakdown strength of up to 1 GV/m at room temperature and 550 MV/m at 200 °C. Meanwhile, its discharged energy density was 1.4 J/cm^3 at 200 °C and 450 MV/m. In addition, Ding et al. [66]. incorporated poly(arylene ether urea) (PEEU) into the PEI matrix to prepare all-organic PEEU/PEI blended films. The PEEU/PEI blended film with 15 wt% PEEU has a high breakdown strength of 672 MV/m at 150 °C. The discharged energy density can reach 4.84 J/cm^3 at 150 °C and 550 MV/m, and the η is higher than 90%.

4.4.2 Polyimide-Based Composite Dielectrics Doped with Organic Semiconductors

Molecular semiconductors usually have high electron affinity energies and can construct traps in polymer dielectrics to capture free electrons, thus suppressing leakage current and increasing the energy density of the composite films (Figure 4.8b). Yuan et al. [67] chose three molecular semiconductors (ITIC, PCBM, and DPDI) as organic fillers, and blended them with PEI, to explore their effects on the high-temperature energy storage properties of PEI. The molecular semiconductors formed deep traps in the composites, which were able to capture the injected and excited electrons by strong electrostatic attraction (Figure 4.8c). The discharged energy density of the composites was up to 3.5 ~ 4.5 J/cm^3 at 150 °C with a η of more than 90%, whereas that of PEI is only 1 J/cm^3 at same temperature (Figure 4.8d). Subsequently, Zhang et al. [68]. used the N-type molecular semiconductor 1,4,5,8-naphthalene tetracarboxylic dianhydride (NTCDA) blended with PEI. The N-type molecular semiconductor NTCDA can form electron traps, which reduces the leakage current, and improves the breakdown strength and energy storage properties of the composites at high temperatures. The breakdown strength of the NTCDA/PEI blended film containing 0.5 vol% NTCDA was as high as 510 MV/m at 200 °C, with a discharged energy density of 3.23 J/cm^3 and a η of 85%.

4.4.3 All-Organic Multilayer Polyimide-Based Composite Dielectrics

Due to the diversity between the electrical properties and molecular structures of heterogeneous polymer, the molecular energy level at the interface of multilayer polymer dielectrics is different, which forms a deep trap and accumulates space

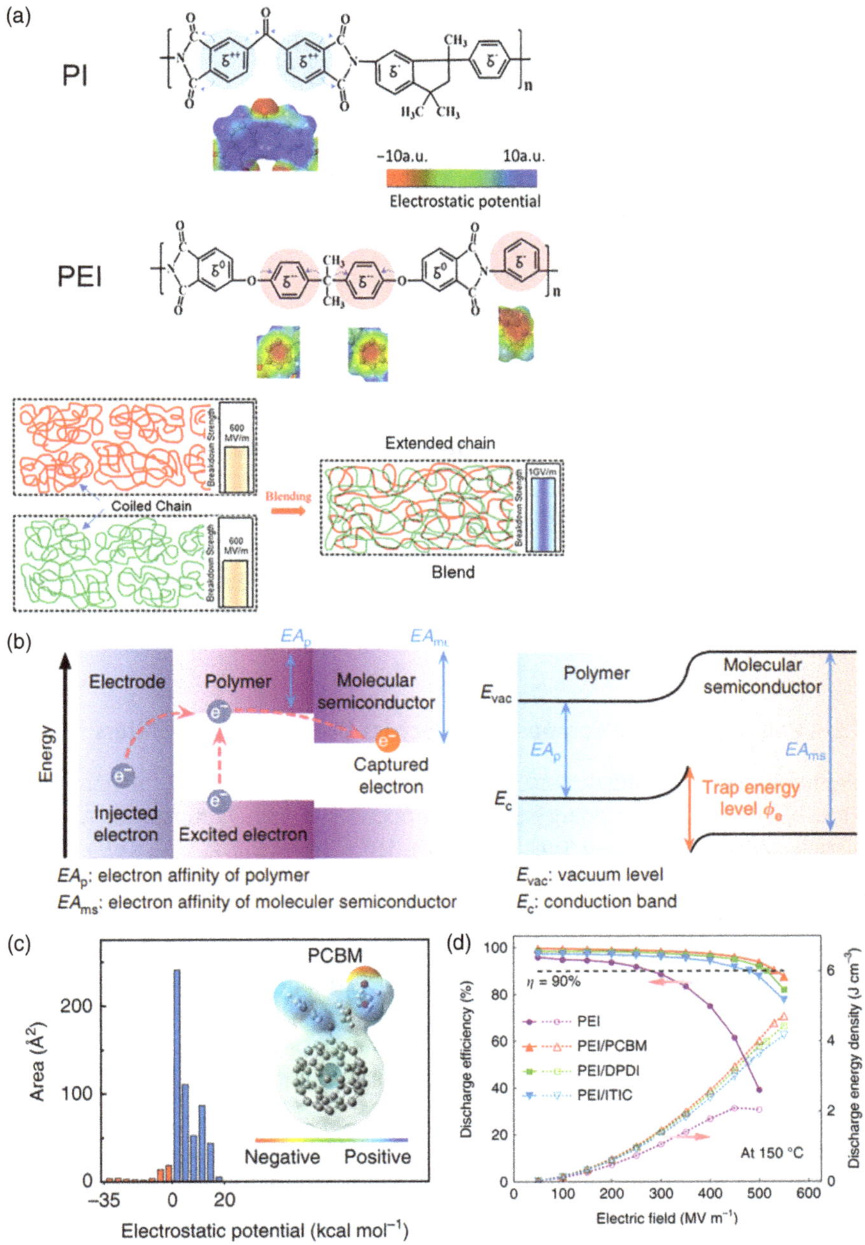

Figure 4.8 (a) Structures of PI and PEI, and schematic diagrams of interchain electrostatic interactions [65]. (b) Charge transfer mechanism in composites and trap energy levels introduced by molecular semiconductors. (c) Surface electrostatic potential distribution of molecular semiconductors. (d) Energy storage properties of polyimide/molecular semiconductor composites at 150 °C [67].

charge at the interface, inhibiting the migration and injection of carriers. Niu et al. [69]. prepared PEI/PI/PEI sandwich films, which were then hot-pressed to improve the adhesion between the interfaces. The changes in polymer structure at the interface were characterized by atomic force microscopy combined with infrared spectroscopy, and it was found that the swelling effect of PEI solution on the surface of the PI film caused a gradient change of molecular content of the two polymers at the interface. This swelling effect increases the deep trap density in the sandwich-structured film and improves the ability to trap electrons. Because the carrier transport is suppressed, the energy storage capacity at high temperatures is improved, and the film can reach U_d of 2 J/cm^3 with η higher than 90% at 200 °C.

4.5 Polyimide-Based Nanocomposite Dielectrics

Improving the high-temperature energy storage properties of PI by exploiting the characteristics of inorganic nanomaterials is an effective way [70]. In this section, filling the polyimide matrix with various nanofillers and designing a multilayer structure will be presented to improve the energy storage properties of polyimide dielectrics.

4.5.1 Polyimide-Based Composite Dielectrics Doped with Nanofillers

To systematically investigate and analyze the effects of ε_r and E_g of nanofillers on the high-temperature energy storage properties of nanocomposites, Ai et al. [71]. selected Al_2O_3 ($\varepsilon_r = 9.5$, $E_g = 8.6$ eV), HfO_2 ($\varepsilon_r = 25$, $E_g = 5.8$ eV), BNNS ($\varepsilon_r = 4$, $E_g = 5.97$ eV), and TiO_2 ($\varepsilon_r = 110$, $E_g = 3.5$ eV) as nanofiller phases to prepare a series of nanocomposites, respectively. It was found that the E_b of the composites increased significantly with the increase of the E_g of the nanofiller (Figure 4.9a), and the nanofiller with higher ε_r exhibited the lowest E_b strength although it could increase the ε_r of the composites more effectively. Similarly, the volume conductivity of PI/TiO_2, PI/HfO_2, PI/BNNS, and PI/Al_2O_3 nanocomposites decreased by 41.1%, 83.2%, 89.9%, and 93.9%, respectively, with the increase of the E_g of the nanofiller at 150 °C compared with that of pure PI. Owing to the wide E_g of Al_2O_3, the conduction loss of PI/Al_2O_3 nanocomposites remains at the lowest level with the increase of the electric field strength and has the highest energy storage efficiency (Figure 4.9b). The thermally stimulated depolarization current (TSDC) test and hopping conduction model further show that the charge-trapping ability is closely related to the E_g of the nanofiller, which is enhanced with the increase of the E_g, which significantly reduces the leakage current of the composites (Figure 4.9c,d).

However, the high content of nanofillers tends to agglomerate in the polymer matrix, resulting in defects within the polymer, which in turn reduces the breakdown strength of the composite. In recent years, researchers have designed a core–shell structure to solve the compatibility problem between the nanofillers

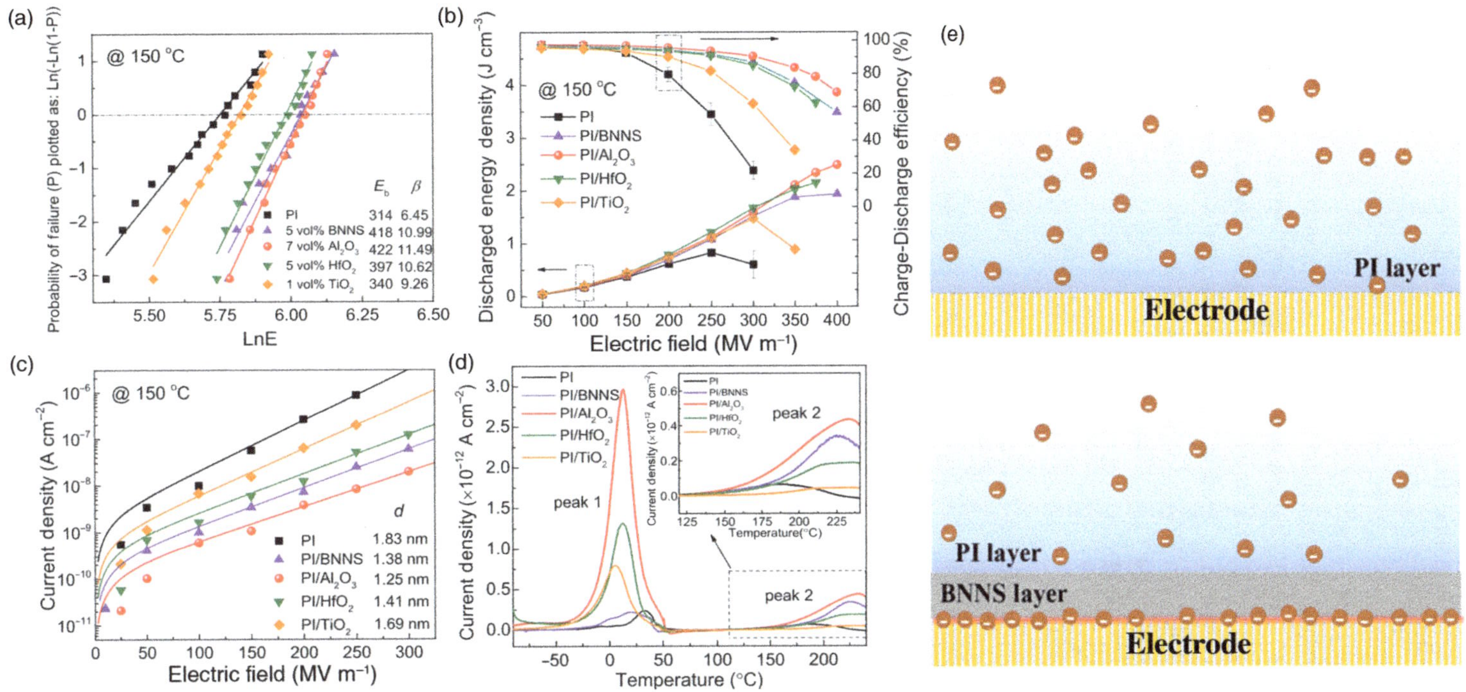

Figure 4.9 (a) Breakdown strength, (b) energy storage properties, (c) current density, and (d) TSDC spectra of PI and PI nanocomposites measured at 150 °C [71]. (e) Schematic diagram of the charge barrier mechanism at the electrode/dielectric interface [72].

and the polymer matrix, and to make the nanofillers uniformly distributed in the polymer matrix, which reduces the occurrence of agglomeration [73]. Yuan et al. [74]. prepared small-sized two-dimensional inorganic ceramic fillers of $BaTiO_3$ by using solid-phase and hydrothermal methods and formed an organic shell layer on the surface of $BaTiO_3$ by chemical self-polymerization of polydopamine (PDA). The organic shell layer solved the interfacial compatibility problem, and the horizontally ordered $BaTiO_3$@PDA was able to provide tortuous breakdown paths, construct an effective conduction barrier, and limit the charge migration and the growth of electric trees. The good interfacial bonding between $BaTiO_3$@PDA and PI matrix promoted the interfacial polarization of the nanocomposites, which led to the improvement of the energy storage properties of the nanocomposites. As a result, the PI/ $BaTiO_3$@PDA nanocomposites had a U_d reached 3.34 J/cm^3 with a η of 83.68% at 150 °C.

4.5.2 Multilayer Polyimide-Based Nanocomposite Dielectrics

The large difference in dielectric properties between the high ε_r filler and the polymer matrix leads to a decrease in breakdown strength. To solve the contradiction between dielectric permittivity and breakdown strength, Wang et al. [75]. designed a bilayer structure of TiO_2-PEI/BNNS-PEI film. The BNNS with wide E_g and TiO_2 nanorod with high ε_r dispersed in different layers can simultaneously enhance the ε_r and E_b of the bilayer composite. Thus, the U_d of TiO_2-PEI/BNNS-PEI bilayer film can reach 5.3 J/cm^3 with an efficiency of 90% at 150 °C.

The incorporation of highly insulating fillers into the polymer matrix can hinder the charge conduction and improve the breakdown field strength and energy density of multilayer composites. In particular, a dense and continuous inorganic layer can more effectively inhibit the development of electrical damage, resulting in higher breakdown field strength and lower energy loss of multilayer composites (Figure 4.9e). Liu et al. [72] prepared BNNS-PI bilayer films by solution casting method. The BNNS acted as a charge-barrier layer, which inhibited charge injection into the electrode/dielectric interface and reduced the leakage current, thus improving the energy storage properties of the composites at high temperatures. The discharged energy density of the optimal BNNS-PI bilayer film was 4.37 J/cm^3 and the η was 92% at 200 °C and 500 MV/m.

Compared to polymer composites made by simply blending polymers and fillers, the advantages of core–shell packings and multilayer composites designed to take into account the micromechanics of composites and interfacial coupling effects are much more pronounced. By optimizing the packing and its distribution, the leakage current can be suppressed and the energy loss reduced, further improving the storage properties of PI at elevated temperatures. However, for organic–inorganic two-phase composites, the realization of large-scale production is a great challenge, and attention should be paid to reducing the filler content to maintain good flexibility and processability under the premise of ensuring good energy storage properties.

4.6 Self-Healing Polyimide Dielectrics

The concept of self-healing can be traced back to the study of metal-oxide-silicon capacitors by Klein [76] and polymer film capacitors by Reed et al. [52]. As examined by Kammermaier et al. [77, 78] in their self-healing studies, the dielectric breakdown can lead to the formation of residual carbon deposits. These residual carbon deposits and other healing scars can limit the recovery of mechanical or electrical strength relative to the original material before the initial breakdown. Zhang et al. [79] demonstrated the deleterious effects of residual defects or scars after self-healing based on a finite element model. The decrease in dielectric strength of the material was attributed to residual healing defects with different dielectric permittivity compared to the pristine polymer. These include voids with lower ε_r than the polymer or conductive carbon deposits, which can be considered to have infinite ε_r under electrostatic conditions. These low or high ε_r defects cause the electric field to concentrate around the defects within the dielectric film, reducing the dielectric strength of the polymer compared to the original material. Therefore, to improve the recovery of dielectric properties after self-healing in dielectric materials, residual defects should be reduced. This can be achieved, as described by Kammermaier et al. [77, 78], through deliberate design to reduce the carbon deposits formed after electrical breakdown, for example, in the choice of organic or electrode materials and thicknesses.

Conventional PI with high carbon content generates large amounts of graphite deposits after electrical breakdown, losing its self-healing properties. Min et al. [80] reported a simple surface engineering strategy to endow metalized PI films with self-healing capabilities. By simply immersing the bare PI film in an epoxy resin (ER) solution and accompanying the curing of the ER, PI films impregnated with ER (P-E films) exhibited excellent self-healing capability after multiple electrical breakdowns, even at high temperatures. For example, PI films impregnated in 1 wt % ER solution (P-1%E) showed a decrease in U_d and η values of less than 6% after 5 breakdown cycles at 150 °C and 350 kV/mm. Subsequently, Song et al. [81]. prepared an alicyclic polyimide, which reduced the carbon content of the polyimide from its inherent molecular composition (Figure 4.10d). As a result, after self-healing, CBDA/DCHM showed only a slight decrease in U_d and η at 150 °C and 600 MV/m, e.g., from 4.61 J/cm^3 and 80.9% for fresh CBDA/DCHM to 4.53 J/cm^3 and 79.5% for self-healing CBDA/DCHM, respectively (Figure 4.10e). Meanwhile, a breakdown hole with a diameter of about 300 μm exists in the CBDA/DCHM film, and three different areas can be identified near the breakdown site (Figure 4.10f). Specifically, area I represents the outside of the breakdown hole, which is completely covered by Au elements, and areas II and III are located inside the breakdown hole, where no Au elements can be seen, revealing the evaporation of Au electrodes during the breakdown process. It should be noted that the appearance of Si elements, which are mainly enriched in zone III, is attributed to the silicone oil residue in the test environment.

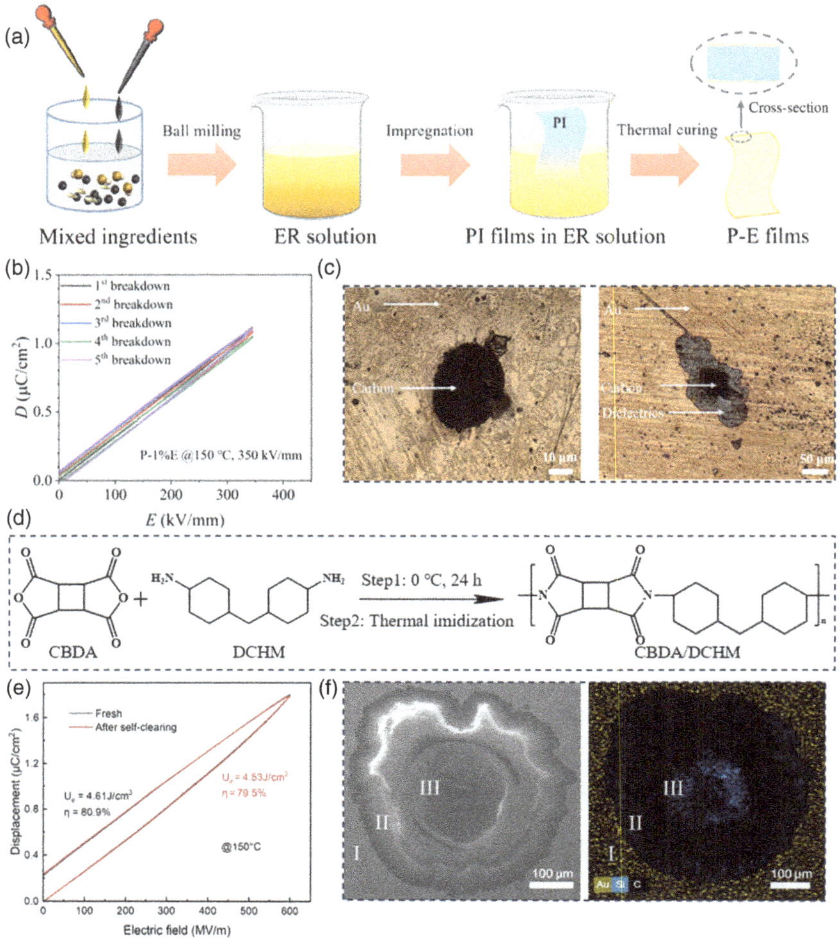

Figure 4.10 (a) Schematic of the fabrication process of impregnated PI films with ER [80]. (b) *D-E* loops of P-1%E films. (c) Optical microscopic images of metalized bare PI film and *P-E* film after electrical breakdown. (d) Molecular structure of CBDA/DCHM. (e) *D-E* loops of fresh and self-healing CBDA/DCHM films. (f) SEM surface morphology and element distribution of metalized CBDA/DCHM film after electrical breakdown [81].

Although the self-healing process is beneficial in terms of extending the dielectric lifetime, the loss of dielectric material results in a degradation of the dielectric properties and mechanical strength of the material [82].

4.7 Conclusion and Perspective

In recent years, multiple structure design approaches for intrinsic PIs and PI composites from the molecular scale to the microscopic scale have made significant progress in significantly improving the high-temperature capacitive properties. Structure design of PI dielectrics should balance dielectric stability, low dielectric loss, and high breakdown strength at high temperatures and high electric fields with the achievement of superior energy and power densities. The high-temperature conduction mechanisms and dielectric and capacitive properties associated with polymer structures should be systematically investigated.

The molecular structure modification of intrinsic PI is a potential method to improve high-temperature energy storage properties. Recently, the design and synthesis of intrinsic PI dielectrics have been focusing on improving ε_r, E_g, and T_g. However, future research should focus on the dielectric loss. Especially in high-temperature environments, a decrease in the tan δ of polymer dielectrics will enhance the efficiency and reliability of capacitors by reducing the possibility of thermal instability. The introduction of highly polarized groups into the polymer backbone or side chains allows for an increase in dipole polarization while maintaining a low tan δ. To clearly understand the structure–property relationship of the polymer dielectric, a synergistic design methodology can be utilized to select a reasonable polymer molecular structure. The methodology includes high-throughput calculations, material synthesis, measurement, and evaluation. The combination of theory and experimentation can greatly improve the efficiency of research. Screening conditions for polymers should not be limited to ε_r and E_g. These properties, together with high thermal stability, low tan δ, film-forming, and mechanical properties, should be considered together to improve the overall properties. In addition, the molecular-structure-modified novel PIs are currently in the laboratory preparation stage, and the reaction processes and yields need to be continuously optimized. Complex organic synthesis methods and precisely controlled film-manufacturing processes are still far from being able to be produced on a large scale. Consequently, new film fabrication methods need to be explored.

All-organic composite dielectrics with appropriate dielectric and mechanical properties are more suitable for scalable, continuous, and large-scale industrial preparation. The key factors to consider are that the selected multiphase/multicomponent material should interact well with the polymer matrix and have a similar or higher temperature resistance than the polymer film. In order to improve the energy storage properties of all-organic polymer capacitors at high temperatures, multiphase/multicomponent high-temperature dielectric polymers, such as blended polymers and multilayer polymers, should be further investigated.

Although the incorporation of wide E_g or high thermal conductivity nanofillers into polymers can reduce the energy loss of polymers at high temperatures, the improvement of dielectric and energy storage properties is limited. The core–shell strategy offers more possibilities to optimize the microstructure and even the dielectric properties of nanocomposites. Designing polymer shells with good compatibility with the polymer matrix is essential to significantly improve the comprehensive properties of nanocomposites. In addition, how to achieve uniform dispersion of nanofillers in industry while further reducing the amount of nanofillers added remains a key technical challenge in the long term. Subsequently, the interfacial effects of nanocomposites and multilayer composites at high temperatures and high electric fields and their mechanisms are still unclear, and it is urgent to seek new characterization techniques for observation and validation study.

Whether the multilayer structures are all-organic composites or nanocomposites, the electric field distribution can be efficiently regulated by a combination of dielectric and insulating layers. These composites obtain high ε_r while maintaining good insulating strength, thus increasing the energy density. High T_g materials promising for high-temperature dielectric applications have been investigated to determine the effect of multilayer structure on the properties, but a systematic exploration of their high-temperature dielectric characteristics, especially the temperature-dependent breakdown strength, is still lacking. In addition, challenges related to the optimal design of structures and the development of scalable processing methods for multilayer-structured films still need to be solved.

A high T_g or T_m value is not sufficient to prevent polymer failure due to long-term heating. Thermal conductivity is an important parameter that should not be neglected in high-temperature capacitor films. The thermal conductivity of polymers can be improved by controlling the stacking and orientation of the chains, and it can also be enhanced by adding thermally conductive fillers to create conductive pathways. Further study should consider the effect of thermal conductivity on the properties of high-temperature capacitors.

Novel polymer dielectric materials generally have high requirements for the preparation process, and it is difficult to achieve large-scale production. Therefore, how to achieve cost-effective and large-scale preparation of polymer films in industry is a hot research topic in the future.

More importantly, future research should focus on the self-healing mechanism of polymer dielectrics at ambient and high temperatures. The self-healing mechanism has received less attention in the current research on polymer dielectrics. The excellent self-healing behavior of all novel polymer dielectrics plays a crucial role in their suitability for practical energy applications.

Rapid advances in a highly interdisciplinary field such as polymer dielectrics are to be expected only through collaborative efforts across the boundaries of chemistry, materials science and engineering, electrical engineering and applied physics.

References

1 Zha, J.-W., Xiao, M., Wan, B. et al. (2023). Polymer dielectrics for high-temperature energy storage: constructing carrier traps. *Progress in Materials Science* 140: 101208.

2 Zha, J.-W., Tian, Y., Zheng, M.-S. et al. (2023). High-temperature energy storage polyimide dielectric materials: polymer multiple-structure design. *Materials Today Energy* 31: 101217.

3 Rabuffi, M. and Picci, G. (2002). Status quo and future prospects for metallized polypropylene energy storage capacitors. *IEEE Transactions on Plasma Science* 30: 1939–1942.

4 Liu, X.-J., Zheng, M.-S., Wang, G. et al. (2022). High energy density of polyimide films employing an imidization reaction kinetics strategy at elevated temperature. *Journal of Materials Chemistry A* 10: 10950–10959.

5 Liu, X.-J., Cheng, M., Zhang, Y. et al. (2024). High-temperature polymer dielectric films with excellent energy storage performance utilizing inorganic outerlayers. *Composites Science and Technology* 245: 110305.

6 Zhou, Y., Zhu, Y., Xu, W. et al. (2023). Molecular trap engineering enables superior high-temperature capacitive energy storage performance in all-organic composite at 200 °C. *Advanced Energy Materials* 13: 2203961.

7 Luo, S., Yu, J., Ansari, T.Q. et al. (2020). Elaborately fabricated polytetrafluoroethylene film exhibiting superior high-temperature energy storage performance. *Applied Materials Today* 21: 100882.

8 Tian, Y., Zheng, M.-S., Li, Y. et al. (2023). Intrinsic-designed polyimide dielectric materials with large energy storage density and discharge efficiency at harsh ultra-high temperatures. *Materials Horizons* 10: 5835–5846.

9 Xiao, M., Wan, B., Wang, X. et al. (2023). Improved high-temperature energy storage of polyetherimide by energy level structure regulation. *Polymer* 286: 126397.

10 Dong, X., Wan, B., Zheng, M.-S. et al. (2024). Dual-effect coupling for superior dielectric and thermal conductivity of polyimide composite films featuring "crystal-like phase" structure. *Advanced Materials* 36: 2307804.

11 Wan, B., Yang, X., Dong, X. et al. (2023). Dynamic sustainable polyimide film combining hardness with softness via a "Mimosa-like" bionic strategy. *Advanced Materials* 35: 2207451.

12 Ma, R., Baldwin, A.F., Wang, C. et al. (2014). Rationally designed polyimides for high-energy density capacitor applications. *ACS Applied Materials & Interfaces* 6: 10445–10451.

13 Chen, J., Zhou, Y., Huang, X. et al. (2023). Ladderphane copolymers for high-temperature capacitive energy storage. *Nature* 615: 62–66.

14 Zheng, W., Li, Z., Chen, K. et al. (2022). Temperature-resistant intrinsic high dielectric constant polyimides: more flexibility of the dipoles. *Larger Permittivity of the Materials. Molecules* 27: 6337.

15 Tong, H., Fu, J., Ahmad, A. et al. (2019). Sulfonyl-containing polyimide dielectrics with advanced heat resistance and dielectric properties for high-temperature capacitor applications. *Macromolecular Materials and Engineering* 304: 1800709.

16 Liu, M., Song, J., Qin, H. et al. (2024). Significant enhancement in dielectric properties of polyimide alloys through a two-phase interlocking structure. *Advanced Functional Materials* 34: 2313258.

17 Feng, M., Feng, Y., Zhang, C. et al. (2024). Enhanced high-temperature energy storage performance of all-organic composite dielectric via constructing Fiber-reinforced structure. *Energy & Environmental Materials* 7: e12571.

18 Zhang, K., Ma, Z., Deng, H. et al. (2022). Improving high-temperature energy storage performance of PI dielectric capacitor films through boron nitride interlayer. *Advanced Composites and Hybrid Materials* 5: 238–249.

19 Yang, M., Zhao, Y., Wang, Z. et al. (2024). Surface ion-activated polymer composite dielectrics for superior high-temperature capacitive energy storage. *Energy & Environmental Science* 17: 1592–1602.

20 Ran, Z., Wang, R., Fu, J. et al. (2023). Spiral-Structured Dielectric Polymers Exhibiting Ultrahigh Energy Density and Charge–Discharge Efficiency at High Temperatures. *Advanced Materials* 35: 2303849.

21 Liu, X.-J., Zheng, M.-S., Chi, Q. et al. (2022). High-temperature energy storage performances of "isomer-like" polyimide and its thermoplastic polyurethane blending system. *Journal of Materials Chemistry C* 10: 17326–17335.

22 Yang, M., Zhou, L., Li, X. et al. (2023). Polyimides physically crosslinked by aromatic molecules exhibit ultrahigh energy density at 200 °C. *Advanced Materials* 35: 2302392.

23 Feng, M., Feng, Y., Zhang, C. et al. (2022). Ultrahigh energy storage performance of all-organic dielectrics at high-temperature by tuning the density and location of traps. *Materials Horizons* 9: 3002–3012.

24 Chen, J., Ren, F., Yin, N. et al. (2024). Significant enhancement of high-temperature capacitive energy storage in dielectric films through surface self-assembly of BNNS coatings. *Chemical Engineering Journal* 479: 147581.

25 Ma, Y., Lin, Y., Zhang, Y. et al. (2024). PEI-based all-organic composite films with simultaneous excellent energy storage density and high efficiency. *Journal of Materials Chemistry A* 12: 12112–12118.

26 Liu, X.-J., Zheng, M.-S., Chen, G. et al. (2022). High-temperature polyimide dielectric materials for energy storage: theory, design, preparation and properties. *Energy & Environmental Science* 15: 56–81.

27 Zha, J.-W., Zheng, M.-S., Fan, B.-H. et al. (2021). Polymer-based dielectrics with high permittivity for electric energy storage: a review. *Nano Energy* 89: 106438.

28 Park, S., Kim, C.-H., Lee, W.-J. et al. (2017). Sol-gel metal oxide dielectrics for all-solution-processed electronics. *Materials Science and Engineering: R: Reports* 114: 1–22.

29 Yao, Z., Song, Z., Hao, H. et al. (2017). Homogeneous/inhomogeneous-structured dielectrics and their energy-storage performances. *Advanced Materials* 29: 1601727.

30 Palneedi, H., Peddigari, M., Hwang, G.-T. et al. (2018). High-performance dielectric ceramic films for energy storage capacitors: progress and outlook. *Advanced Functional Materials* 28: 1803665.

31 Feng, Q.-K., Zhong, S.-L., Pei, J.-Y. et al. (2022). Recent progress and future prospects on all-organic polymer dielectrics for energy storage capacitors. *Chemical Reviews* 122: 3820–3878.

32 Fan, B., Zhou, M., Zhang, C. et al. (2019). Polymer-based materials for achieving high energy density film capacitors. *Progress in Polymer Science* 97: 101143.

33 Zhang, H., Marwat, M.A., Xie, B. et al. (2020). Polymer matrix nanocomposites with 1D ceramic nanofillers for energy storage capacitor applications. *ACS Applied Materials & Interfaces* 12: 1–37.

34 Nan, C.-W., Shen, Y., and Ma, J. (2010). Physical properties of composites near percolation. *Annual Review of Materials Research* 40: 131–151.

35 Ramesh, S., Shutzberg, B.A., Huang, C. et al. (2003). Dielectric nanocomposites for integral thin film capacitors: materials design, fabrication and integration issues. *IEEE Transactions on Advanced Packaging* 26: 17–24.

36 Amin, M., Samy, H., and Küpper, J. (2019). Robust and accurate computational estimation of the polarizability tensors of macromolecules. *The Journal of Physical Chemistry Letters* 10: 2938–2943.

37 Liu, F., Li, Q., Cui, J. et al. (2017). High-energy-density dielectric polymer nanocomposites with trilayered architecture. *Advanced Functional Materials* 27: 1606292.

38 Tan, D.Q. (2020). The search for enhanced dielectric strength of polymer-based dielectrics: a focused review on polymer nanocomposites. *Journal of Applied Polymer Science* 137: 49379.

39 Yang, L., Kong, X., Li, F. et al. (2019). Perovskite lead-free dielectrics for energy storage applications. *Progress in Materials Science* 102: 72–108.

40 Huang, X., Sun, B., Zhu, Y. et al. (2019). High-k polymer nanocomposites with 1D filler for dielectric and energy storage applications. *Progress in Materials Science* 100: 187–225.

41 Laghari, J.R. and Sarjeant, W.J. (1992). Energy-storage pulsed-power capacitor technology. *IEEE Transactions on Power Electronics* 7: 251–257.

42 Li, Q., Yao, F.-Z., Liu, Y. et al. (2018). High-temperature dielectric materials for electrical energy storage. *Annual Review of Materials Research* 48: 219–243.

43 Pei, J.-Y., Yin, L.-J., Zhong, S.-L. et al. (2023). Suppressing the loss of polymer-based dielectrics for high power energy storage. *Advanced Materials* 35: 2203623.

44 Zhang, Z., Wang, D.H., Litt, M.H. et al. (2018). High-temperature and high-energy-density dipolar glass polymers based on sulfonylated poly(2,6-dimethyl-1,4-phenylene oxide). *Angewandte Chemie International Edition* 57: 1528–1531.

45 Chiu, F.-C. (2014). A review on conduction mechanisms in dielectric films. *Advances in Materials Science and Engineering* 2014: 578168.

46 Liu, A., Zhu, H., Sun, H. et al. (2018). Solution processed metal oxide high-κ dielectrics for emerging transistors and circuits. *Advanced Materials* 30: 1706364.

47 Sarjeant, W.J., Zirnheld, J., and MacDougall, F.W. (1998). Capacitors. *IEEE Transactions on Plasma Science* 26: 1368–1392.

48 Zhou, Z., Mackey, M., Carr, J. et al. (2012). Multilayered polycarbonate/poly(vinylidene fluoride-co-hexafluoropropylene) for high energy density capacitors with enhanced lifetime. *Journal of Polymer Science Part B: Polymer Physics* 50: 993–1003.

49 Heywang, H. (1976). Physikalische und chemische Vorgänge in selbstheilenden Kunststoff-Kondensatoren. *Colloid and Polymer Science* 254: 139–147.

50 Ho, J.S. and Greenbaum, S.G. (2018). Polymer capacitor dielectrics for high temperature applications. *ACS Applied Materials & Interfaces* 10: 29189–29218.

51 Chen, S., Meng, G., Kong, B. et al. (2020). Asymmetric alicyclic amine-polyether amine molecular chain structure for improved energy storage density of high-temperature crosslinked polymer capacitor. *Chemical Engineering Journal* 387: 123662.

52 Reed, C.W. and Cichanowskil, S.W. (1994). The fundamentals of aging in HV polymer-film capacitors. *IEEE Transactions on Dielectrics and Electrical Insulation* 1: 904–922.

53 Zhu, Y.-F., Zhang, Z., Litt, M.H. et al. (2018). High dielectric constant sulfonyl-containing dipolar glass polymers with enhanced Orientational polarization. *Macromolecules* 51: 6257–6266.

54 Wang, Q. and Zhu, L. (2011). Polymer nanocomposites for electrical energy storage. *Journal of Polymer Science Part B: Polymer Physics* 49: 1421–1429.

55 Zhu, T., Yu, Q., Zheng, W. et al. (2021). Intrinsic high-k–low-loss dielectric polyimides containing ortho-position aromatic nitrile moieties: reconsideration on Clausius–Mossotti equation. *Polymer Chemistry* 12: 2481–2489.

56 Zheng, W., Yang, T., Qu, L. et al. (2022). Temperature resistant amorphous polyimides with high intrinsic permittivity for electronic applications. *Chemical Engineering Journal* 436: 135060.

57 Pei, Z., Liu, Y., Zhao, W. et al. (2024). Wide bandgap Heterostructured dielectric polymers by rapid photo-crosslinking for high-temperature capacitive energy storage. *Advanced Functional Materials* 34: 2307639.

58 Wu, C., Deshmukh, A.A., Li, Z. et al. (2020). Flexible temperature-invariant polymer dielectrics with large bandgap. *Advanced Materials* 32: 2000499.

59 Deshmukh, A.A., Wu, C., Yassin, O. et al. (2022). Flexible polyolefin dielectric by strategic design of organic modules for harsh condition electrification. *Energy & Environmental Science* 15: 1307–1314.

60 Wu, C., Deshmukh, A.A., Yassin, O. et al. (2021). Flexible cyclic-olefin with enhanced dipolar relaxation for harsh condition electrification. *Proceedings of the National Academy of Sciences* 118: e2115367118.

61 Wang, R., Zhu, Y., Fu, J. et al. (2023). Designing tailored combinations of structural units in polymer dielectrics for high-temperature capacitive energy storage. *Nature Communications* 14: 2406.

62 Ren, W., Yang, M., Zhou, L. et al. (2022). Scalable ultrathin all-organic polymer dielectric films for high-temperature capacitive energy storage. *Advanced Materials* 34: 2207421.

63 Yu, S., Zhou, J., Xu, A. et al. (2023). The scalable and high performance polyimide dielectrics containing alicyclic structures for high-temperature capacitive energy storage. *Chemical Engineering Journal* 469: 143803.

64 Dong, J., Li, L., Niu, Y. et al. (2024). Scalable high-permittivity polyimide copolymer with ultrahigh high-temperature capacitive performance enabled by molecular engineering. *Advanced Energy Materials* 14: 2303732.

65 Zhang, Q., Chen, X., Zhang, B. et al. (2021). High-temperature polymers with record-high breakdown strength enabled by rationally designed chain-packing behavior in blends. *Matter* 4: 2448–2459.

66 Ding, S., Bao, Z., Wang, Y. et al. (2023). Excellent high-temperature dielectric energy storage of flexible all-organic polyetherimide/poly(arylene ether urea) polymer blend films. *Journal of Power Sources* 570: 233053.

67 Yuan, C., Zhou, Y., Zhu, Y. et al. (2020). Polymer/molecular semiconductor all-organic composites for high-temperature dielectric energy storage. *Nature Communications* 11: 3919.

68 Zhang, B., Chen, X.-M., Pan, Z. et al. (2023). Superior high-temperature energy density in molecular semiconductor/polymer all-organic composites. *Advanced Functional Materials* 33: 2210050.

69 Niu, Y., Dong, J., He, Y. et al. (2022). Significantly enhancing the discharge efficiency of sandwich-structured polymer dielectrics at elevated temperature by building carrier blocking interface. *Nano Energy* 97: 107215.

70 Wang, P., Guo, Y., Zhou, D. et al. (2022). High-temperature flexible nanocomposites with ultra-high energy storage density by nanostructured MgO fillers. *Advanced Functional Materials* 32: 2204155.

71 Ai, D., Li, H., Zhou, Y. et al. (2020). Tuning nanofillers in in situ prepared polyimide nanocomposites for high-temperature capacitive energy storage. *Advanced Energy Materials* 10: 1903881.

72 Liu, X.-J., Zhong, S.-L., Zheng, M.-S. et al. (2023). Enhanced high-temperature capacitive performance of a bilayer-structured composite film employing a charge blocking layer. *ACS Applied Materials & Interfaces* 15: 1105–1114.

73 Xie, Z., Le, K., Li, H. et al. Interfacial engineering using covalent organic frameworks in polymer composites for high-temperature electrostatic energy storage. *Advanced Functional Materials* 34: 2314910.

74 Yuan, P., Xue, R., Wang, Y. et al. (2024). Horizontally-oriented barium titanate@polydomine/polyimide nanocomposite films for high-temperature energy storage. *Journal of Colloid and Interface Science* 662: 1052–1062.

75 Wang, P., Yao, L., Pan, Z. et al. (2021). Ultrahigh energy storage performance of layered polymer nanocomposites over a broad temperature range. *Advanced Materials* 33: 2103338.

76 Klein, N. (1966). The mechanism of self-healing electrical breakdown in MOS structures. *IEEE Transactions on Electron Devices* ED-13: 788–805.

77 Kammermaier, J. (1987). Chemical prpcesses during electrical breakdown in an organic dielectric with evaporated thin electrodes. *IEEE Transactions on Electrical Insulation* EI-22: 145–149.

78 Kammermaier, J., Rittmayer, G., and Birkle, S. (1989). Modeling of plasma-induced self-healing in organic dielectrics. *Journal of Applied Physics* 66: 1594–1609.

79 Zhang, Y., Ellingford, C., Zhang, R. et al. (2019). Electrical and mechanical self-healing in high-performance dielectric elastomer actuator materials. *Advanced Functional Materials* 29: 1808431.

80 Min, Y., Yu, J., Xu, P. et al. (2022). Polyimide films impregnated with epoxy resin demonstrating superior self-healing properties for thermally stable energy storage capacitors. *ACS Applied Materials & Interfaces* 14: 51265–51274.

81 Song, J., Qin, H., Qin, S. et al. (2023). Alicyclic polyimides with large band gaps exhibit superior high-temperature capacitive energy storage. *Materials Horizons* 10: 2139–2148.

82 Hunt, S., McKay, T.G., and Anderson, I.A. (2014). A self-healing dielectric elastomer actuator. *Applied Physics Letters* 104: 113701.

5

Polyimide Dielectrics with Low Dielectric Permittivity

Xiaodi Dong and Jun-Wei Zha

University of Science and Technology Beijing, School of Chemistry and Biological Engineering, Department of Chemistry, No.30 Xueyuan Road, Beijing 100083, China

5.1 Fundamental Principles for Low Dielectric Permittivity (Low-*k*)

The rapid development of integration and miniaturization of modern electronic and electrical equipment has led to serious capacitive-resistive effects between metal layers. This gives rise to problems such as signal transmission lag and increased losses, which cannot meet the needs of the microelectronics industry and the communications field [1–3]. According to the basic equations affecting the signal transmission speed (v) and delay time (T_d) Eqs. (5.1) and (5.2), decreasing the permittivity of the interlayer insulating material will effectively improve the quality of signal transmission [4, 5]. Therefore, investigation of the low dielectric properties of interlayer electrolyte materials is essential.

$$T_d = 2\rho k k_0 \left[\frac{4L^2}{P^2} + \frac{L^2}{D^2}\right] \tag{5.1}$$

$$v = \frac{c}{\sqrt{k}} \tag{5.2}$$

where ρ is the specific resistance of the conductor, k_0 is the permittivity of the vacuum, L and D are the length and thickness of the conductor, respectively, P is the distance between the two wires, and c is the speed of light.

Polyimide, as a bottleneck material limiting the development of electronics and electric power fields, has received much attention from researchers in recent years [6–10]. According to the basic factors affecting the dielectric property, a significant decrease in the permittivity can be achieved by adjusting the polarization rate and the density of polarized molecules. As illustrated in Figure 5.1, approaches to reduce polarization include introducing units of low polarity or adjusting the molecular structure to restrict the movement of polarized molecules. On the other hand, ways to decrease the density of polarized molecules include reducing the number of polarized molecules per unit volume by constructing a large free volume of space, or introducing a porous structure in a dense system.

Functional Polyimide Dielectrics: Structure, Properties, and Applications, First Edition.
Edited by Jun-Wei Zha.

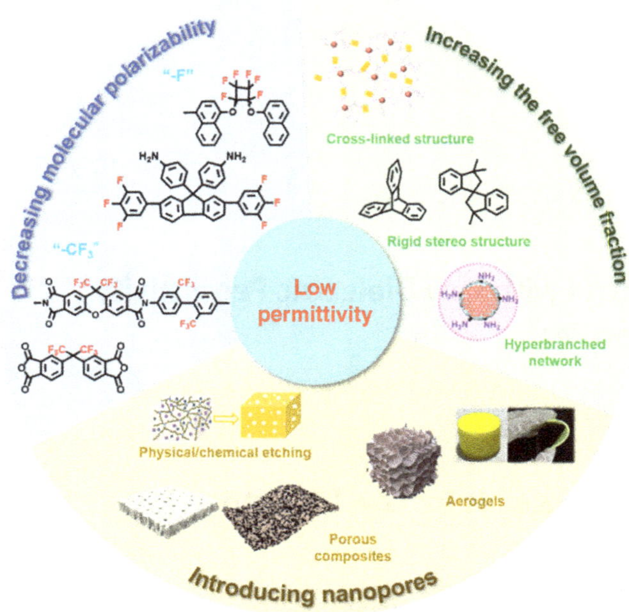

Figure 5.1 Basic strategies for reducing dielectric permittivity.

In this chapter, research progress of intrinsic low-*k* PIs, composite low-*k* PIs, and porous low-*k* PIs is comprehensively presented from the perspectives of molecular structure design, introduction of functional fillers, and induction of porous structure. The process design, property variation, and intrinsic mechanism of different types of low-*k* PIs are discussed. Moreover, researches on emerging multifunctional low-*k* PI materials, such as low-*k* and high thermal conductivity PIs, low-*k* and electromagnetic shielding PIs, and low-*k* and low thermal expansion PIs, are also presented in detail. Finally, the current problems and challenges facing the research on low-*k* PIs are summarized, and insights into the future development of low-*k* materials are also provided.

5.2 Intrinsic Low-*k* PI

The PI molecular structure can be effectively modulated by changing the structure of the synthetic monomer dianhydride or diamine. Generally, the introduction of highly electronegative atoms, bulky alicyclic rings, or non-planar rigid structures into the monomer structure will regulate the polarizability and polarization density of PIs and achieve a reduction of the permittivity. In this subsection, the researches related to intrinsic low-*k* PIs will be presented as represented by the introduction of fluorine (F) atom, alicyclic ring, and special rigid structures such as anthracene groups and crown ethers.

5.2.1 Introduction of Fluorinated Structures

Due to the strong electronegativity of the F atom, the —C—F— bond possesses a low dipole polarization. Therefore, the fabricated fluorinated PI (FPI) molecules exhibit a lower polarization rate. Additionally, the strong repulsion between the F atoms hinders the stacking between the molecular chain segments, leading to the formation of a larger loose structure inside the system [11, 12]. This increases the free volume fraction (FFV) of the PI molecule, which further decreases the permittivity of PI. There are two strategies for preparing FPI, direct fluorine substitution, and the introduction of trifluoromethyl groups (—CF_3) with a large free volume.

Recently, there have been various researches on FPI, and the permittivity of general FPI can be kept below 3.0. Wang et al. [13] investigated the variation of the permittivity of FPI with —CF_3 content, and an FPI with a permittivity of 2.74 was achieved at 1 MHz. Bong et al. [14] successfully synthesized asymmetric diamine monomers containing —CF_3 and polymerized them with fluorinated dianhydride (6FDA). FPIs with permittivity as low as 2.633 were obtained. In particular, Nagella et al. [15] synthesized a new type of diamine (BPBTAB) using bisphenol-P having diisopropyl benzene units, and a series of FPIs with ultralow permittivity ($k = 1.13$–1.84) were produced by in situ polymerization with different types of dianhydrides. Meanwhile, these FPIs maintained good thermal stability, tensile modulus, optical properties, and hydrophobicity (moisture absorption less than 0.63%), showing great potential for applications in high-frequency 5G and 6G mobile communications.

Notably, the permittivity of PI does not always decrease with increasing fluorine content. Excessive fluorine content causes entanglement between molecular chains, which reduces the free volume of the structure and increases the packing density of molecules and thus causes the permittivity to rise. Therefore, it is crucial to specify the optimal fluorine content for effective regulation of the dielectric properties of FPI. Fan et al. [16] collected the structural parameters of 78 PIs containing different molecular structures through a combination of quantum chemical and group contribution methods. The optimal fluorine content introduced into PIs was accurately predicted by constructing two quantitative constitutive relationship models for permittivity (with a relative error of less than 5%). It was shown that the optimal fluoride content was between 0.25 and 0.37.

5.2.2 Introduction of Alicyclic Rings or Steric Rigid Structures

On the one hand, the incorporation of aliphatic structures will separate the rigid conjugated structures on the polymer backbone, which will fundamentally disrupt the formation of conjugated charge transfer complexes of the molecules and inhibit intermolecular charge transfer [17–19]. On the other hand, the presence of sterically rigid groups (such as phenylpropyl cyclobutene, anthracenyl, and biphenyl groups) will reduce the stacking density between the polymer chains, increase the molar volume of the repeating units in the molecular chains, and reduce the polarized molecular density [20–22]. In practical investigations, researchers tend to combine the introduction of fluorinated structures and alicyclic/rigid structures to promote a steady decrease in permittivity.

Zhuang et al. [23] prepared partially alicyclic polyimides (AcPI-TBs) via Troger bases (TBs), displaying good mechanical strength (72.5–102.3 MPa) and low permittivity ($k = 2.66$). Specifically, the films exhibited complete colorlessness and transparency, and also showed bifluorescence and phosphorescence emission at low temperatures. It has a high potential for applications in micro-optical devices such as displays and lighting. Yu et al. [24] produced a colorless copolymerized PI film with an o-phenylene aliphatic ring structure. The permittivity of this semi-aliphatic PI was in the range of 2.15–2.47, and the dielectric loss was less than 0.023.

Moreover, the introduction of an adamantane or benzolipid ring structure in the main chain of PI not only reduces the permittivity, but also improves the mechanical properties and heat resistance. Yoo et al. [25] developed a new type of diamine, bis(4-aminophenyl)-1-adamantane amide phosphine oxide (DAAPO). The permittivity of PI prepared based on DAAPO and 6FDA could be reduced from 3.03 to 2.49. Meanwhile, the thermal decomposition temperature of 5% ($T_{d5\%}$) of PI was stable at 500 °C.

For the synthesis of low-k PI molecules containing steric rigid structures, Li et al. [26] introduced —CF_3 and crown ether (CE) during the synthesis of PI to form a similar host–guest inclusion complex. It was found that the permittivity of this PI films gradually decreased when lower CE content was introduced, and the k value at 10 GHz could be as low as 2.3. This was attributed to the molecular chains that could be uniformly wrapped with a low content of cyclic macromolecular crown ethers, which increased the FFV of the PI molecules and reduced the polarized molecular density. Furthermore, complemented by the low polarity of —CF_3, the PI films also exhibited a low dielectric loss ($\tan\delta = 0.00337$). As shown in Figure 5.2, Lee et al. [27]

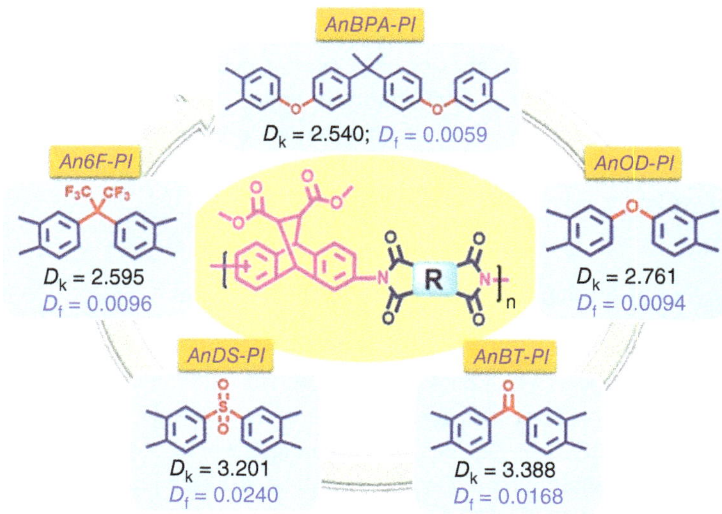

Figure 5.2 Low-k PIs with rigid anthracene-based structures. Source: Ref. [27]/American Chemical Society.

synthesized a series of PIs with low permittivity using ester-functionalized anthracene diamine monomers and four commercial dianhydrides.

In view of the safety and reliability of dielectric materials in practical applications, dielectric materials must always exhibit low permittivity over a wide range of temperatures. Chen et al. [28] prepared two diamines containing different biphenyl side groups and polymerized them with 6FDA to obtain two different PIs, i.e., 6FDA-GA and 6FDA-GB. The variation patterns of the permittivity with temperature and frequency were explored. The 6FDA-GB type PI was found to exhibit lower dielectric properties ($k = 2.7$, $\tan \delta = 0.0022$). It was because the shorter side groups in GB not only created site barriers between the molecules to prevent chain stacking, but also increased the torsional rigidity of the main chain, which allowed the PI to obtain a larger free volume. More importantly, the low dielectric properties of this PI remain stable over a temperature range of 50–200 °C. This was attributed to the presence of biphenyl side groups inhibiting the rotation of the polar ring at different temperatures.

5.2.3 Introduction of Other Special Structures

In addition to the introduction of simple alicyclic or rigid side groups into PI molecules as described above, researchers have attempted to construct more complex crosslinked or micro-crosslinked structures in FPIs to further reduce the permittivity of PIs [28–31]. This mainly stems from the fact that, firstly, the crosslinked network can reduce the dipole orientation by limiting the mobility of the polymer chains, thus effectively preventing the polarization of the electron cloud in the PI. Then, the rigid crosslinked structure prevents the stacking of molecules and increases the FFV of PI, which reduces the number of polarized molecules per unit volume.

Huang et al. [32] introduced triallyl isocyanurate (TAIC) as a crosslinking agent for PAA oligomers, and the synthesized crosslinked PI (CPI) exhibited permittivity as low as 2.34 and improved thermodynamic properties. This was mainly because of the construction of crosslinking structure induces the formation of microporous structure in the PI system, which will reduce the intrinsic k value of PI. Remarkably, the decrease in permittivity by increasing the proportion of crosslinking agent is limited. When the polymer chains are drawn too close together, the presence of the crosslinking agent instead slightly increases the permittivity of PI. It has been shown that the maximum percentage of crosslinking agent is about 20%.

Density functional theory (DFT) and molecular dynamics (MD) simulations become more universal methods in analyzing the mechanism of the effect of crosslinked or micro-crosslinked structures on the permittivity of PI. Furthermore, it could achieve closed-loop research between experimental investigation and theoretical simulation. Peng et al. [33] grafted polyhedral oligomeric silsesquioxane (POSS) onto PI molecules to obtain PI molecules with a micro-crosslinked structure. Figure 5.3a displays the electrostatic potential (ESP) of three different modified POSS/PI, phenyl-POSS/PI, isobutyl-POSS/PI, and $8NH_2$-POSS/PI. The ESP around the aminopropyl heptyl group in $8NH_2$-POSS showed negative

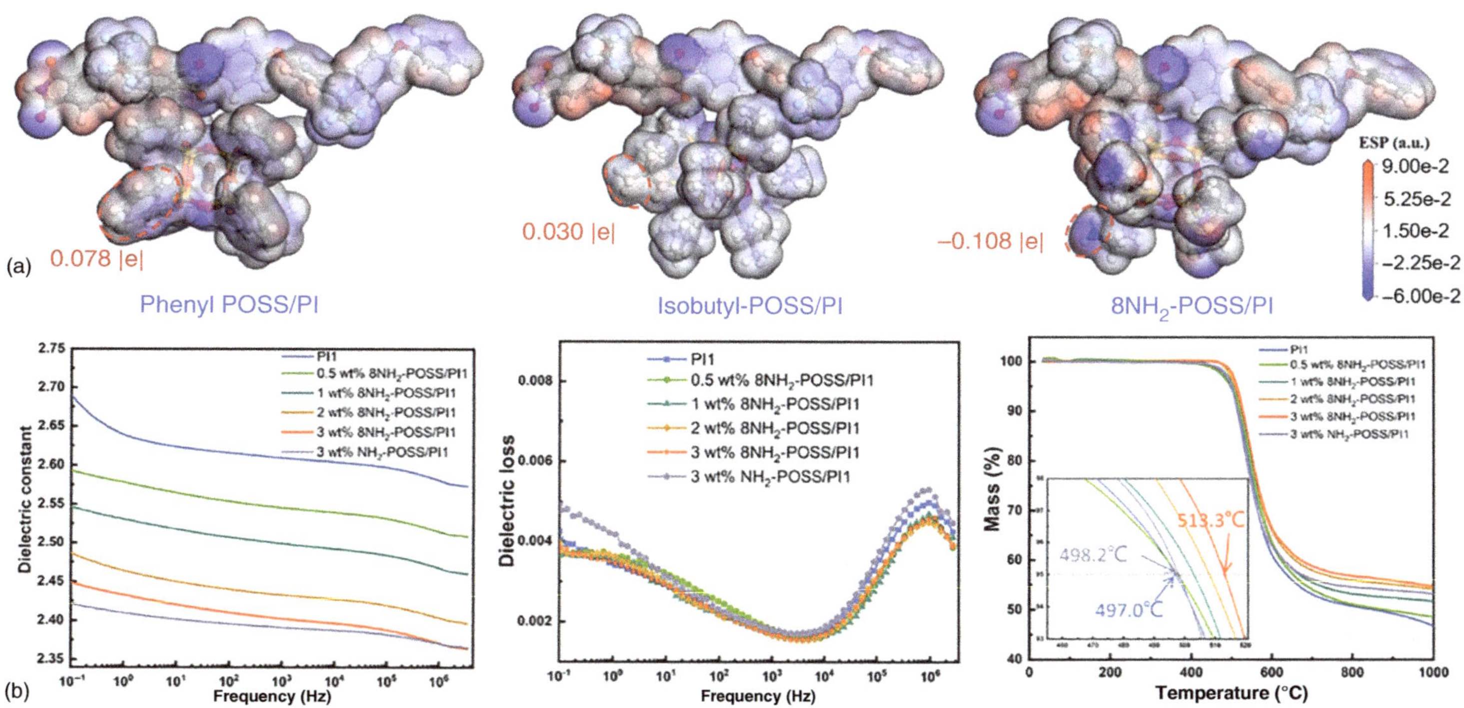

Figure 5.3 (a) Schematic diagram of electrostatic potential of phenyl POSS/PI, isobutyl POSS/PI, and $8NH_2$-POSS/PI. (b) Dielectric and thermal properties of $8NH_2$-POSS/PI. Source: Ref. [33]/Springer Nature.

potentials, suggesting the deepest trapping effect at the interfacial site. This can significantly suppress the injection and migration of charge carriers, contributing to the decrease in dielectric constant. Thus, 3 wt% $8NH_2$-POSS/PI exhibited the lowest permittivity and dielectric loss (k = 2.4 and tan δ = 0.0017) as illustrated in Figure 5.3b. Moreover, this micro-crosslinked structure also gave the PI better thermal stability (513.3 °C), lower coefficient of thermal expansion, and acid resistance for better application in printed circuits.

PDMS has excellent hydrophobicity and large free volume. It can be copolymerized with PI by modification of its terminal amino group to obtain PDMA-PI, which has been shown to be a better low-k copolymer material. For instance, Qi et al. [34] designed and synthesized a symmetric disiloxane-linked alkyl diamine, bis(aminopropyl)tetramethyldisiloxane (BATMS). The PDMS-PI copolymer possessed the lowest dielectric property when the molar percentage of BATMS was 75% (k = 2.48, tan δ = 0.014).

Currently, researchers have achieved productive results in regulating the dielectric properties of PIs at the molecular structure level. A more mature and systematic theory has been established regarding the correlation between the molecular structure and the low permittivity of PI. However, there is still a relative paucity of researches on the permittivity and dielectric loss of PI at high-frequency (GHz) level.

5.3 Low-*k* PI Composites

The acquisition of intrinsic low-k PI depends more on the selection and construction of special monomer structure or molecular structure, which involves high cost and complicated synthesis process. It is more common to prepare low-k PI composites by introducing functional fillers. Typical fillers include fluorinated graphene (FG), modified graphene oxide, and fluorinated ethylene propylene (FEP). However, due to the differences in composition, properties, and structure between the multiphases, there is a certain interfacial effect between the PI matrix and fillers. This will affect the interfacial polarization of the PI composite and thus directly affect the dielectric properties. Therefore, optimizing the interactions within the composite system is crucial for the preparation of high-performance low-k PI composites. In this section, an overview of the research on low-k PI composites is presented in terms of the introduction of inorganic and organic fillers.

5.3.1 Introduction of Inorganic Functional Filler

FG is one of the most promising dielectric fillers available nowadays, inheriting the unique properties of graphene and fluorinated materials and exhibiting more novel chemical properties. It has been demonstrated that a high fluorine/carbon (F/C) ratio can lead to a wider band gap, greater hydrophobicity, and thermal stability of FG sheets. The FG surface can be chemically modified to improve interfacial compatibility with PI substrates [35, 36].

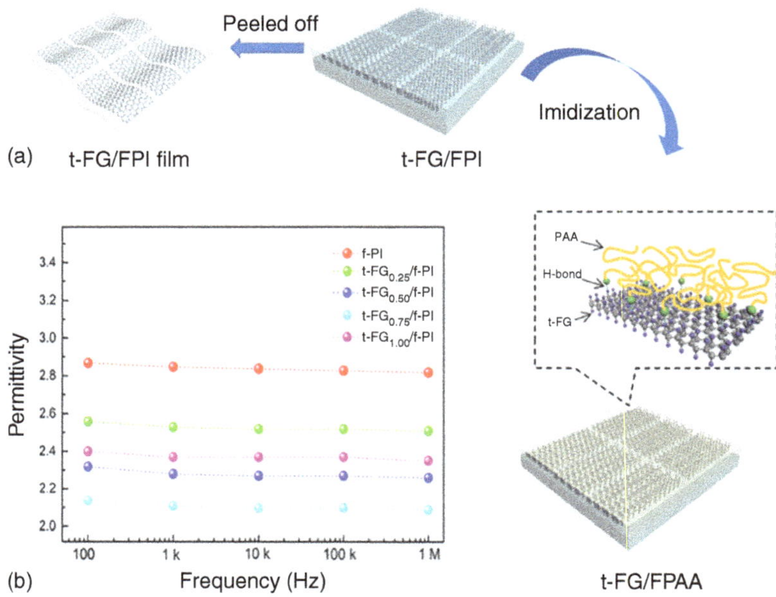

Figure 5.4 (a) Schematic of the preparation of t-FG/FPI films. (b) Variation of t-FG content on permittivity of t-FG/FPI. Source: Ref. [37]/Royal Society of Chemistry.

As depicted in Figure 5.4, Yin et al. [37] grafted FG using 2,2′ bis(trifluoromethyl)-(1,1′biphenyl)-4,4′amine (TFDB), and the obtained grafted FG (t-FG) had sites on the surface that could be reacted with fluorinated PAA (FPAA). The transparent, homogeneous t-FG0.75/FPI film was obtained with permittivity as low as 2.09 and dielectric losses as low as 0.0019. It was primarily due to the effective adsorption of monolayer t-FG nanosheets with high aspect ratios on the PI molecular chains, which hindered the movement of the dipoles of the FPI molecules and limited the dipole relaxation. However, when the loading of t-FG is too high, large aggregation and phase separation will occur within the composite system and induce interfacial polarization. Therefore, there is an inverse increase in the permittivity of the composite system.

Graphene oxide (GO) is one of the derivatives of graphene, and the abundance of oxygen-containing functional groups on its substrate surface, as well as carbonyl and carboxyl groups at the edges of the lamellae, will enhance the activity of GO [38, 39]. The surface modification of GO is required when it is used as a functional filler for low-*k* research. For example, Wang et al [40] synthesized aminoquinoline-functionalized graphene oxide (AQL-GO). This PI composite could only be fully imidized at 200 °C due to the presence of quinoline. And the permittivity of composite was as low as 2.96 at the AQL-GO content of 0.5 wt%. The low-temperature imidization was achieved mainly owing to the many curing acceleration points on the surface of AQL-GO. Similar to FG, there have been some investigations on FGO/PI composites [41, 42]. He et al. [41] prepared a low-*k* filler by a simple hydrothermal reaction between GO dispersion and hydrofluoric acid (HF). The permittivity of FGO-PI decreased to 2.28 at 1.0 wt% of FGO.

Moreover, Wang et al. [43] synthesized a series of fluorinated sponge graphene (FSG) sheets that maintained the pristine alveolar-like structure using sponge graphene as the raw material. 1 wt% FSG-3/PI was able to exhibit good low dielectric properties with $k = 2.1$ and $\tan\delta < 0.03$. This was attributed to the high F/C ratio of FSG which may lower the discrete electronic energy levels in the composite system and reduces the electronic polarization. Meanwhile, the hydrophobicity of the FSG/PI composite films was improved, which could prevent the composites from moisture during application. The combination of fluorinated graphene and boron nitride nanosheets also confers multifunctionality to low-*k* polyimides [44].

Besides, inorganic materials such as titanium dioxide (TiO_2) [45], montmorillonite [46], zeolite [47], and fluorine-containing graphene quantum dots (FGQDs) [48] have been developed in the field of low-*k* PIs. The obtained inorganic–organic low-*k* PI composites have permittivity in the range of 2.5–3.0.

5.3.2 Introduction of Organic Functional Fillers

The advantage of all-organic PI films lies in the good similarity and compatibility between organic fillers and PI, which enables the PI films to be more flexible. Presently, there are fewer organic fillers that can be applied to low-*k* PIs, and it has been reported that FEP and fluoroelastomers can significantly reduce the permittivity of PI composite systems. Cheng et al. [49] prepared an FEP/PI composite film with low permittivity and high bonding strength. The PI composite containing 60 wt% FEP showed low permittivity ($k = 2.69$) and ultralow dielectric loss ($\tan\delta = 0.006$) at 10 GHz. The incorporation of this flexible resin also increased the elongation at break of the PI composite films to 13.7%. Dong et al. [50] fabricated all-organic PI composite films with ultralow-*k* by incorporating fluoroelastomers (FEMs) into the PI matrix for the first time. Based on molecular dynamics simulations, it was verified that the "hydrogen bonding" between FEM and PI limits the dipole polarization of the composite system. 7 wt% FEM/PI composites exhibited a k value of 1.21, ultra-low water absorption of 0.44%, and high tensile strength of 136.8 MPa.

5.4 Porous Low-*k* PIs

Ultralow-*k* materials require the permittivity below 2.0, and the dense PIs require more demanding synthesis conditions if they intend to achieve ultralow-*k* levels. Air has the lowest permittivity. When a large amount of air is present in the polymer system, the density of polarized molecules is greatly reduced, giving more feasibility and versatility in obtaining ultralow permittivity. Porous PI will combine the high thermal stability of PI itself with the high specific surface area and ultralow dielectric properties of porous materials. It has a wide range of applications and great commercial value in fields such as adsorption, aerospace, and gas separation.

Strategies for the preparation of porous PIs include intrinsic microporosity, thermal degradation (physical) porosity, chemical etching porosity, introduction of porous fillers, and preparation of aerogels. Among them, the realization of intrinsically microporous PIs has a more demanding choice of monomers, higher process costs, and limited improvement. Therefore, in this chapter, the process synthesis, performance investigations, and industrial applications of representative porous PIs are described in detail from four aspects, including phase separation porogenesis, physical/chemical porogenesis, introduction of porous fillers, and preparation of aerogels.

5.4.1 Phase-separation-induced Pores

The main methods for inducing the formation of porous structures by phase separation are non-solvent-induced phase separation (NIPS) [51], water-vapor-induced phase separation (WVIPS) [52, 53], and thermally induced phase separation (TIPS). NIPS is a commonly used approach to induce porous PI films by immersing a homogeneous polymer solution in a differential solvent and utilizing the principle of mass transfer between the solvent and the non-solvent to obtain a porous structure [54, 55]. Khim et al. [56] grafted amino-functionalized mesoporous silica (AMS) onto PAA molecules and prepared porous PI films with multimodal porous network structure based on NIPS. It was showed that the permittivity and dielectric loss of 1 wt% AMS-PI were 1.84 and 0.0018, respectively. Meanwhile, the film maintained excellent ultralow-k stability over a wide temperature range from −20 °C to 200 °C.

However, NIPS cannot effectively control the porous structure and its properties due to the shortcomings of various fabrication variables such as liquid–liquid separation rate, polymer concentration, solvent fluctuation, and temperature [57]. TIPS is essentially a simpler and more general method with fewer variables to control than NIPS. This method first induces the separation of a homogeneous polymer solution by temperature change to create a multiphase system and subsequently removes the split-phase material by solvent extraction, leaving porous structure in the polymer. Hwang et al. [58] introduced dibutyl phthalate (DBP) into NH_2-POSS-PAA solution, and the homogeneous transparent DBP/ NH_2-POSS-PAA solution formed was subjected to cast film laid on glass plates. As the thermal imidization process proceeded, the transparent film gradually became opaque, i.e., phase separation was observed. After the completion of the thermal imidization, the DBP phase was then extracted with acetone to form a porous PI film. The permittivity of this porous PI film was as low as 2.08, and the dielectric loss was only 0.005. This was mainly attributed to the pores generated by the TIPS and the porous structure generated by the nanocaged POSS, which promoted the reduction of the polarized molecular density inside the PI system. Thus, this porogenic approach could lead to the development of promising ultralow-k dielectrics for efficient microelectronic dives.

5.4.2 Physically or Chemically Induced Pores

Physical porogenesis is mainly the removal of thermally unstable small molecules by thermal degradation, thus leaving a rich porous structure inside the polymer matrix [59, 60]. A commonly used template for thermal degradation is polyethylene glycol (PEG). Lv et al. [61] successfully prepared PI with nanopores using PEG as the porogenic template, and the permittivity of porous film decreased from 2.80 to 2.12 with the increase of PEG content. Moreover, the permittivity of 15 wt% PEG-induced porous PI films was stable at around 2.46 in the temperature range of $-150\,°C \sim 150\,°C$. Besides, Jin et al. [62] used triblock copolymer F127 as thermal degradation template in PAA. After removing the particle template during solvent evaporation, porous structures with the diameter of less than 50 nm can be formed in the polymer matrix. The minimum permittivity of this porous PI film was 1.92 when the molar ratios of F127 to PMDA (R) was 0.005.

Chemical-solvent-induced porosity is mainly achieved by etching the nanofillers with the help of highly corrosive solvents (e.g., hydrofluoric acid (HF) and potassium permanganate ($KMnO_4$)) [63]. Jiang et al [64] obtained nanoporous PIs with pore sizes ranging from 20 nm to 120 nm by etching silica (SiO_2) nanoparticles with HF. The permittivity of this film can be as low as 1.18 with increasing porosity. Huang et al. [65] prepared hollow SiO_2 microspheres by using $KMnO_4$ as an etchant, which were subsequently modified by fluorination to obtain fluorinated hollow SiO_2 (FH-SiO_2). It was shown that the k value of the porous film was significantly reduced to 2.61 at only 2 wt% FH-SiO_2 filling without sacrificing thermal stability.

Although the method of preparing porous PI films by sacrificing templates through physical or chemical method is easier to operate, there are certain problems. High temperature or the use of corrosive reagents will have certain operational risks.

When the introduced template is not completely decomposed or etched, it can also lead to an increase in the dielectric properties of the materials.

5.4.3 Introduction of Porous Fillers

5.4.3.1 Porous Inorganic Fillers

Inorganic fillers possess good electrical properties and dimensional stability. Nowadays, the most representative porous inorganic fillers applied include porous SiO_2, POSS, and porous nanotubes.

Mesoporous SiO_2 possesses an ordered and regular spatial structure and is widely used in molecular sieves, catalysts, and adsorbents [66, 67]. The earlier development of mesoporous SiO_2 belongs to SiO_2 mesoporous molecular sieve (MCM-41), which were first successfully synthesized by Kresge et al in 1992 [68]. So far, reports on low-k MCM-41/PI porous composites have become more mature [69, 70]. Additionally, novel MCM-41 hybrid fillers can be prepared to further reduce the permittivity of PI. Zhao et al. [69] designed a novel hybrid filler, POSS-functionalized MCM-41 (MCM-41@POSS). 3 wt% MCM-41@POSS/PI showed the lowest permittivity of 1.88 at 10^6 Hz and maintained a low dielectric loss ($\tan\delta = 0.01$). Here, the modifying effect of POSS was to prevent the penetration of PI molecular chains into the mesopores of MCM-41 in order to retain more air inside the PI system. This led to a significantly better permittivity of POSS@MCM-41/PI than that of the porous PI film containing only MCM-41.

Furthermore, the porous SiO_2 family also includes SiO_2 hollow spheres (SHS) and hollow tubes. Lee et al. [71] prepared SHS with a diameter of 200 nm using polystyrene (PS) as the template. When the content of SiO_2 hollow spheres was 5 wt%, the permittivity of porous PI composite film was reduced to 2.24, while its thermal stability was improved. Zhou et al [72] obtained the SHS by sol–gel method. The permittivity of the PI composite with 10 wt% SHS was reduced from 3.41 to 2.09.

Polyhedral oligomeric silsesquioxane (POSS) is the most representative class of materials in the siloxane family. Due to its characteristic high porosity, rigidity, and high thermal stability of the internal inorganic cage structure, polymer composites containing POSS usually exhibit excellent thermal stability and low permittivity [73–75]. And, it has also been mentioned in the above sections that POSS as part of the hybridized filler can assist in the synthesis of ultralow-*k* porous PI films. There are two forms of POSS presence in the PI matrix, i.e., embedded and filled. For the embedded POSS-PI system, the more typical representative is single active site modified POSS [76] and multi-active site modified POSS, such as octaaminophenylsilsesquioxane (OAPS) [77] as shown in Figure 5.5. This POSS-diamine can be involved in the synthesis of PI through covalent interaction, which not only promotes the reduction of dielectric properties, but also obtains controllable thermodynamic performance [2]. Leu et al. [78] obtained POSS-CPI by in situ polymerization of different ratios of NH_2-POSS monomer with PAA solution. The permittivity of POSS-CPI was about 2.32 and the thermal decomposition temperature was slightly decreased, but still remained at about 530 °C when the molar percentage of the POSS-diamine in the PI molecule was 16 mol%. But due to the increase in the FFV of POSS-CPI molecules, the coefficient of thermal expansion

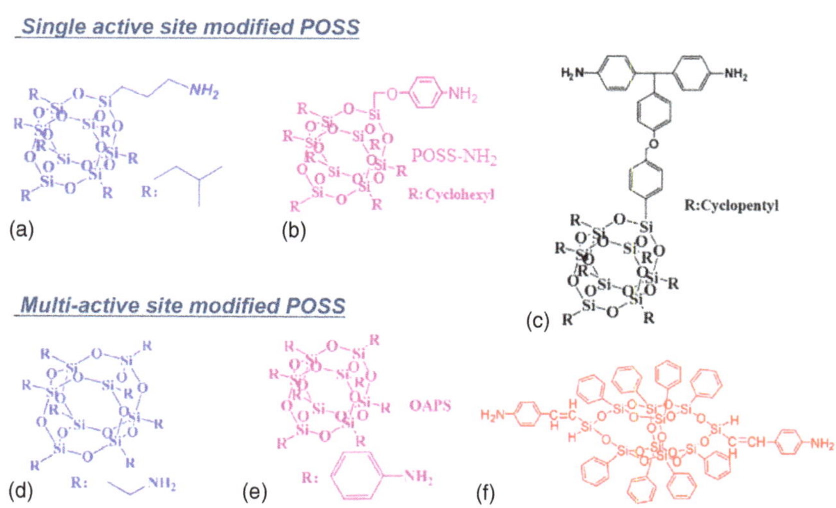

Figure 5.5 Different types of POSS-diamines: (a) aminopropylisobutyl-POSS (Source: Ref. [78]/American Chemical Society), (b, c) NH_2-POSS [78, 79], (d) octa (aminopropylsilsesquioxane) POSS-NH_2 (Source: Ref. [80]/with permission of Elsevier), (e) OAPS (Source: Ref. [77]/John Wiley & Sons) (f) 3,13-divinyloctaphenyl double decker silsesquioxane (Source: Ref. [81]/The Royal Society of Chemistry).

(CTE) of the composites increased with increasing POSS content. Therefore, Liu et al. [82] designed a crosslinked network and aminopropyl-isobutyl-modified POSS to jointly inhibit the enhancement of the CTE. The 14.6 wt% POSS-PI showed a decrease in the dielectric properties (k = 2.67, tan δ = 0.007) while achieving a good low thermal expansion (CTE = 24 ppm/K).

For the research of doping POSS into PI system as a separate filler, Chen et al. [83] formed nanoscale POSS aggregate based on maleimide isobutyl POSS (MIPOSS). The permittivity of MIPOSS/PI composite could be reduced to 2.09 when the concentration of POSS aggregates was 2.8 wt%. Shi et al. [84] synthesized CE6-PAA molecules featuring a necklace-like interlocking structure and then introduced highly rigid OAPS. This OAPS/CE6-PI nanocomposite with permittivity as low as 2.43 was obtained.

Furthermore, based on the idea of cross integration of multiple fields, researchers are also committed to exploring other porous inorganic fillers suitable for low-k fields.

Im et al. [85] obtained PI films with the k value of 1.9 by adding porous activated carbon to the PI matrix. Yin et al. [86] synthesized five-layer films of poly (amic acid) salt (PAAs) and poly (allylamine hydrochloride) (PAH) by electrostatic assembly, and subsequently immersed the films in an aqueous solution containing imogolite nanotubes. During thermal imidization, the water inside the multilayer film evaporates, i.e., nanopores are introduced into the PI film. The permittivity of this porous PI film is 2.78. Polyoxymetal (POM) is a class of inorganic metal oxide cluster compounds with special properties and structure, which has received increasing attention due to its nanoporous structure, controllable porosity, and simple synthesis method. However, Pang et al. [5] mentioned that POM is poorly compatible with PI and thus requires surface modification for effective filling. Tan et al. [87] prepared POM/PI composite films by organic modification of POM using silane coupling agents. The permittivity of the POM/PI films decreased to 2.05 when the loading of POM was 20 wt%. It was shown that the large free volume fraction and the intrinsic porosity of POM were the main reasons for the decrease in the permittivity.

However, it is worth noting that the incorporation of inorganic nanoporous fillers inevitably weakens the mechanical properties of PIs. Meanwhile, surface modification of the fillers is usually required to provide interfacial compatibility between the polyphases, which increases the complexity and cost of preparation [88].

5.4.3.2 Porous Organic Fillers

Covalent organic framework (COF) is a new type of porous crystalline material composed of light elements, which is characterized by customizable surface functional groups, ordered channel structures, and good thermal and chemical stability [89–91]. In 2021, Evans et al. [92] reported that 2D COF has an ultralow permittivity of 1.6, making it a promising ultralow-k interlayer dielectric material. Then, Zhao et al. [88] successfully introduced COF as a functional filler into a CPI matrix. The synthesized COF surface is rich in —NH_2 functional groups, which can crosslink with PAA to a certain extent (Figure 5.6). Therefore, thanks to the synergistic effect of

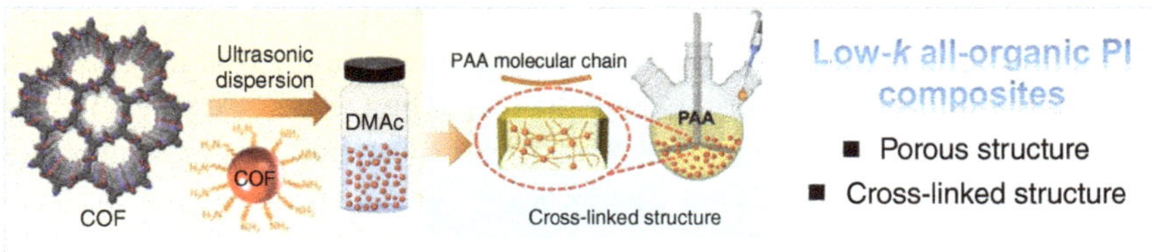

Figure 5.6 Schematic diagram of the spatial structure of COF and the crosslinked structure formed with PAA. Source: Ref. [88]/with permission of Elsevier.

the inherent porous of COF and crosslinked structures, the permittivity of the 2 wt% COF/PI composite film was reduced to 2.72 at 1 MHz, with a dielectric loss of 0.0077.

Moreover, Zhao et al. [93] synthesized a novel hollow-polydopamine microsphere (H-PDA) using polystyrene as a sacrificial template. The permittivity of 2 wt% H-PDA/PI was 1.96, and the dielectric loss was 0.0123. Notably, the significant decrease in dielectric property partly originated from the abundant amino groups on the surface of H-PDA which can react with the dianhydride monomer to form a crosslinked structure, producing a dielectric confinement effect and increasing the FFV of the PI composites.

Hypercrosslinked polymer (HCP) is a porous organic polymer composed of covalently crosslinked light elements (C, O, N, H) with high specific surface area, excellent thermal and chemical stability [94]. Therefore, the HCP is widely used in energy and environmental applications such as gas storage, carbon capture, and catalysis [95]. Sun et al. [94] investigated the effect of the introduction of HCP on the permittivity of PI (6FDA-ODA) at high frequencies. It was shown that an ultralow k value of 1.7 was obtained for 10 wt% HCP/PI at 8.2–9.6 GHz.

5.4.3.3 PI Aerogel

Aerogel is a lightweight mesoporous material obtained by removing the solvent from hydrogel, which is widely used in the fields of adsorption, dielectric property, thermal insulation, sensing, etc. [96–98] Organic polymer aerogels have better intrinsic flexibility than brittle inorganic aerogels (like silica aerogels). PI aerogels stand out among the many polymer aerogels by exhibiting stronger mechanical properties and temperature resistance [99]. Strategies for the preparation of PI aerogels include supercritical CO_2 drying involving sol–gel conversion, freeze drying, and ambient pressure drying.

The preparation of PI aerogels by supercritical CO_2 drying is one of the most common ways and has undergone mature research in recent years. Conventional linear PI structures have poor volume stability during drying and imidization, which can easily lead to volume shrinkage and structural collapse of the aerogel in the synthesis process. Therefore, Meador et al. [100] have successively reported on the synthesis of ultralow-k crosslinked PI aerogels. A crosslinked PI aerogel (CPA) with a k value of 1.26 was obtained by synthesizing PAA oligomers based on BPDA, 2,2′-dimethylbenzidine (DMBZ) and by using 1,3,5-tris(4-aminophenoxy) benzene (TAB) as the crosslinking agent. When this PI aerogel is used as an antenna substrate, it outperforms ordinary commercial antenna substrates and has great potential for application in fields such as aerospace [100]. Furthermore, CPA synthesized on the basis of ODA-6FDA-TAB exhibited a lower permittivity ($k = 1.08$). This was mainly attributed to the presence of isopropyl fluoride leading to a larger internal volume and lower shrinkage [101].

Recently, more in-depth investigations have been carried out on CPAs prepared by supercritical CO_2 drying [99, 102]. For example, Li et al. [99] investigated the effect of single and double crosslinking on the dielectric properties of CPA. In Figure 5.7, the permittivity of single crosslinked PI aerogel was in the range of 1.34–1.53, and the dielectric loss was in the range of 0.008–0.020 at 12 GHz. While the permittivity of double crosslinked PI aerogel varied from 1.10 to 1.17, the dielectric loss was in the range of 0.003–0.006. It was because the double crosslinked network gave the CPA a lower shrinkage and density. Meanwhile, the double CPAs also exhibited superior thermal insulation properties. Dong et al. [103] prepared a series of CPAs with good compression properties and tunable permittivity by freeze-drying using tris(4-aminophenylamine) (TPA) as a crosslinking agent. Owing to the bridging effect of TPA, a "spider-like" spatial structure was supported inside the CPAs. This not only enabled the CPAs to exhibit low dielectric properties, with a minimum k value of 1.09 and dielectric loss of 0.03, but also imparted good compressive resilience to the CPAs. This will further expand the application potential of PI aerogels in the field of flexible electronics.

However, supercritical CO_2 drying involves extremely high pressure (critical pressure needs to be above 7.4 MPa), while freeze-drying requires low temperature (−10–60 °C) and vacuum (30–10 Pa) environments, both of which inevitably increase the cost and safety of aerogel membrane preparation. The ambient pressure drying method offers a promising application for the safe and efficient preparation of porous PI aerogels [104]. However, it should be noted that fully imidized PI aerogels are difficult to dry under ambient pressure. Ding et al. [105] obtained PI aerogels in atmospheric pressure drying environment by a partial pre-imidization process and solvent substitution. Specifically, γ-aminopropyltriethoxysilane (APTES) was introduced into PAA oligomers to obtain APTES-modified PAA-PI sols. Then it was sequentially substituted with ethanol and hexane, and finally chemically imidized to obtain PI aerogels. Here, the presence of APTES inhibits irreversible

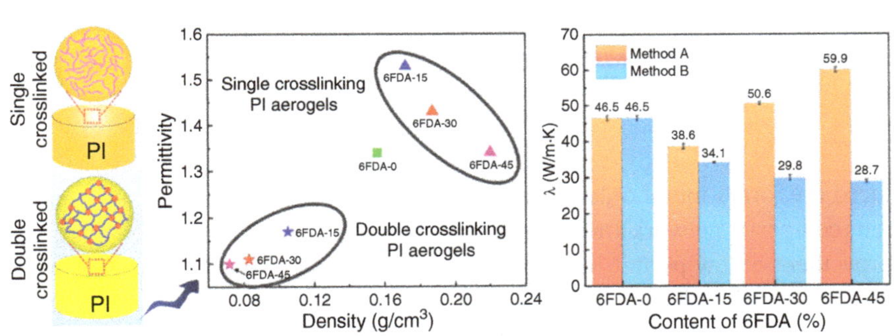

Figure 5.7 (a) Internal network structures of single and double crosslinked PI aerogel and their permittivity and thermal conductivity. Source: [99]/with permission of Elsevier.

shrinkage of the gels during the drying process. The PI aerogel possessed a low permittivity of 1.414 and good thermal insulation properties.

5.5 Multifunctional Low-*k* PIs

In the field of electronics and electricity, the diversification of the functionality of low-*k* materials is also required for specific applications. For example, low-*k* and high thermal conductive PIs for microelectronic packages, low-*k* PIs with high electromagnetic shielding for smart electromagnetic devices, low-*k* PIs with low thermal expansion for printed circuits, and low-*k* and high flame-retardant PIs for aerospace applications. In this subsection, researches on multifunctional low-*k* PIs will be presented and discussed with several specific application scenarios as representatives.

The increasing integration and high power of microelectronic devices make the heat flow generated by the devices tend to show periodic and non-uniform aggregation; thus, the rapid increase in temperature due to ultrahigh heat flow density has become a bottleneck restricting the development of the devices [106–109]. Therefore, there have been several researches carried out for synergistic improvement of thermal conductivity and low-*k* of dielectric materials. Zhao et al. [110] prepared alkylated boron nitride nanosheets (BNNSs) by exfoliation modification of hexagonal boron nitride (h-BN) using hexanol. It was found that the obtained alkyl-BNNSs/PI composites had high thermal conductivity of 6.21 W/(m·K) and relatively low permittivity of 3.23 at 20 wt% loading. Similarly, Zhang et al. [111] utilized alkyl-BNNSs as a functional filler and combined with the hot-pressing process to further order the orientation of alkyl-BNNSs in FPI matrix. The thermal conductivity of 4.99 wt% alkyl-BNNSs/PI raised to 2.36 W/(m·K), the permittivity was as low as 1.76, and the dielectric loss was 0.0065 (1 MHz). The significant reduction in dielectric properties was attributed to the introduction of low loading of alkyl-BNNS uniformly dispersed in FPI to optimize the multiphase interface, suppress the electrical conductivity, and inhibit the interfacial polarization. Nevertheless, it is difficult to achieve higher thermal conductivity and stable low permittivity of PI composites only by introducing modified functional fillers. Dong et al. [112] proposed a strategy to construct a "3D porous composite network" to develop a unique low-*k* and high thermal conductive porous PI composites (Figure 5.8a). The first use of calcium fluoride (CaF_2) in combination with ionic liquids induced pores led to the construction of 3D porous composite networks in PI matrix. Interestingly, the charge transfer interaction between CaF_2 and PI induced the formation of ordered "crystal-like" structures. The combination of 3D porous composite network and "crystal-like" structure contributes to the synergistic improvement of low permittivity and high thermal conductivity of PI composites. The 10 vol% IL-CaF_2/PI porous composite film possessed the lowest permittivity of 2.14 and the highest thermal conductivity of 7.22 W/(m·K).

Additionally, wireless communication electronic devices also generate serious electromagnetic radiation during operation, which not only threatens human

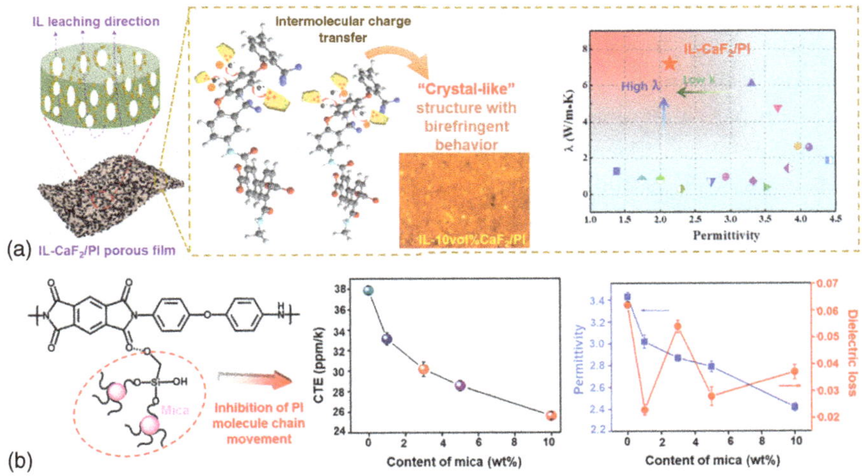

Figure 5.8 (a) Schematic molecular and spatial structure of IL-CaF_2/PI composites and their permittivity (Inset is a polarized microscope (POM) image of the IL-10 vol% CaF_2/PI film). Source: Ref. [112]/John Wiley & Sons. (b) CTE and permittivity of mica/PI. Source: Adapted from [113].

health, but also causes damage to sensitive electromagnetic equipment [114, 115]. Therefore, low-k materials with both high electromagnetic shielding are receiving more and more attention to shield undesirable electromagnetic waves and prevent signal interference. Considering the application of equipment in extreme environments, electromagnetic shielding materials also need to provide good thermal stability and mechanical properties. Zhou et al. [116] developed an aramid nanofiber (ANF)/reduced graphene oxide (rGO)/PI composite aerogel. The ANF/rGO/PI-15 maintained low permittivity (k = 2.0–2.3) in the high-frequency range (8.2–12.4 GHz) and exhibited the strongest effective absorption capacity for electromagnetic waves. Generally, 90% of EM wave will be effectively absorbed when the reflection loss value is less than 10 dB [117]. The reflection loss minimum (RL_{min}) of this composite aerogel reached −41.0 dB at 10.8 GHz. This was mainly due to the fact that the graphene and porous structure inside the aerogel triggers the electromagnetic wave to undergo multiple reflections and scattering, increasing the propagation path and dissipation intensity of the electromagnetic wave.

Printed circuit boards will involve the lamination of PI films with copper foils. Due to the structural differences between polymers and metals, the CTE of aromatic PIs (66 ~ 67 ppm/K) is much higher than that of copper foils (17 ~ 18 ppm/K). Therefore, the design and development of low-k PIs with low CTE can effectively solve the quality problems of warpage, fracture and delamination that are prone to occur in copper foil laminates during operation [118, 119]. The reduction of CTE can be achieved by blocking the movement of PI molecular chains by enhancing the interactions between the molecular chains in the PI system [113, 120]. Zhao et al. [113] prepared mica/PI composite films by modifying mica with a coupling agent (γ-aminopropyltriethoxysilane, APTS). The variation rules of low permittivity and thermal expansion coefficient of the synergistically modulated mica/PI composite films were thoroughly explored. As shown in Figure 5.8b, the CTE and permittivity of the 10 wt% mica/PI composite films decreased to 25.36 ppm/k and 2.42, respectively, which exhibited excellent high-temperature dimensional stability.

Low-*k* PIs with high-temperature thermally insulating and flame resistance also have a wide market in aerospace, construction, and fire protection. The limiting oxygen index (LOI) is utilized to assess the flame retardancy of materials [121]. When the LOI value is greater than 26, the material has self-extinguishing behavior and high flame-retardant properties. Devaraju et al [122] obtained OAPS/PI composite films by in situ co-polymerization of OAPS, bisphenol A ether diamine, and PMDA. The *k* value of PI film decreased from 3.34 to 2.68 at 15 wt% of OAPS, and it also exhibited excellent flame-retardant properties with LOI ranging from 34.9 to 37.7. Although there have been some studies on polymers with low-*k* and high flame-retardant properties [123–125], there are still relatively few researches on low-*k* and high flame-retardant PI.

5.6 Conclusion and Prospects

In conclusion, this chapter describes the basic schemes to reduce the permittivity of PI and summarizes the research progress of low-*k* PI materials at home and abroad in recent years. Intrinsic low-*k* PI can achieve the reduction of permittivity by the choice of special monomer structure or increasing the FFV of molecules on the basis of maintaining the good mechanical properties of PI. But the disadvantage is that the level of permittivity reduction is limited, and the monomer molecules containing special structures are synthesized at a high cost and complex process. PI composites and porous PI are currently more popular mainstream solutions for the synthesis of ultralow-*k* PI. However, the pursuit of lowering the permittivity will inevitably reduce the intrinsic thermal stability and mechanical strength of PI. Therefore, although some research results on the synthesis of low-*k* PI materials have been achieved, the exploration of low-*k* PI materials still needs to be further developed and faces big challenges.

To address the future direction and challenges of low-*k* PIs, we summarize several perspectives here:

1) Although more mature low-*k* or ultralow-*k* PIs are now available at the laboratory stage, the industrial-scale production process for low-*k* or ultralow-*k* PIs at suitable molecular weights is not yet clear.
2) The lifetime of PI films under complex operating conditions needs to be further focused and evaluated in the future to improve the reliability of the films for applications in extreme environments.
3) The presence of complex operating conditions (vibration, compression, and bending) can lead to changes in the gaps between material interfaces. Interlayer insulating materials with good thermal conductivity and dielectric property also need to be resistant to deformation in order to maintain the continued effectiveness of interfacial heat transfer.

4) The cumulative increase in e-waste makes incineration of PI an urgent problem. Besides its dielectric, mechanical, and thermal conductivity properties, PI should be highly valued for their recyclability. The design of low-*k* dynamic PIs will create more opportunities for the upgrading of smart electronic and electrical equipment.

References

1 Volksen, W., Miller, R.D., and Dubois, G. (2010). Low dielectric constant materials. *Chemical Reviews* 110 (1): 56–110.

2 Miao, L., Zhan, L.L., Liao, S.L. et al. (2024). The recent advances of polymer-POSS nanocomposites with low dielectric constant. *Macromolecular Rapid Communications* 45 (5): 2300601.

3 Hatton, B.D., Landskron, K., Hunks, W.J. et al. (2006). Materials chemistry for Low-k materials. *Materials Today* 9 (3): 22–31.

4 Maier, G. (2001). Low dielectric constant polymers for microelectronics. *Progress in Polymer Science* 26 (1): 3–65.

5 Pang, Z.J., Sun, H.C., Guo, Y. et al. (2023). Research advances of porous polyimide-based composites with low dielectric constant. *Polymers* 15 (16): 3341.

6 Zha, J.W., Dang, Z.M., Zhou, T. et al. (2010). Electrical properties of TiO_2-filled polyimide nanocomposite films prepared via an in-situ polymerization process. *Synthetic Metals* 160 (23): 2670–2674.

7 Wan, B.Q., Dong, X.D., Yang, X. et al. (2022). Rising of dynamic polyimide materials: a versatile dielectric for electrical and electronic applications. *Advanced Materials* 35 (39): 2301185.

8 Liu, X.J., Zheng, M.S., Chen, G. et al. (2022). High-temperature polyimide dielectric materials for energy storage: theory, design, preparation and properties. *Energy & Environmental Science* 15 (1): 56–81.

9 Dong, X.D., Zheng, M.S., Wan, B.Q. et al. (2021). Low-permittivity copolymerized polyimides with fluorene rigid conjugated structure. *Materials* 14 (21): 6266.

10 Zhang, Y.L., Ruan, K.P., Zhou, K. et al. (2023). Controlled distributed $Ti_3C_2T_x$ hollow microspheres on thermally conductive polyimide composite films for excellent electromagnetic interference shielding. *Advanced Materials* 35 (16): 2211642.

11 Li, Y.H., Sun, G.H., Zhou, Y. et al. (2022). Progress in low dielectric polyimide film - a review. *Progress in Organic Coatings* 172: 107103.

12 Yamaguchi, H., Ueda, T., Miura, K. et al. (2012). 11.7-Inch flexible AMOLED display driven by a-IGZO TFTs on plastic substrate. *SID Symposium Digest of Technical Papers* 43: 1002–1005.

13 Wang, C.Y., Chen, W.T., Chen, Y.Y. et al. (2014). New fluorinated poly (ether sulfone imide) s with high thermal stability and low dielectric constant. *Materials Chemistry and Physics* 143 (2): 773–778.

14 Bong, S., Yeo, H., Ku, B. et al. (2018). Highly soluble polyimide based on asymmetric diamines containing trifluoromethyl group for high performance dielectric material. *Macromolecular Research* 26 (1): 85–91.

15 Nagell, S.R., Park, S.S., Chitumall, R.K. et al. (2024). Intrinsic low-dielectric-onstant polyimides based on a novel diamine having -CF_3 groups and kinkable 1,4-diisopropylbenzene units. *Polymer International* 73 (3): 238–247.

16 Fan, F.G., Liu, S.W., Chi, Z.G. et al. (2021). Construction and study of quantitative structure-property relationship model for intrinsic polyimide dielectric constant. *Acta Polymerica Sinica* 52: 750–761.

17 Chisca, S., Musteata, V.E., Stoica, I. et al. (2013). Effect of the chemical structure of aromatic-cycloaliphatic copolyimide films on their surface morphology, relaxation behavior and dielectric properties. *Journal of Polymer Research* 20: 111–121.

18 He, J.H., Wu, X.L., and Cheng, Y.R. (2023). Low dielectric post-cured benzocyclobutene-functionalized fluorine-containing polyimide material. *European Polymer Journal* 196: 112334.

19 Bei, R.X., Qian, C., Zhang, Y. et al. (2017). Intrinsic low dielectric constant polyimides: relationship between molecular structure and dielectric properties. *Journal of Materials Chemistry C* 5 (48): 12807–12815.

20 Liu, Y.W., Zhang, Y., Lan, Q. et al. (2013). Synthesis and properties of high-performance functional polyimides containing rigid nonplanar conjugated tetraphenylethylene moieties. *Journal of Polymer Science Part A: Polymer Chemistry* 51 (6): 1302–1314.

21 Chen, W.X., Zhou, Z.X., Yang, T.T. et al. (2016). Synthesis and properties of highly organosoluble and low dielectric constant polyimides containing non-polar bulky triphenyl methane moiety. *Reactive and Functional Polymers* 108: 71–77.

22 Liu, Y.W., Qian, C., Qu, L.J. et al. (2015). A bulk dielectric polymer film with intrinsic ultralow dielectric constant and outstanding comprehensive properties. *Chemistry of Materials* 27 (19): 6543–6549.

23 Zhuang, Y.B., Orita, R., Fujiwara, E. et al. (2019). Colorless partially alicyclic polyimides based on Tröger's base exhibiting good solubility and dual fluorescence/phosphorescence emission. *Macromolecules* 52 (10): 3813–3824.

24 Yu, Y.F., Zhu, W.J., Li, H.T. et al. (2024). Transparent semi-aliphatic polyimide films with excellent thermal stabilities for applications in the display technology. *European Polymer Journal* 203: 112653.

25 Yoo, D.K., Lim, S.K., Yoon, T.H. et al. (2003). Physical properties and stress analysis of low dielectric polyimide films containing adamantane pendant group. *Polymer Journal* 35: 697–703.

26 Li, H.M., Wang, X.M., Gong, Y.Z. et al. (2023). Polyimide/crown ether composite film with low dielectric constant and low dielectric loss for high signal transmission. *RSC Advances* 13 (11): 7585–7596.

27 Lee, Y.H., Lin, J.W., Huang, H.C. et al. (2023). High thermally stable anthracene-based polyimides with low dielectric constants and dissipation factors. *ACS Applied Polymer Materials* 5 (12): 10196–10204.

28 Chen, Y., Chen, X.Y., Wang, Y. et al. (2023). Increasing twist rigidity to prepare polyimides with low dielectric constants and dissipation factors over a wide temperature range. *Macromolecules* 56 (23): 9379–9388.

29 Kratochvil, A.M. and Koros, W.J. (2008). Decarboxylation-induced cross-linking of a polyimide for enhanced CO_2 plasticization resistance. *Macromolecules* 41 (21): 7920–7927.

30 Zhou, H., Lei, H., Wang, J. et al. (2019). Breaking the mutual restraint between low permittivity and low thermal expansion in polyimide films via a branched crosslink structure. *Polymer* 162: 116–120.

31 Han, S.H., Li, Y.N., Hao, F.Y. et al. (2021). Ultra-low dielectric constant polyimides: combined efforts of fluorination and micro-branched crosslink structure. *European Polymer Journal* 143: 110206.

32 Huang, Y.J., Lin, J.W., Lee, Y.H. et al. (2023). Monomer ratio-controlled polyimides with enhanced dielectric properties and thermal stabilities through crosslinking network. *Journal of Polymer Science* 62 (6): 1–11.

33 Peng, Z.Y., Ye, A., Zhang, L. et al. (2024). Micro-crosslinked polyimide nanocomposites with low dielectric constant and low dielectric loss for microwave antenna with molecular dynamics. *Composites Communications* 46: 101804.

34 Qi, H.X., Wang, X.L., Zhu, T.S. et al. (2019). Low dielectric poly (imide siloxane) films enabled by a well-defined disiloxane-linked alkyl diamine. *ACS Omega* 4 (26): 22143–22151.

35 Pu, L.Y., Huang, X., Wang, W. et al. (2019). Strategy to achieve ultralow dielectric constant for polyimide: introduction of fluorinated blocks and fluorographene nanosheets by in situ polymerization. *Journal of Materials Science: Materials in Electronics* 30: 14679–14686.

36 Kumar, S., Arumugham, H., Roy, D. et al. (2022). Synthesis and characterization of fluorine functionalized graphene oxide dispersed quinoline-based polyimide composites having low-k and UV shielding properties. *Polymers for Advanced Technologies* 33: 427–439.

37 Yin, X.D., Feng, Y.F., Zhao, Q. et al. (2018). Highly transparent, strong, and flexible fluorographene/fluorinated polyimide nanocomposite films with low dielectric constant. *Journal of Materials Chemistry C* 6 (24): 6378–6384.

38 Kovtyukhova, N.I., Ollivier, P.J., Martin, B.R. et al. (1999). Layer-by-layer assembly of ultrathin composite films from micron-sized graphite oxide sheets and polycations. *Chemistry of Materials* 11 (3): 771–778.

39 Lerf, A., He, H.Y., Forster, M. et al. (1998). Structure of graphite oxide revisited. *The Journal of Physical Chemistry B* 102 (23): 4477–4482.

40 Wang, T., Li, J.H., Niu, F.F. et al. (2022). Low-temperature curable and low-dielectric polyimide nanocomposites using aminoquinoline-functionalized graphene oxide nanosheets. *Composites Part B: Engineering* 228: 109412.

41 He, D.Q., Wang, Z.G., Long, J.P. et al. (2018). Preparation of fluorinated graphene oxide/polyimide composites with low dielectric constant and moisture resistance. *Nano* 13: 1850098–1850109.

42 Chen, X., Huang, H.H., Shu, X. et al. (2017). Preparation and properties of a novel graphene fluoroxide/polyimide nanocomposite film with a low dielectric constant. *RSC Advances* 7 (4): 1956–1965.

43 Wang, X., Dai, Y.Y., Wang, W.M. et al. (2014). Fluorographene with high fluorine/carbon ratio: a nanofiller for preparing low-k polyimide hybrid films. *ACS Applied Materials & Interfaces* 6 (18): 16182–16188.

44 Zhang, Y., Wang, J.W., Ma, Y.J. et al. (2024). Polyimide composites with fluorinated graphene and functionalized boron nitride nanosheets for heat dissipation. *ACS Applied Nano Materials* 7 (5): 5009–5018.

45 Zhao, G.Q., Mu, X., Ma, D.D. et al. (2023). Dielectric and mechanical properties of TiO_2/polyimide composites with low dielectric constant. *Polymer Engineering and Science* 63 (7): 1953–1960.

46 Zhang, Y.H., Dang, Z.M., Fu, S.Y. et al. (2005). Dielectric and dynamic mechanical properties of polyimide-clay nanocomposite films. *Chemical Physics Letters* 401 (4): 553–557.

47 Huang, Z.X., Liu, S.M., Yuan, C.Y. et al. (2015). High-performance fluorinated polyimide/pure silica zeolite nanocrystal hybrid films with a low dielectric constant. *RSC Advances* 5 (93): 76476–76482.

48 Zhang, J.L., Lai, X.W., Li, J.H. et al. (2023). Comprehensive properties study of polyimide with fluorinated nanofillers. *Polymer Composites* 44: 5058–5069.

49 Cheng, T.J., Lv, G.P., Li, Y.T. et al. (2021). Low dielectric polyimide/fluorinated ethylene propylene (PI/FEP) nanocomposite film for high-frequency flexible circuit board application. *Macromolecular Materials and Engineering* 306 (7): 2100086.

50 Dong, X.D., Wan, B.Q., Feng, Y. et al. (2021). Ultra-low-permittivity, high hydrophobic, and excellent thermally stable fluoroelastomer/polyimide composite films employing dielectric reduction. *European Polymer Journal* 181: 111667.

51 Kim, S., Son, J., Park, H. et al. (2022). Polymer concentration and liquid-liquid demixing time correlation with porous structure of low dielectric polyimide in diffusion-driven phase separation. *Polymers* 14 (7): 1425.

52 Dong, F.P., Li, H.W., Lu, L.Y. et al. (2019). Superhydrophobic and low-k polyimide film with porous interior structure and hierarchical surface morphology. *Macromolecular Materials and Engineering* 304: 1900252–1900256.

53 Ren, Y.X. and Lam, D.C.C. (2008). Properties and microstructures of low-temperature processable ultralow-dielectric porous polyimide films. *Journal of Electronic Materials* 37: 955–961.

54 Guillen, G.R., Pan, Y., Li, M. et al. (2011). Preparation and characterization of membranes formed by nonsolvent induced phase separation: a review. *Industrial and Engineering Chemistry Research* 50 (7): 3798–3817.

55 Kahrs, C. and Schwellenbach, J. (2020). Membrane formation via non-solvent induced phase separation using sustainable solvents: A comparative study. *Polymer* 186: 122071.

56 Khim, S., Hwang, Y.C., Choi, J. et al. (2023). Temperature-invariant large broadband polyimide dielectrics with multimodal porous networks. *ACS Applied Polymer Materials* 5 (6): 4159–4169.

57 Li, Z., Zou, H., and Liu, P. (2015). Morphology and properties of porous polyimide films prepared through thermally induced phase separation. *RSC Advances* 5: 37837–37842.

58 Hwang, Y.C., Khim, S., Sohn, J.M. et al. (2023). Controllable growth of porous morphology in low dielectric polyimides via thermal-driven spontaneous phase separation. *European Polymer Journal* 195: 112195.

59 Zhang, C., Zhai, T.L., Wang, J.J. et al. (2014). Triptycene-based microporous polyimides: synthesis and their high selectivity for CO_2 capture. *Polymer* 55: 3642–3647.

60 Yang, Z.H., Guo, H.Q., Gao, L.X. (2020). Structural design of polyimide and its high frequency dielectric properties. *Proceedings of the 21st China CCL Technology Symposium* 241–245.

61 Lv, P.X., Dong, Z.X., Dai, X.M. et al. (2017). Low-dielectric polyimide nanofoams derived from 4,4′ -(hexafluoroisopropylidene) diphthalicanhydride and 2,2′-bis (trifluoromethyl)benzidine. *RSC Advances* 8: 4848–4854.

62 Jin, Y.X., Tang, J., Hu, J. et al. (2011). One-step fabrication of ultralow dielectric polyimide films consisting of size-controlled mesoporous nanoparticles. *Colloids and Surfaces A: Physicochemical and Engineering Aspects* 392 (1): 178–186.

63 Huang, B.L., Li, K., Peng, M.Y. et al. (2022). Polyimide/fluorinated silica composite films with low dielectric constant and low water absorption. *High Performance Polymers* 34 (4): 434–443.

64 Jiang, L.Z., Liu, J.G., Wu, D.Z. et al. (2006). Methodology for the preparation of nonporous polyimide films with low dielectric constants. *Thin Solid Films* 510 (1): 241–246.

65 Chen, W.C., Lin, R.C., Tseng, S.M. et al. (2018). Minimizing the strong screening effect of polyhedral oligomeric silsesquioxane nanoparticles in hydrogen-bonded random copolymers. *Polymers* 10 (3): 303.

66 Yang, C.M., Cho, A.T., Pan, F.M. et al. (2001). Spin-on mesoporous silica films with ultralow dielectric constants, ordered pore structures, and hydrophobic surfaces. *Advanced Materials* 13 (14): 1099–1102.

67 Li, Y. and Shi, J. (2014). Hollow-structured mesoporous materials: chemical synthesis, functionalization and applications. *Advanced Materials* 26 (20): 3176–3205.

68 Kresge, C.T., Leonowicz, M.E., Roth, W.J. et al. (1992). Ordered mesoporous molecular sieves synthesized by a liquid crystal template mechanism. *Nature* 395: 710–712.

69 Zhao, W.J., Lu, C.H., Zhao, H. et al. (2022). Achieving hydrophobic ultralow dielectric constant polyimide composites: combined efforts of fluorination and porous fillers. *Macromolecular Materials and Engineering* 307 (9): 2200291.

70 Dang, Z.M., Ma, L.J., Zha, J.W. et al. (2009). Origin of ultralow permittivity in polyimide/mesoporous silicate nanohybrid films with high resistivity and high breakdown strength. *Journal of Applied Physics* 105 (4): 044104.

71 Lee, S., Choi, M., Park, S. et al. (2011). Synthesis and characterization of hybrid films of polyimide and silica hollow spheres. *Macromolecular Research* 19: 599–607.

72 Zhou, H., Wei, D.Y., Fan, Y. et al. (2016). Dielectric properties of polyimide/ SiO_2 hollow spheres composite films with ultralow dielectric constant. *Materials Science and Engineering B* 203: 13–18.

73 Wu, J. and Mathur, P.T. (2009). POSS polymers: physical properties and biomaterials applications. *Polymer Reviews* 49 (1): 25–63.

74 Xu, Z.G., Zhao, Y., Wang, X.G. et al. (2013). A thermally healable polyhedral oligomeric silsesquioxane (POSS) nanocomposite based on Diels-Alder chemistry. *Chemical Communications* 49 (60): 6755–6757.

75 Ye, Q., Zhou, H., and Xu, J. (2016). Cubic polyhedral oligomeric silsesquioxane based functional materials: synthesis, assembly, and applications. *Chemistry - An Asian Journal* 11 (9): 1322.

76 Huang, C., Li, J.H., Xie, G.X. et al. (2019). Low-dielectric constant and low-temperature curable polyimide/POSS nanocomposites. *Macromolecular Materials and Engineering* 304 (12): 1900505.

77 Govindaraj, B., Sundararajan, P., and Sarojadevi, M. (2012). Synthesis and characterization of polyimide/polyhedral oligomeric silsesquioxane nanocomposites containing quinolyl moiety. *Polymer International* 61 (8): 1344.

78 Leu, C.M., Chang, Y.T., and Wei, K.H. (2003). Synthesis and dielectric properties of polyimide-tethered polyhedral oligomeric silsesquioxane (POSS) nanocomposites via POSS-diamine. *Macromolecules* 36 (24): 9122–9127.

79 Leu, C.M., Reddy, G.M., Wei, K.H. et al. (2003). Synthesis and dielectric properties of polyimide-chain end tethered polyhedral oligomeric silsesquioxane nanocomposites. *Chemistry of Materials* 15 (11): 2261–2265.

80 Seckin, T., Köytepe, S., and Adıgüzel, H.I. (2008). Molecular design of POSS core star polyimides as a route to low-k dielectric materials. *Materials Chemistry and Physics* 112 (3): 1040–1046.

81 Zheng, S., Wang, L., and Liu, N. (2016). Organic-inorganic polyimides with double decker silsesquioxane in the main chains. *Polymer Chemistry* 7 (5): 1158–1167.

82 Liu, X., Zhou, J.J., Zhou, Y.B. et al. (2022). Chemically crosslinked polyimide-POSS hybrid: a dielectric material with improved dimensional stability and dielectric properties. *European Polymer Journal* 173: 111315.

83 Chen, Z.G., Zhou, Y.B., Wu, Y.C. et al. (2019). Fluorinated polyimide with polyhedral oligomeric silsesquioxane aggregates: toward low dielectric constant and high toughness. *Composites Science and Technology* 181: 107700.

84 Shi, C.Q., Liu, S.M., Li, Y. et al. (2017). Imparting low dielectric constant and high modulus to polyimides via synergy between coupled silsesquioxane and crown ethers. *Composites Science and Technology* 142: 117–123.

85 Im, J.S., Bae, T.S., Lee, S.K. et al. (2010). Effects of porous carbon additives and induced fluorine on low dielectric constant polyimide synthesized with an e-beam. *Materials Research Bulletin* 45 (11): 1641–1647.

86 Yin, L.Y., He, Y.Q., Guo, W.H. et al. (2023). Low dielectric constant ultrathin polyimide films alternated by poly (amic acid) salt/poly (allylamine hydrochloride)/imogolite through layer by layer deposition. *Advanced Composites and Hybrid Materials* 6: 212.

87 Tan, L., Liu, S., Zeng, F. et al. (2011). A low dielectric constant polyimide/polyoxometalate composite. *Polymers for Advanced Technologies* 22 (2): 209–214.

88 Zhao, W.J., Wei, Z.Y., Lu, C.H. et al. (2023). Construction of all-organic low dielectric polyimide hybrids via synergistic effect between covalent organic framework and cross-linking structure. *Nano Materials Science* 5 (4): 429–438.

89 Liu, X.L., Pang, H.W., Liu, X.W. et al. (2021). Orderly porous covalent organic frameworks-based materials: superior adsorbents for pollutants removal from aqueous solutions. *The Innovation* 2 (1): 100076.

90 Kou, X.X., Tong, L.J., Huang, S.M. et al. (2021). Recent advances of covalent organic frameworks and their application in sample preparation of biological analysis. *Trends in Analytical Chemistry* 136: 116182.

91 Xu, S. and Zhang, Q. (2021). Recent progress in covalent organic frameworks as light-emitting materials. *Materials Today Energy* 20: 100635.

92 Evans, A.M., Giri, A., Sangwan, V.K. et al. (2021). Thermally conductive ultra-low-k dielectric layers based on two-dimensional covalent organic frameworks. *Nature Materials* 20: 1142–1148.

93 Zhao, W.J., Tong, Y.Z., Lu, C.H. et al. (2024). Bioinspired construction of size adjustable hollow polydopamine microspheres for ultra-low dielectric polyimide composites. *Applied Surface Science* 643: 158714.

94 Sun, W.K., Yang, M., Peng, K. et al. (2023). All-organic hyper-crosslinked polymer/polyimide composite films with ultralow high-frequency dielectric constant. *Macromolecular Rapid Communications* 44 (12): 2200956.

95 Tan, L.X. and Tan, B. (2017). Hypercrosslinked porous polymer materials: design, synthesis, and applications. *Chemical Society Reviews* 46 (11): 3322.

96 Wang, Y.J., Cui, Y., Shao, Z.Y. et al. (2020). Multifunctional polyimide aerogel textile inspired by polar bear hair for thermoregulation in extreme environments. *Chemical Engineering Journal* 390: 124623.

97 Li, C., Zhou, C., Lv, J. et al. (2019). Bio-molecule adenine building block effectively enhances electromagnetic interference shielding performance of polyimide-derived carbon foam. *Carbon* 149: 190–202.

98 Zhu, C.Y., Xue, T.T., Ma, Z.C. et al. (2023). Mechanically strong and thermally insulating polyimide aerogel fibers reinforced by prefabricated long polyimide fibers. *ACS Applied Materials & Interfaces* 15 (9): 12443–12452.

99 Li, Y., Ma, B., Zhang, R.B. et al. (2022). Mechanically strong, thermal-insulated, and ultralow dielectric polyimide aerogels with adjustable crosslinking methods. *Polymer* 253: 125035.

100 Meador, M.A.B., Wright, S., Sandberg, A. et al. (2012). Low dielectric polyimide aerogels as substrates for lightweight patch antennas. *ACS Applied Materials & Interfaces* 4: 6346–6353.

101 Meador, M.A.B., Mcmillon, E., Sandberg, A. et al. (2014). Dielectric and other properties of polyimide aerogels containing fluorinated blocks. *ACS Applied Materials & Interfaces* 6: 6062–6068.

102 Li, D.D., Lu, Z.H., Ke, Z. et al. (2024). Moisture resistant polyimide aerogel membranes with low dielectric constant and super thermal insulation for electronic device under harsh environment. *Polymer* 290: 126478.

103 Dong, X.D., Wan, B.Q., Zheng, M.S. et al. (2023). Versatile spider-web-like cross-linked polyimide aerogel with tunable dielectric permittivity for highly sensitive flexible sensors. *Chemical Engineering Journal* 465: 143034.

104 Zhu, Z., Yao, H., Dong, J. et al. (2019). High-mechanical-strength polyimide aerogels crosslinked with 4, 4′-oxydianiline-functionalized carbon nanotubes. *Carbon* 144: 24–31.

105 Ding, C.C., Li, R.Y., Yu, J. et al. (2022). Synthesis of porous polyimide films with low dielectric constant and excellent mechanical properties by ambient pressure drying. *Journal of Materials Science* 57: 9480–9492.

106 Wang, S.L., Cheng, Y., Wang, R.R. et al. (2014). Highly thermal conductive copper nanowire composites with ultralow loading: toward applications as thermal interface materials. *ACS Applied Materials & Interfaces* 6 (9): 6481–6486.

107 Zhang, D.L., Zha, J.W., Li, C.Q. et al. (2017). High thermal conductivity and excellent electrical insulation performance in double-percolated three-phase polymer nanocomposites. *Composites Science and Technology* 144: 36–42.

108 Ma, J. and He, Z.M. (2004). Thermal, electrical, and mechanical properties of layered substrates for microelectronic applications. *Thin Solid Films* 462: 477–480.

109 Liu, J.C., Guo, Y.F., Weng, C.X. et al. (2020). High thermal conductive epoxy based composites fabricated by multi-material direct ink writing. *Composites Part A Applied Science and Manufacturing* 129: 105684.

110 Zhao, K., Wei, S.Y., Wang, M. et al. (2023). Alkylated modified boron nitride nanosheets/polyimide composite films with advanced thermal conductivity and low dielectric constant. *Ceramics International* 49 (20): 32577–32587.

111 Zhang, Y., Wang, J.W., Ma, Y.J. et al. (2014). Multifunctional boron nitride nanosheets/fluorinated-polyimide composites with ultra-low dielectric constant and high thermal conductivity. *Polymer Composites* 45 (8): 1–12.

112 Dong, X.D., Wan, B.Q., Zheng, M.S. et al. (2024). Dual-effect coupling for superior dielectric and thermal conductivity of polyimide composite films featuring "crystal-like phase" structure. *Advanced Materials* 36: 2307804.

113 Zhao, W., Xu, Y., Song, C.R. et al. (2019). Polyimide/mica hybrid films with low coefficient of thermal expansion and low dielectric constant. *E-Polymers* 19: 181–189.

114 Yan, F., Kang, J.Y., Zhang, S. et al. (2018). Enhanced electromagnetic wave absorption induced by void spaces in hollow nanoparticles. *Nanoscale* 10 (39): 18742.

115 Sun, Y., Zhang, J.W., Zong, Y. et al. (2019). Crystalline–amorphous permalloy @ iron oxide core–shell nanoparticles decorated on graphene as high-efficiency, lightweight, and hydrophobic microwave absorbents. *ACS Applied Materials & Interfaces* 11 (6): 6374.

116 Zhou, Y., Wang, S.J., Li, D.S. et al. (2021). Lightweight and recoverable ANF/rGO/PI composite aerogels for broad and high-performance microwave absorption. *Composites Part B: Engineering* 213: 108701.

117 Miao, P., Zhou, R., Chen, K.J. et al. (2020). Tunable electromagnetic wave absorption of supramolecular isomer-derived nanocomposites with different morphology. *Advanced Materials Interfaces* 7 (4): 1901820.

118 Li, D.W., Li, D.D., Ke, Z. et al. (2023). Synthesis of colorless polyimides with high T_g and low coefficient of thermal expansion from benzimidazole diamine containing biamide. *Journal of Polymer Science* 61 (9): 818–828.

119 Lian, M., Lu, X.M., and Lu, Q.H. (2018). Synthesis of superheat-resistant polyimides with high T_g and low coefficient of thermal expansion by introduction of strong intermolecular interaction. *Macromolecules* 51: 10127–10135.

120 Li, X.S., Hao, J.M., Jiang, Q.Y. et al. (2015). Phosphorus-containing polyhedral oligomeric silsesquioxane/polyimides hybrid materials with low dielectric constant and low coefficients of thermal expansion. *Journal of Applied Polymer Science* 132 (39): 42611.

121 Lu, J., Zhang, Y., Li, J. et al. (2023). Tröger's Base (TB)-based polyimides as promising heat-insulating and low-k dielectric materials. *Macromolecules* 56 (5): 2164–2174.

122 Devaraju, S., Vengatesan, M.R., Selvi, M. et al. (2012). Synthesis and characterization of bisphenol-a ether diamine-based polyimide POSS nanocomposite. *High Performance Polymers* 24 (2): 1–12.

123 Zhang, S.Y., Fu, T., Gong, Y. et al. (2023). Design and synthesis of liquid crystal copolyesters with high-frequency low dielectric loss and inherent flame retardancy. *Chinese Chemical Letters* 34 (5): 107615.

124 Zhang, S., Sun, H.Y., Lan, T.G. et al. (2021). Polybenzoxazine/boron nitride foam: a promising low-k, flame-retardant and robust material. *Journal of Materials Science* 56: 18749–18761.

125 Zhao, L., Yuan, L., Liang, G.Z. et al. (2015). Novel tough and thermally stable cyanate ester resins with high flame retardancy, low dielectric loss and constant based on a phenolphthalein type polyarylether sulfone. *RSC Advances* 5 (73): 58989–59002.

6

High Thermally Conductive Polyimide Dielectrics

Fan Wang, Gengbo Liu, and Jun-Wei Zha

University of Science and Technology Beijing, School of Chemistry and Biological Engineering, Department of Chemistry, No. 30 Xueyuan Road, Beijing 100083, P. R. China

6.1 Introduction

Under the wave of the digital era, intelligence, personalization, lightweight, portability, and wearability have become important driving forces for the innovation of electronic devices. With the improvement of integrated circuit chip technology and battery technology, the number of functional components is increasing. The size of electronic components keeps shrinking, and electrons move faster and faster in silicon circuits, which brings about a rapid increase in power density and heat. If the accumulated heat cannot be dissipated in time, it will seriously affect the reliability and service life of the electronic equipment, and even bring the risk of fire [1, 2]. In addition, when electronic devices operate in environments with excessively high temperatures, the insulating properties, mechanical properties, and thermal stability of the materials are also affected. Therefore, in the context of the current social environment, the need for heat dissipation of electronic devices has become increasingly prominent. The heat dissipation problem has become a technical challenge that hinders further innovation and development of electronic devices [3, 4]. Thermally conductive materials usually have high thermal conductivity, low thermal expansion coefficient, and good thermal stability. It has a wide range of applications in electronic information, energy batteries, transportation, and aerospace [5–7]. However, polymers usually exhibit low thermal conductivity due to the heat transfer mode within the polymer, the irregular entanglement of the polymer chain segments, and their own amorphous morphology [8].

Polymer thermally conductive composites can be divided into intrinsic type and filling type. The intrinsic thermally conductive polymer material maintains the insulation properties, mechanical properties, and processability of the polymer itself due to the absence of external thermally conductive fillers. Generally, the molecular chain structure is designed to change the aggregation state structure, increase the microscopic orderliness of the segments, and build an ordered path for phonon transmission, thus improving the intrinsic thermal conductivity of the polymer.

Functional Polyimide Dielectrics: Structure, Properties, and Applications, First Edition.
Edited by Jun-Wei Zha.

However, due to the complex preparation process and high cost of the intrinsic thermally conductive polymer, its application in the commercial field is limited. The thermally conductive composite material of the filling type has become a hot topic of current research and attention due to its advantages such as low preparation cost, strong filler selectivity, significant improvement in thermal conductivity, and easy processing and molding. By adding fillers with high thermal conductivity to the polymer matrix, heat transfer paths can be constructed through the interconnection and stacking of thermally conductive fillers. However, it is often necessary to add a large number of fillers in order to realize the effective enhancement of thermal conductivity of the composite system, but this also causes the mechanical properties of the composite to drop sharply, which cannot meet the application requirements under actual working conditions. More importantly, interfacial thermal resistance is a key factor that hinders the composites from exerting the thermal conductivity of the filler itself. How to reduce the interfacial thermal resistance and achieve high thermal conductivity with low filling content is a technical difficulty and development bottleneck of thermally conductive composites.

Polyimide (PI) has excellent high-temperature resistance, insulation properties, and mechanical properties, and is widely used as a flexible printed circuit board substrate and high-temperature-resistant electrical insulation material for motors [9, 10]. However, PI belongs to an amorphous polymer, and the irregular arrangement of molecules inside causes severe phonon scattering. In recent years, the research and application of PI high thermal conductivity composite materials have made great progress. Researchers have been committed to improving thermal conductivity by regulating the molecular structure of PI and the directional arrangement of thermally conductive fillers, but interfacial thermal resistance remains the main challenge restricting the improvement of thermal conductivity of composites. This chapter mainly introduces the intrinsic thermally conductive PI and filled thermally conductive PI, focusing on the effective strategies to enhance the thermal conductivity of PI from three aspects: directional structure design, non-directional structure design, and multi-layer structure design. Given the universality of research and the widespread application of thermally conductive PI, it also emphasizes the multifunctionality of thermally conductive PI, providing some references for the research of high-performance thermally conductive PI.

6.2 Intrinsic Thermally Conductive Polyimide

Heat transfer mainly depends on the collision of particles such as molecules, atoms, and free electrons, which is closely related to the collision, diffusion, and transfer of microscopic particles [11]. The main carriers of heat transfer are molecules, electrons, and phonons. Most polymers are saturated systems in which there are no freely moving electrons and no violent collisions between electrons. The heat transfer of the polymer is mainly accomplished by phonons. Polymer molecular chains vibrate when heated. Heat is transferred to neighboring molecules or atoms in turn. However, the way atoms are arranged is not as regular and ordered

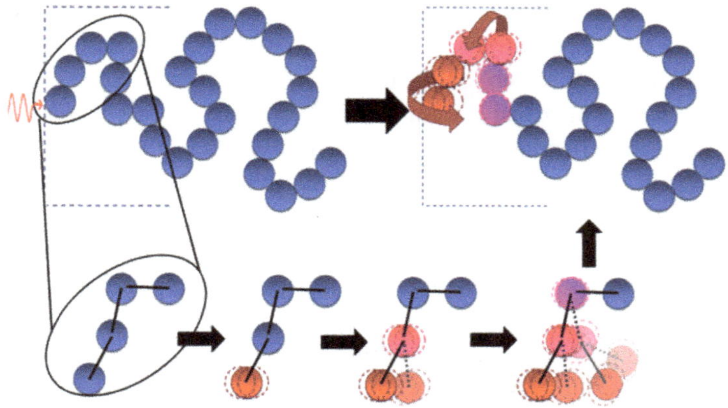

Figure 6.1 Heat transfer mechanism of polymers. Source: Ref. [11]/with permission of Elsevier.

as in crystals. The atoms on the molecular chain generate disorderly vibration and rotation at their equilibrium positions, resulting in slower heat transfer and diffusion speeds between atoms. The heat transfer mechanism of polymers is shown in Figure 6.1. The heat transfer efficiency of phonons is determined by the mean speed and mean free path of phonons. The mean free range is all the paths that the phonons travel through before the scattering phenomenon occurs, and its magnitude is related to the geometrical scattering of phonons and the collisional scattering between phonons in the crystal point structure [12].

However, phonon collisions and phonon scattering due to structural defects, molecular vibrations, and anharmonic lattice vibrations in the polymer itself significantly shorten the mean free range of phonon motion. Additionally, defects, voids, interfaces in the polymer crystal structure as well as disordered parts in the amorphous structure will cause phonon scattering, resulting in a generally low thermal conductivity of polymers of only 0.2–0.5 W/(m·K) [13, 14]. Polyimide has been widely used in flexible electronics, biomedicine, aerospace, and other fields because of its excellent mechanical properties, insulation properties, and high temperature resistance. Phonon scattering can be suppressed by altering the aggregated state structure of the polymer molecular chain segments and enhancing the ordering of the structure, thereby facilitating heat transfer and increasing the intrinsic thermal conductivity of the polymer [9, 15]. Currently, the enhancement of the intrinsic thermal conductivity of PIs is mainly realized by regulating the chain structure of PIs, constructing molecular chain orientations and intermolecular interactions.

6.2.1 Molecular Chain Structure

The structure and length of the main chain, side chain, and branch chain of polymer molecule chain have an effect on its thermal conductivity. Therefore, reducing the rotation of single chains can increase the intrinsic thermal conductivity of polymers [16, 17]. In the main chain, the rigid backbone and π-π conjugate structure can effectively inhibit the rotation inside the chain links, increase the phonon group velocity, and prevent the temperature-induced phase transition, thus enhancing the thermal conductivity and thermal stability of the material. The chain structure mainly consists of a primary proximal structure and a secondary remote structure. The primary structure can be regulated by introducing flexible groups such as ether, ester, amide, or phthalimide rigid units into the main chain of the molecule to regulate the chemical composition of the structural units in the PI molecular chain [18]. Introducing a phenylalkynyl group at the end and chemically cross-linking it to maintain the linkage order of the PI chain [19]. Altering the stereo configuration of the PI molecular chain to improve the structural ordering of the aggregates of the molecular chain segments [20]. The main means of regulating the secondary structure are: simulation of the conformation of the PI molecular chain to study the state in which the PI molecular chain shows a regular and orderly arrangement and maintains the lowest energy state. Adding terminators and choosing different preparation processes to regulate the size of molecular weight, so as to make the curled molecular chain show a stretching state, thus improving the thermal conductivity of the material [21, 22].

The intrinsic thermal conductivity of PI is largely hindered by the conformation of random crimped and entangled chains. Flexible functional groups can be introduced into the primary structure of the main chain to regulate the thermal conductivity of PI. Liu et al. [23] synchronously controlled the intra-chain and inter-chain hopping behaviors through an innovative copolymerization strategy. 2,4,5,7-tetraamino-1,8-dihydroxyanthracene-9,10-dione (4NADA) was introduced into the PI matrix as shown in Figure 6.2. The introduction of 4NADA enhanced the rigidity of the main chain and altered the intramolecular heat transport. Moreover, the coplanar morphology of 4NADA inhibits chain distortion and increases the thickness of π-stacking. As a result, effective heat transfer channels are formed in the copolymerized PI. When the loading of 4NADA is 10%, the thermal conductivity of PI can reach 0.58 W/(m·K), which is about 3 times of that of pure PI.

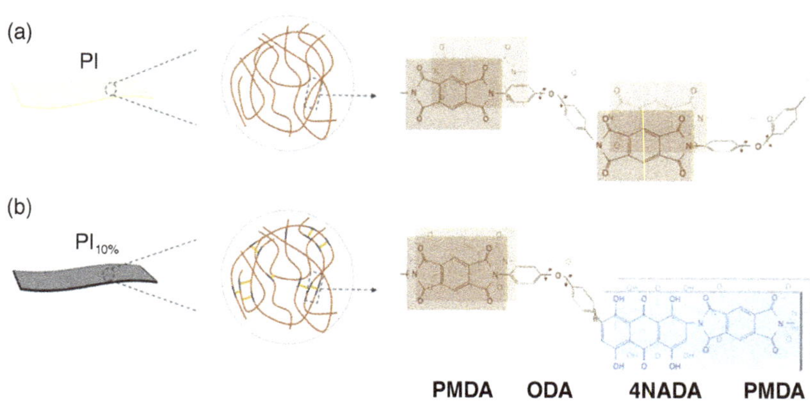

Figure 6.2 Schematic diagram of molecular chains and molecular structure: (a) pure PI film; (b) copolymerized PI film with 10% 4NADA. Source: Ref. [23]/with permission of Elsevier.

Different internal structures of molecules correspond to different aggregation patterns and thus produce different thermal conductivity values. The thermal conductivity of PI is influenced by the crystalline, amorphous, liquid crystal, and other aggregate structures. Ruan et al. [24] prepared a liquid crystal polyimide (LC-PI) film with high thermal conductivity by using phthalimide groups as rigid mesocrystalline units and ether bonds as flexible groups. The molecular chains can maintain a perfectly ordered structure in the liquid crystal temperature range. The in-plane and out-of-plane thermal conductivities of LC-PI films are 2.11 and 0.32 W/(m·K), respectively, which are significantly higher than those of conventional PI films. LC-PI also retains excellent mechanical and thermal properties, which have potential applications in highly integrated electronics.

6.2.2 Molecular Chain Orientation

The molecular chains are aligned along the direction of orientation. The amorphous part of the molecular structure is ordered, and the crystallinity is therefore increased. The crystal structure can provide high-speed channel for phonon transfer. Chain orientation can be considered as the most effective factor to improve the intrinsic thermal conductivity of polymers [25, 26]. Due to the complex and amorphous chain structure of PI, studies on its orientation have mainly focused on PI composites. The current studies on the molecular chain orientation on the intrinsic thermal conductivity of polymers are mainly on crystalline or semi-crystalline polymers. Shen et al. [27] successfully prepared super-stretched polyethylene nanofibers with thermal conductivity up to 104 W/(m·K). This high thermal conductivity is attributed to the stretching effect in the nanofibers, which promotes the reorganization of polyethylene molecular chains to form single-crystal fibers close to the "ideal" state. For polyethylene, which is inherently crystalline, the stretching orientation process may result in the formation of needle-like crystals consisting of a large number of oriented molecular chains, which significantly enhances the thermal conductivity. This also opens up the possibility of improving the thermal conductivity of PI by stretching.

Another strategy to enhance the orientation of the molecular chains is the introduction of specific monomers, such as liquid crystal motifs, during the polymerization process. The rigid or semi-rigid chain segments in liquid crystal polymers give the molecular chains an ordered structure and the molecular chains exhibit periodic orientation. Liquid crystal polymers usually exhibit microstructural anisotropy and macrostructural isotropy.

Ruan et al. [28] used the positive liquid crystal 4′-heptyl-4-biphenylcarbonitrile (7CB) for surface functionalization of carbon nanotubes (LC-CNT) to prepare LC-CNT/LC-PI composite films with enhanced thermal conductivity, as shown in Figure 6.3a. Under the effect of AC electric field, LC-CNT can be uniformly distributed in the bright yellow spots of LC-PI to achieve good convergence (Figure 6.3b). When the filling amount of LC-CNT is 15 wt%, the in-plane and out-of-plane thermal conductivities of the LC-CNT/LC-PI composite films are enhanced to 4.02 W/(m·K) and 0.55 W/(m·K), respectively. Compared to the intrinsically thermally conductive LC-PI films, the improvements are 90.5% and

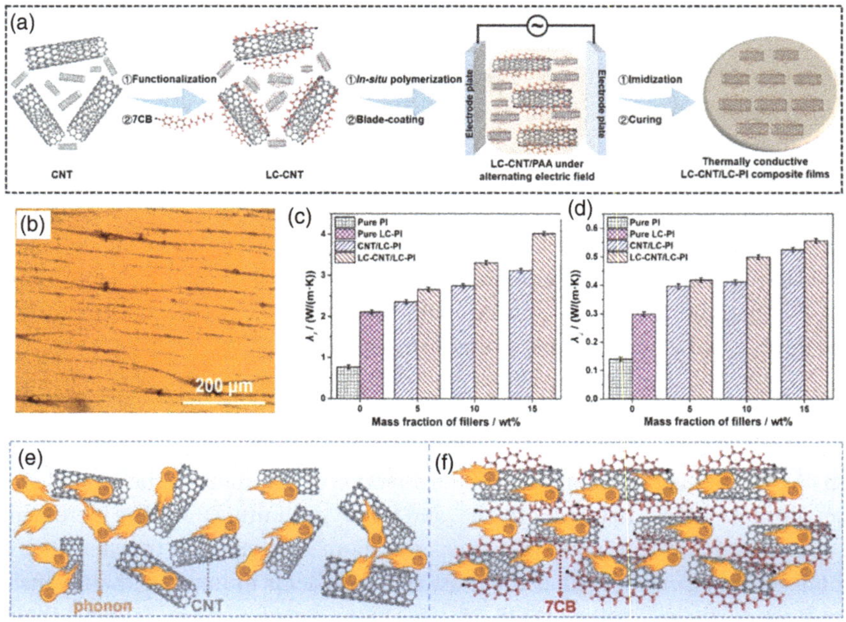

Figure 6.3 (a) Preparation schematic, (b) polarizing microscope image, (c) in-plane thermal conductivity and (d) out-of-plane thermal conductivity of LC-CNT/LC-PI composite films; (e) schematic of the thermal conduction mechanism of CNT/LC-PI, and (f) LC-CNT/LC-PI composite films. Source: Ref. [28]/John Wiley & Sons.

71.9%, respectively, as shown in Figure 6.3c,d. In the CNT/LC-PI system, CNTs are randomly distributed in the LC-PI matrix, which cannot form an effective CNT-CNT heat conduction pathway. In contrast, in the LC-CNT/LC-PI system, the LC-CNT is highly ordered along the in-plane direction of the composite film, forming multiple heat conduction pathways, which results in a significant enhancement of the thermal conductivity of LC-PI. The heat transfer mechanisms are shown in Figure 6.3e,f.

Shoji et al. [29] prepared cross-linked liquid crystalline PI films containing siloxane spacer units by introducing acetylene end-groups at the end of the PI molecular chain. The relationship between the liquid crystal morphology and the thermal diffusivity of the PI films was also investigated. The results show that the thermal diffusivity in the direction perpendicular to the plane increases with the degree of cross-linking. The PI molecular chains are arranged vertically along the thickness direction, and this chain orientation plays an important role in phonon conduction. Additionally, they [30] synthesized a new semi-fatty liquid crystal PI containing siloxane in order to determine the effect of the type and position of substituents on the reduction of liquid crystal PI transition temperature. The results show that the chlorine and fluorine substituents can promote the formation of liquid crystal phase, but reduce the liquid crystal transition temperature significantly. The methyl substituents can seriously damage the stability of liquid crystal phase.

The key to preparing highly thermally conductive liquid crystal PIs lies in the introduction of liquid crystal units with rigid structures or high aspect ratios into the PI network. These liquid crystal units are assembled with the PI matrix to form an ordered microstructure, which promotes phonon transport along the ordered thermal conductivity network. In addition, by regulating external factors such as temperature, pressure, electric or magnetic fields, the microstructures are induced to be arranged in an ordered state, resulting in higher thermal conductivity. Compared with conventional PIs, intrinsically thermally conductive liquid crystal PIs can retain excellent thermo-mechanical and electrical properties. This unique structure facilitates the development of new thermally conductive PIs and provides ideas for solving the problem of low thermal conductivity of PIs [31].

6.2.3 Intermolecular Interaction

Intermolecular interactions mainly include electrostatic forces, van der Waals forces, and hydrogen bonding. The construction of intermolecular interactions leads to the formation of crystal-like structures, which promotes the orderly arrangement of molecular chains. These interactions contribute to the construction of long-range ordered structures, providing a smoother path for heat transfer. Among them, hydrogen bonding and van der Waals forces mainly enhance the thermal conductivity of the material by promoting crystallization and limiting the twisting of molecular chains [32]. The presence of imide groups in the PI structure tends to introduce hydrogen bonding, which enhances intermolecular interactions and forms a stable thermally conductive network structure. Ran et al. [33] systematically investigated the effect of hydrogen bonding on intrinsic thermal conductivity in ordered polymers with lamellar structures by molecular dynamics simulations. Hydrogen bonding provides an ideal structure in heat transfer. Hydrogen bonding interactions lead to enhanced intermolecular van der Waals force interactions, which significantly reduce the interfacial thermal resistance of layered polymers. Also, the effect of introducing hydrogen bonding on the chain morphology is a factor to be considered for improving the thermal conductivity.

Nicholls et al. [34] took thermoplastic PI-polyurea as the research system and combined urea bond into PI to enhance hydrogen bond interaction. Kim et al. [35] investigated the effect of several different hydrogen bonds on thermal conductivity from the point of view of intermolecular chain interactions. Achieving high thermal conductivity of polymers requires strong intermolecular bonds instead of weak interactions. The intermolecular bonds must be as tightly connected to the polymer backbone as possible, and the concentration of certain hydrogen bonds must exceed the percolation threshold.

Currently, the research on the introduction of interactions such as hydrogen bonding in PIs focuses on filled thermally conductive PIs. The non-oriented structure of thermally conductive fillers is designed to have reactive functional groups on their surfaces, which are prone to hydrogen bonding with PIs. The thermal conductivity of PI is highly dependent on its microstructure. The enhancement of thermal conductivity is limited by a number of factors such as the chemical structure of the

PI, the orientation of the molecular chains, the degree of crystallization, and the intermolecular interactions.

6.3 Filled Thermally Conductive Polyimide

The thermal conductivity of filled composites is the result of the joint action of the polymer matrix and the thermally conductive filler. The introduction of fillers creates contact interfaces with the polymer matrix, and overlap between fillers creates interfaces. When heat passes through the interface, there is an interfacial thermal resistance, which greatly reduces the efficiency of heat transfer. For this reason, the microstructure of the filler is the key to constructing the heat transfer path and reducing the interfacial thermal resistance between the polymer and filler to achieve high thermal conductivity. This section summarizes the effects of directional structure design, non-directional structure design, and multi-layer structure design on the thermal conductivity of PI, which provides a reference for the development of new high thermal conductivity PI composites.

6.3.1 Thermal Conductivity Mechanism of Polymer Composites

The theory of thermally conductive pathways is commonly used to analyze and elucidate the thermal conductivity mechanism of filled composites. The theory suggests that the filler paths formed by thermally conductive fillers lapping over each other in the polymer matrix are the paths with the lowest thermal resistance in the heat conduction process. As shown in Figure 6.4, the thermally conductive fillers form effective thermal channels through close contact with the polymer matrix. As the filler content increases, the thermally conductive fillers pass through separation from each other to the formation of localized thermally conductive chains to the formation of a complete thermally conductive network. The heat flow is transferred in the matrix along the paths lapped by the thermally conductive fillers [4]. The theory provides an important guiding direction for the design of polymer-based thermally conductive materials, which is the key to the preparation of high-performance polymer-based composites.

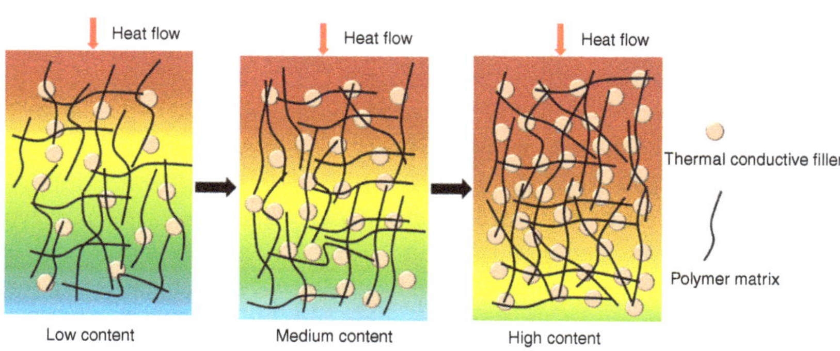

Figure 6.4 Schematic diagram of the thermal conductivity network.

The theory of thermally conductive pathways is used to describe systems in which conventional polymers are blended with thermally conductive fillers. The formation of a continuous thermally conductive network requires highly loaded fillers, which in turn leads to an increase in thermal conductivity at the expense of the mechanical properties of the composite. Moreover, the random arrangement and dense stacking of thermally conductive fillers generate high interfacial thermal resistance, which seriously affects the further improvement of thermal conductivity of composites. Researchers are committed to reducing the repetitive stacking between fillers by regulating the arrangement of thermally conductive fillers in the matrix to achieve high thermal conductivity at low filler loadings, thus preserving the excellent properties of the fillers and the polymer matrix itself. Therefore, elucidating the relationship between thermal conductive paths and interfacial thermal resistance, and establishing the deep thermal conductivity mechanism of composites are of scientific value and practical significance for the development of high-performance thermally conductive composites.

6.3.2 Effective Strategies for Improving Thermal Conductivity of Polyimide

The heat conduction of composites is affected by the properties of the interface. The difference in polarity between the filler and the polymer matrix leads to poor interfacial compatibility between them. For nanofillers with high surface energy, the fillers themselves are prone to agglomeration. In addition, defects and pores appearing at the interface can hinder the thermal conductivity of composites. For example, the theoretical thermal conductivity of CNTs can reach up to 3000 W/(m·K). However, by dispersing 1 vol% of CNTs into the polymer, the thermal conductivity of the composite is only 10 W/(m·K). The thermal conductivity of the composites is much lower than that of the filler itself, which is mainly due to the fact that the simple blending of CNTs with polymers produces a large amount of interfacial thermal resistance in the composites, which greatly affects the interfacial heat transport [36]. Therefore, in-depth understanding of interfacial heat transfer is important for the effective regulation of interfacial structure and the preparation of high thermal conductivity composites. Interfacial reinforcement is beneficial to improve the interfacial compatibility between matrix and filler, and interfacial regulation strategies are widely used to improve the thermal conductivity of composites. In this section, the importance of interface regulation to improve the thermal conductivity of PI is illustrated from the aspects of directional structure design, non-directional structure design, and multi-layer structure design.

6.3.2.1 Directional Structure Design

In the design of oriented structures, the thermally conductive filler is oriented by external forces to achieve the ordering of the internal microstructure, which results in the directional conduction of heat flow. Oriented structures usually require specific processes to realize, such as electrostatic spinning, mechanical stretching, magnetic field orientation, and hot pressing. The pathways for heat transfer are formed under external forces to reduce disordered stacking between fillers and maximize the advantages of thermally conductive fillers.

Guo et al. [37] used electrostatic spinning combined with hot pressing technique to improve the thermal conductivity of PI by adding chemically modified graphene oxide (CMG) to the PI matrix. The thermal conductivity and thermal index of the PI composites increased with the filler loading. The thermal conductivity of 5 wt% CMG/PI composite film is increased to 1.05 W/(m·K) which is 4 times that of pure PI film. This process allows the CMG to be uniformly dispersed and oriented in the PI matrix. Compared with randomly dispersed fillers, the oriented arrangement of fillers can reduce the interfacial thermal barrier due to the disordered overlap between fillers, thus forming an efficient heat conduction path. The modified CMG has good interfacial compatibility with the PI matrix, which reduces the interfacial thermal barrier to form an effective thermal conductive network. Zhang et al. [38] prepared a PI/BNNS/CNT@αPVA composite film with a dual thermal conductivity network structure (Figure 6.5a). The network structure of interpenetrating CNT@αPVA was constructed in the PI/BNNS film using the co-electrostatic spinning technique. CNT@αPVA acts as a continuous thermally conductive network connecting two-dimensional BNNS to reduce the voids between BNNS. These two types of fibers are uniformly interwoven in the film to form a unique dual thermally conductive network structure, as shown in Figure 6.5b,c. When the filling amount of CNT and BNNS is 0.3 wt% and 30 wt%, respectively, the thermal conductivity of PI/BNNS/CNT@αPVA composite film is as high as 8.40 W/(m·K), about twice that of PI/BNNS composite film (Figure 6.5d). In addition, the PI composite film also has good insulation properties, its volume resistivity is about 10^{15} Ω cm, which

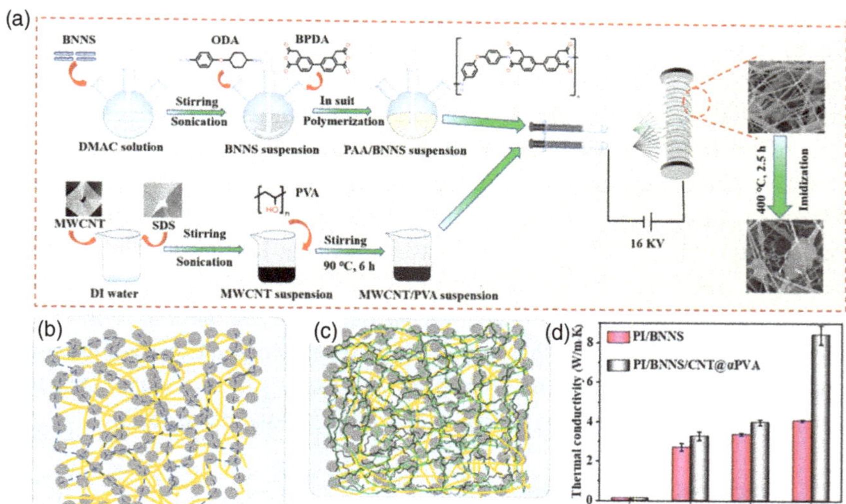

Figure 6.5 (a) Preparation diagram of PI/BNNS/CNT@αPVA composite film; (b) schematic diagram of heat conduction network of PI/BNNS film; (c) schematic diagram of dual thermal conductivity network of PI/BNNS/CNT@αPVA membrane; (d) thermal conductivity of PI/BNNS film and PI/BNNS/CNT@αPVA film. Source: Ref. [38]/Springer Nature.

provides an effective method for constructing a continuous thermal conductivity network in the PI matrix.

Liu et al. [39] prepared GF-BN/PI(MF) composite films with polydopamine (PDA)-modified reduced graphene and magnetic needle iron oxide (GF) as fillers by a moving magnetic field induction strategy (MF). The GF is disordered in PI matrix without applied magnetic field. After the magnetic field is applied, the original disordered distribution becomes directional arrangement. In the PI/BN composite film, the oriented GF nanoparticles constructed an efficient thermal conductivity network with BN in both in-plane and out-of-plane directions under the effect of magnetic field induction. The moving magnetic field induction and heat conduction mechanism of the composite film are shown in Figure 6.6a. The in-plane thermal conductivity of 30 wt% GF-BN/PI (MF) composite film is as high as 2.532 W/(m·K), which is 1233%, 27%, and 25% higher than that of pure PI, 30 wt% BN/PI, and 30 wt% GF-BN/PI composite films, respectively (Figure 6.6b). Also, the GF-BN/PI (MF) composite film also has the highest out-of-plane thermal conductivity (0.452 W/(m·K)), which is 150% higher than that of the pure PI film,

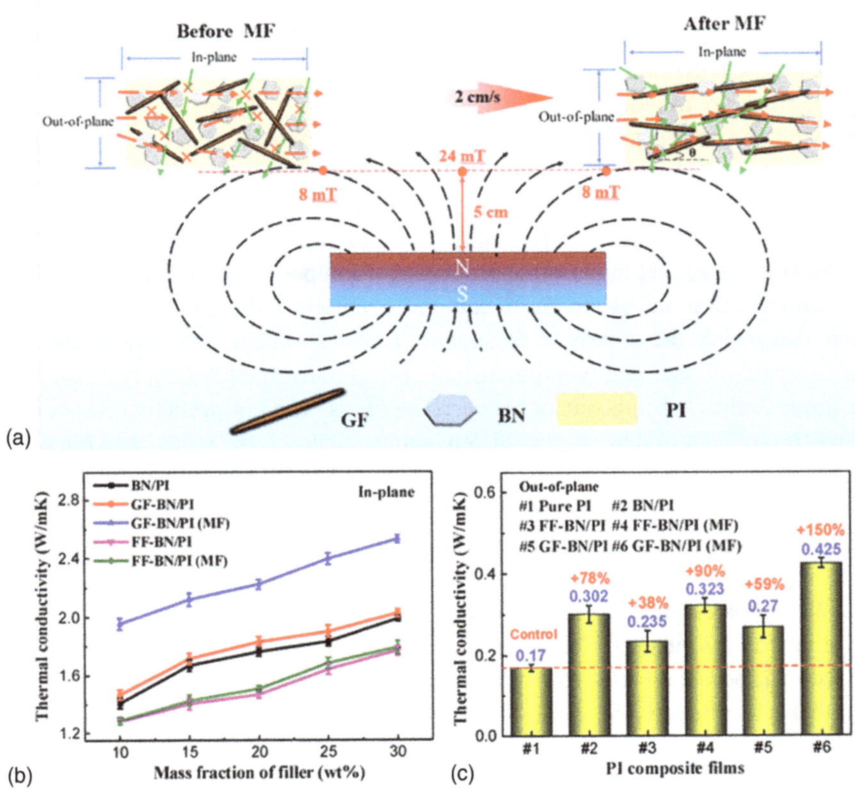

Figure 6.6 (a) Motion magnetic field induction and heat conduction mechanism, (b) in-plane, and (c) out-of-plane thermal conductivity of PI composite films. Source: Ref. [39]/with permission Elsevier.

as shown in Figure 6.6c. This indicates that the magnetic field induction changes the anisotropy of the PI composites, resulting in a certain degree of improvement in the out-of-plane thermal conductivity. Orientation structure is the key to design highly oriented thermal conductive materials. One-dimensional fillers such as wires, tubes, and needles and flaky two-dimensional materials are easily processed and oriented to give high thermal conductivity to the composites.

Ding et al. [40] prepared h-BN/PI composites with well-aligned structures using a simple "vacuum filtration-hot pressing" process. The modification by PDA enhanced the interfacial interaction between PI matrix and h-BN. The vacuum filtration process assists the directional alignment of h-BN in the matrix and provides a pathway for heat transfer. The in-plane thermal conductivity of the 20 vol% h-BN/PI composite film is 3.01 W/(m·K). The vacuum filtration combined with hot-pressing process can press the two-dimensional h-BN into a tight stack, forming an ordered layered structure along the horizontal direction, thus obtaining high in-plane thermal conductivity. However, the films prepared by this process have a rougher surface and a weak layer-to-layer interface. In addition, it is difficult to realize the directional alignment in the out-of-plane direction, which limits the application in scenarios with high out-of-plane thermal conductivity. Comparatively, electrospinning and magnetic field orientation can make the filler obtain any orientation, and the connection between the filler and the matrix is closer, which is a more practical method at present.

In contrast to random dispersion of fillers in the polymer matrix, uniform dispersion and orderly arrangement of fillers in the polymer matrix can be achieved by structural design. Effective structural design can reduce the contact interfaces between fillers or between fillers and matrix and build a continuous thermal conductivity path. Meanwhile, interface improvement and good interfacial interaction are also key factors to enhance other properties of composites. Based on the preparation method and molding process of composite materials, the directional structure design can realize the orderly arrangement of internal fillers. The easy processing characteristics of one- and two-dimensional fillers make the fillers fully connected and improve the thermal contact between the fillers. The hot-pressing process can reduce the voids caused by the insufficient contact between the matrix and the filler and improve the interfacial interaction between the two. Moreover, by controlling the temperature and pressure, it is also possible to change the arrangement of fillers inside the composite material to form a continuous thermal conductive network.

6.3.2.2 Non-directional Structural Design

The directional structural design of composites allows the internal fillers to be aligned in a specific direction to optimize the thermal conduction path, resulting in high thermal conductivity. Non-directional structural design of thermally conductive fillers from the point of view of lowering interfacial thermal resistance can also achieve an effective increase in thermal conductivity. Non-directional structural design usually builds covalent or non-covalent interactions, such as hydrogen bonding, π-π interactions, cation-π, and covalent bonding in the molecular structure, so as to build a continuous heat conduction network. The interfacial behavior between fillers or between fillers and matrix is regulated from the perspective of intermolecular interactions, and novel microstructures are constructed to form a unique heat transfer network, which promotes the effective transfer of heat along the thermal conductivity path. Due to the high surface energy of the thermal conductive filler, its dispersion in the polymer matrix is the first problem to be considered in the preparation of the polymer thermal conductive composite. In order to minimize the agglomeration of the filler and reduce the scattering of phonons at the interface, the non-directional structural design of the thermally conductive filler is an effective strategy to enhance the thermal conductivity. Fillers with microstructural design are easy to form thermal conductivity paths, and the unique thermal conductivity network makes the composites have excellent thermal conductivity.

Enhancing the interaction force between matrix and filler can effectively reduce the interfacial thermal resistance. In order to significantly improve the interfacial thermal conductivity, current research hotspots focus on enhancing the interaction between polymer matrix and thermally conductive fillers. The formation of covalent bonds is one of the effective strategies. The covalent bond usually means that after the two atomic orbitals effectively overlap, the electrons in the atomic orbitals appear between the two atomic nuclei, and the electrons are shared by the two atoms. Choy et al. [41] mentioned that heat transfer along the direction of covalent bond formation on the polymer molecular chain contributes to the thermal conductivity of the polymer. The formation of covalent bonds is equivalent to phonons building bridges at the interface between the matrix and the thermally conductive filler. At the atomic scale, the chemical bond strength can be analogized to a spring. When the chemical bond is strong enough, this spring-like elastic wave resonant motion frequency will effectively couple the phonons between different materials for a smoother transport. The researchers systematically studied the effect of covalent bond strength and density on the interfacial thermal conductivity [42–45]. The results show that covalent interactions affect the interactions and spatial configuration of molecular chains, which help to build long-range ordered structures and promote the formation of pathways for phonon transport.

The covalent bond between the polymer matrix and the thermal conductive filler can improve the dispersion of the filler in the matrix, which is conducive to phonon coupling and heat transfer. Wang et al. [46] prepared highly thermally conductive PI composites with in-plane orientation using carbon nitride nanosheets (CNNS) as thermally conductive filler and PI as matrix. The amino group ($—NH_2$) on the surface of CNNS can form a strong covalent interaction with the carboxyl group (—COOH) in polyamide acid (PAA), which promotes the self-orientation of CNNS along the in-plane in the PI matrix. The in-plane orientation is realized with low filler content and the thermal conductivity is improved. The in-plane thermal conductivity of PI/CNNS composites is 2.04 W/(m·K) when the CNNS filler content is 20 wt%. Wan et al. [47] prepared m-BN/PI composite films by surface activation and modification of h-BN by in situ polymerization. The m-BN/PI is attached to the PI molecular chain by participating in the polymerization of PI, which promotes the orientation of m-BN along the in-plane direction. The covalent bonding between the m-BN and PI molecular chains in the m-BN/PI composites effectively reduces the interfacial thermal resistance. The thermal conductivity of 30 wt% m-BN/PI reaches 4.43 W/(m·K), which is about 12 times that of pure PI film.

Lan et al. [48] performed non-destructive functional group modification on graphene. The crystal defects generated by surface modification are reduced while maintaining the high thermal conductivity of graphene. The interfacial structure between fillers and between fillers and matrix is regulated by covalently grafting maleimide groups to the graphene surface through Diels–Alder(D–A) reaction. The modified graphene nanosheets (M@CNS) are reacted with PAA and thermally imidized to obtain PI composites with multidirectional covalent bonds. Figure 6.7a shows the preparation process of M@GNS$_x$/PI. The reduction of contact thermal resistance between GNS and the interfacial thermal resistance between GNS and PI effectively promotes the in-plane heat transfer. The thermal conduction path forms between the multidirectional covalent bonds minimizes the interfacial thermal resistance. The thermal conductivity of the M@GNS$_{15}$/PI can reach up to 16.10 W/(m·K) (Figure 6.7b). As shown in Figure 6.7c, the M@GNS/PI composite also exhibits good thermal conductivity stability, which remains 15.38 W/(m·K) at 100 °C. The interfacial thermal resistance (R_c) between GNS and the interfacial thermal resistance (R_i) between GNS and PI is calculated to illustrate the role of maleimide in interfacial regulation. The R_c and R_i of M@GNS/PI composite films decrease to 3.29×10^{-9} m^2·K·W^{-1} and 4.74×10^{-9} m^2·K·W^{-1}, respectively, as shown

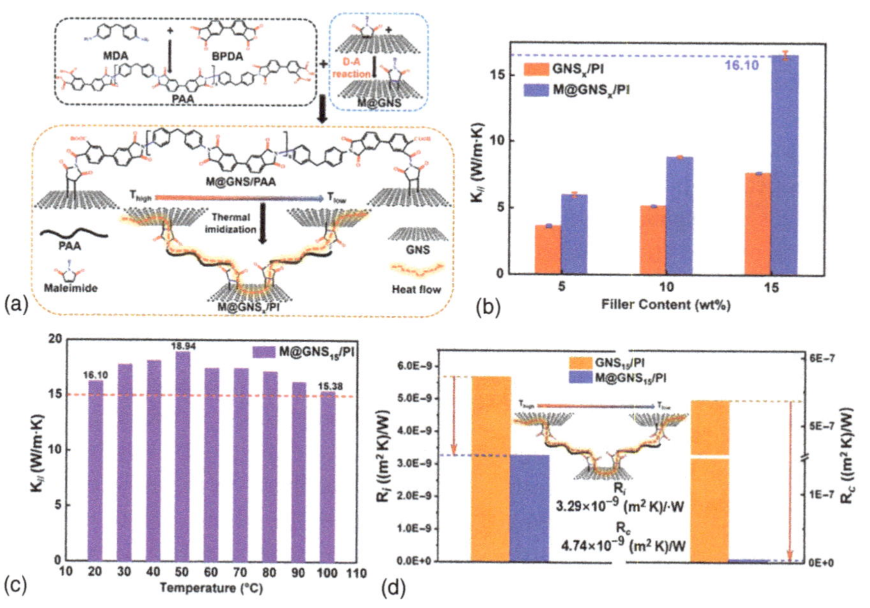

Figure 6.7 (a) Schematic of the preparation of M@GNS/PI composite films; (b) in-plane thermal conductivity of GNS/PI and M@GNS/PI composite films; (c) in-plane thermal conductivity versus temperature; and (d) R_i and R_c of GNS/PI and M@GNS/PI composite films. Source: Ref. [48]/The Royal Society of Chemistry.

in Figure 6.7d. Maleimide is not only involved in the construction of "bridge" with GNS, but also in the formation of PI. The covalent bonding provides an additional transport channel for heat transfer in the M@GNS/PI composite films.

The covalent bond formation may be accompanied by the introduction of heteroatoms or functional groups, which can disrupt the integrity of the filler's crystallization and result in the aggravation of phonon non-harmonic vibration. Chemical modification may also lead to the formation of defects that disrupt the perfect structure of the filler surface itself, thus reducing the mean free range of the phonon, and the resulting composite does not have as high a thermal conductivity as expected. Therefore, the key to surface functional modification of fillers from the perspective of constructing covalent bonds lies in introducing functional groups while preserving the integrity of the thermally conductive fillers, and also considering the influence of the functionalization rate on thermal conductivity [49, 50].

Unlike covalent bonds, non-covalent interactions do not share electrons when they occur. It mainly includes electrostatic interaction, π-conjugation effect, van der Waals force, and hydrogen bond. The intermolecular force of electrostatic interaction and van der Waals force is weak, resulting in a low coupling coefficient between phonons, which has no obvious effect on the thermal conductivity of reinforced polymer composites. The theoretical in-plane thermal conductivity of graphene can reach 100–400 W/(m·K), but the layer-to-layer force of graphene is van der Waals force. The interlayer thermal conductivity is less than 1 W/(m·K). Huxtable et al. [51] used sodium dodecyl benzene sulfonate (SDS) to simulate polymer molecules, SDS-coated CNTs, and dispersed in aqueous solution. According to the curve of light absorption of CNTs solution with time, the thermal conductivity of the interface between CNTs and SDS is 12 W/(m·K). The SDS molecules and CNTs do not form chemical bonds, but interact with each other in the form of weak van der Waals forces. Therefore, CNTs can only interact with SDS molecules through phonon–phonon coupling in the low-frequency range. Further study found that after replacing the molecules on the surface of CNTs, the main force between them is van der Waals force. The thermal conductivity of the interface does not change much, which proves that the van der Waals force has little effect on the improvement of the thermal conductivity of the composite.

Hydrogen bonding is considered to be a special type of intermolecular or intramolecular interaction that is 10–100 times stronger than van der Waals forces. Hydrogen bonding between the polymer matrix and the thermally conductive filler is constructed to form a strong bond. This bonding helps to enhance the interfacial interaction between the filler and the matrix and improves the dispersion of the filler in the matrix, thus reducing the interfacial thermal resistance of the composite. Cao et al. [52] prepared self-assembled high thermal conductivity PI/BNNS composites using van der Waals interactions between PI matrix and filler. h-BN powder is edge hydroxylated during ultrasonic-assisted liquid phase exfoliation to form O—H polar groups and attach to the surface of BNNS. The abundant carbonyl groups in the imide ring formed hydrogen bonds with the polar O—H, which helps to improve the compatibility between the filler and PI chains. The thermal conductivity of PI/BNNS composite with 12.4 vol% BNNS is 4.25 W/(m·K).

The π-π interaction is a weak interaction exhibited by a π-electron conjugated system satisfying a particular spatial arrangement. It usually exists between two molecules that are relatively electron-rich and electron-deficient, and is a non-covalent bonding interaction that is as important as hydrogen bonding. Xu et al. [53] achieved a simple self-assembly of PI matrix and r-GO filler by using hydrogen bonding effect and π-π stacking interaction, and prepared PI/r-GO composites with core–shell structure to construct a highly ordered three-dimensional graphene network. Graphene oxide is amphiphilic and has hydrophilic functional groups such as hydroxyl, carbon, and carboxyl groups, which can form hydrogen bonds between the carbon groups of PI and enhance the intermolecular interaction between PI microspheres and graphene oxide.

The non-covalent interaction is constructed in the composites, which shows low interfacial thermal resistance. The structurally designed fillers have good dispersion and affinity in the polymer matrix. However, compared with covalent interaction, the strength and stability of bonding force of non-covalent interaction need to be improved. The influence of non-covalent interaction on thermal conductivity of composites is indirect, and its influence degree is restricted by many factors. For example, factors such as the type, content and dispersion of fillers, the type and properties of polymer matrix, and the preparation process may affect the degree of influence of non-covalent interactions on thermal conductivity. Therefore, when designing and preparing composites with high thermal conductivity, various factors need to be considered to achieve optimal thermal conductivity. The key to regulate the thermal conductivity of composites by forming non-covalent interactions is how to ensure the sustainability and stability of the interactions.

6.3.2.3 Multi-Layer Structure Design

The introduction of fillers and the creation of interfaces coexist and influence each other. The interfacial bond strength and interfacial properties are closely related to the molecular structure of filler and matrix as well as the composite configuration. The multilayer structure design can alleviate the contradiction between the thermally conductive filler and the interface, which is an important strategy to improve the thermal conductivity of composite materials. Due to the complementary advantages of different layers, the thermal conductivity composite with multilayer structure is a hot research direction at present. In a multilayer structure, each layer can be optimized for the specific function required. For example, certain layers can be specialized to provide heat transfer, while certain layers can provide the necessary mechanical support or electrical insulation properties through the adjustment of specific layers to meet specific application requirements.

Guo et al. [54] prepared high-performance PI composite materials using the strategy of layered design and assembly. With electrostatically PI fibers as the base layer, Fe_3O_4/PI as the intermediate shielding enhancement layer, and graphene oxide/expanded graphite (GO/EG) as the top layer for thermal conductivity and electromagnetic shielding, the synergistic enhancement of thermal conductivity, mechanical properties, and electromagnetic interference shielding performance has been achieved. The binding of the GO/EG layer to the Fe_3O_4/PI layer mainly relies on hydrogen bonding and π -π interactions. The binding of the Fe_3O_4/PI layer to the PI fiber layer relies on the entanglement of the PI molecular chains. The structure diagram of PI composite film is shown in Figure 6.8a,b. The brittleness of EG layer is solved by introducing GO into EG. Fe_3O_4/PI prepared by in situ polymerization is helpful to enhance the electromagnetic shielding properties of the composites. PI fiber is used as the base layer to maintain the mechanical properties of PI composite. The in-plane thermal conductivity of the PI composite film containing 61.0 wt% GO/EG and 23.8 wt% Fe_3O_4/PI reaches 95.40 W/(m·K), the electromagnetic shielding interference efficacy is 34.0 dB, and the tensile strength is 93.6 MPa (Figure 6.8c,d).

Wang et al. [55] prepared a flexible printed circuit board (G-FPC) with PI as the top and bottom layers and graphene film (M-GF) as the middle layer. The thermal conductivity of the graphene film in the middle layer is up to 739.56 W/(m·K), and the in-plane and inter-plane thermal conductivities of the G-FPC are 222.04 W/(m·K) and 0.34 W/(m·K). G-FPC maintains good flexibility, stability, reliability, and heat dissipation capability, and it has potential applications in flexible and wearable electronics. Chen et al. [56] developed a PI/carbon ring-carbon nitride (C_{ring}-C_3N_4) composite film with a three-layer structure. The upper and lower layers are 20 wt% C_{ring}-C_3N_4/PI as the mechanical support layer. The middle layer C_{ring}-C_3N_4 is filled with 50 wt%. The out-of-plane thermal conductivity of the PI/C_{ring}-C_3N_4 composite film reaches 1.35 W/(m·K) and also shows good photothermal response. Wu et al. [57] reported a composite with large-size graphite nanosheets to enhance the thermal conductivity of PI. The PI microspheres were filled into PDA-modified expanded graphite (EF-PDA) by self-assembly process, and the interlayer space of EG-PDA was decorated by PI resin, and finally densified by hot pressing. This unique layered structure gives PI composites excellent thermal conductivity and electromagnetic shielding properties. In the EG-PDA/PI system, continuous thermally and electrically conductive channels constructed from graphite nanosheets are formed. The large size graphite nanosheets with high orientation and high interconnection reduce the interfacial thermal resistance between fillers. The hydrogen bond interaction between PI and EG-PDA enhances the interface affinity between them. The thermal conductivity of the 60 wt% EG-PDA/PI composite is increased to 65.6 W/(m·K).

The multilayer structural design of thermally conductive composites allows for excellent mechanical and electromagnetic shielding properties. However, it exhibits a contradiction between its own structure and the trans-planar thermal conductivity. The increase in the number of layers also hinders the heat transfer, while phonons propagating in the vertical direction need to cross more and more interfaces, thus limiting the increase in the out-of-plane thermal conductivity. The multilayer structural design combined with the hot-pressing process is able to reduce the defects and voids within the film, allowing for a more pronounced orientation of the filler in the matrix. In addition, the design of the multilayer structure needs to consider the matching of the coefficient of thermal expansion to ensure the stability and reliability of the overall structure. The interfacial thermal resistance between different

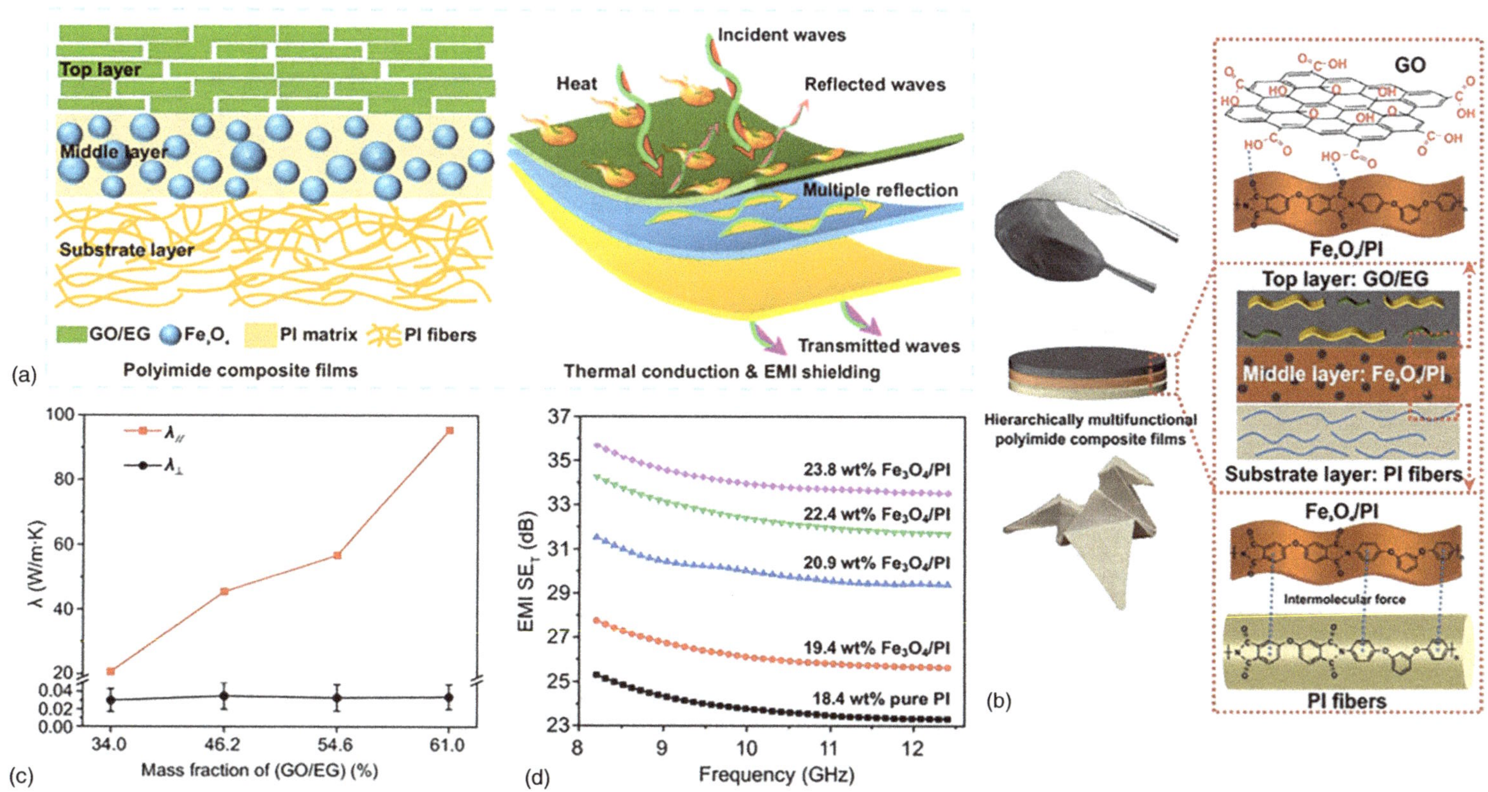

Figure 6.8 (a) Structure diagram of PI composite film with layered structure; (b) schematic diagram of the interaction between layers; (c) thermal conductivity; and (d) electromagnetic shielding efficiency of PI composite films. Source: Ref. [54]/Springer Nature/CC By 4.0.

fillers may affect the overall thermal conductivity, so the selection of fillers and interface design needs to be optimized. Meanwhile, the existence of interfaces hinders the formation of conductive paths, which provides a feasible solution for the preparation of highly thermally conductive and electrically insulating composites.

6.4 Multifunctionalization of Thermally Conductive Polyimides

Driven by the wave of digitization, electronic devices are developing rapidly in the direction of intelligence, personalization, lightweight, portability, and wearability. Faced with increasingly complex and demanding application environments, PI composites are not only required to exhibit excellent insulating, mechanical, and electromagnetic interference shielding (EMI) properties to adapt to high-temperature and high-pressure operating environments. This trend puts forward more stringent and cutting-edge requirements for the versatility of PI thermally conductive composites, which in turn promotes the research and development of thermally conductive and insulating, high-performance, and intelligently responsive PI composites.

6.4.1 Thermal Conductivity and EMI Shielding

In highly integrated electronic systems, the accumulation of heat generated by electronic equipment seriously affects the performance and service life of the equipment. Furthermore, electromagnetic wave radiation is released during the operation of electronic equipment, and this radiation causes interference with neighboring components. More seriously, this electromagnetic wave radiation can also be potentially harmful to human health and other organisms. For this reason, the electromagnetic shielding performance of thermally conductive composites is required at a higher level in order to maintain the normal operation of electronic components and ensure human health [6].

The working mechanism of electromagnetic shielding materials is to block unwanted electromagnetic radiation by reflecting or absorbing incident electromagnetic waves. Conductor materials with low resistance values are usually used, and the basic principle is to utilize the reflection of electromagnetic waves on the surface of the conductor and their absorption inside the conductor, as well as the loss during transmission to produce a shielding effect [58, 59]. Electromagnetic waves are reflected and absorbed through the conductive network, thus weakening or eliminating the propagation of electromagnetic waves.

Zhang et al. [60] prepared (Fe_3O_4/PI)-$Ti_3C_2T_x$-(Fe_3O_4/PI) composite films using poly (methyl methacrylate) (PMMA) microspheres as templates by the sacrificial template method (Figure 6.9a). When the mass ratio of PMMA to $Ti_3C_2T_x$ is 2 : 1, the thermal conductivity and EMI shielding efficacy of the PI composite films with multilayer/porous structure are 3.49 W/(m·K) and 85 dB, respectively, as shown in Figure 6.9b,c. When electromagnetic waves are incident on the surface of the

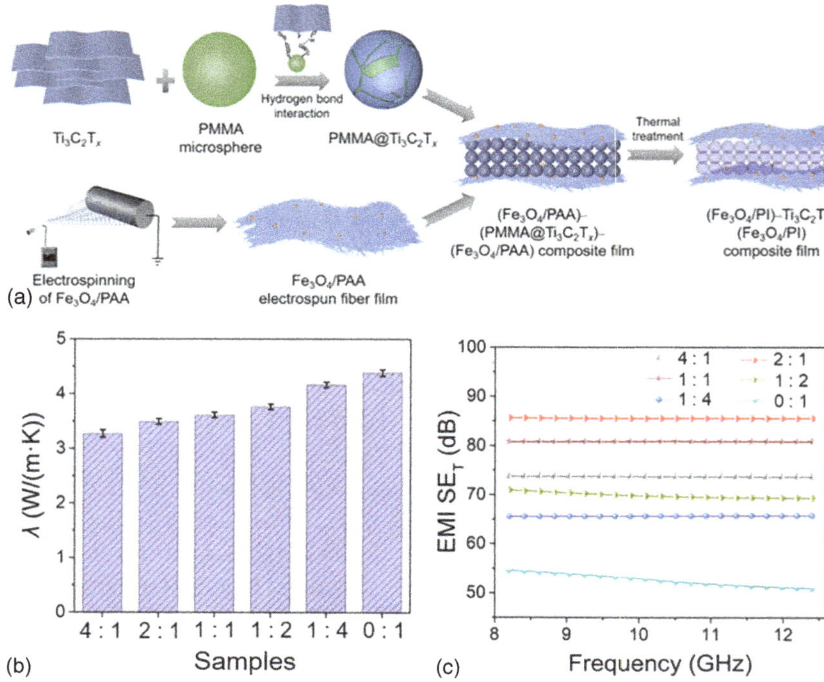

Figure 6.9 (Fe_3O_4/PI)-$Ti_3C_2T_x$-(Fe_3O_4/PI) composite films with (a) preparation schematic, (b) in-plane thermal conductivity, and (c) EMI SE_T. Source: Ref. [60]/John Wiley & Sons.

PI composite film, the electromagnetic waves interact with the Fe_3O_4 magnetic particles in the surface layer, resulting in hysteresis loss. When the electromagnetic wave enters into the hollow microspheres in the middle layer, the internal hollow structure effectively enlarges the inter-sheet spacing, prolongs the transmission path of the electromagnetic wave, and generates a large impedance mismatch. Electromagnetic waves are reflected several times inside the $Ti_3C_2T_x$ hollow microspheres, which effectively improves the absorption ability of the composite film for electromagnetic waves.

The combination design of filler and matrix, the construction of layered structure, and the design of three-dimensional porous structure are the main strategies to improve the thermal conductivity and electromagnetic shielding properties of PI composites. The three-dimensional efficient conductive network is constructed through the structural design of filler and matrix. The interaction between filler and matrix is enhanced by chemical bonding and other means, which improves the thermal conductivity and electromagnetic shielding performance of the composites. The layered structure can precisely control the transfer path of heat flow and the reflection, absorption, and scattering behaviors of electromagnetic waves inside the material, so as to optimize the thermal conductivity and electromagnetic shielding properties. The 3D porous structure increases the specific surface area and interface effect of the composites by introducing a large number of internal pores and channels. This structure not only contributes to the rapid transfer of heat inside the material, but also effectively reduces the transmission and leakage of electromagnetic waves through multiple reflection and scattering effects [61, 62]. The fine design and optimization of the layered structure and three-dimensional porous structure of the composites help to develop PI composites with high thermal conductivity and electromagnetic shielding properties to meet the increasing performance demands.

6.4.2 Thermal Conductivity and Electrical Insulation

The application of thermal conductive composites in the field of electronic packaging often requires good electrical insulation properties. When graphite and graphene, which have high thermal conductivity, are used as thermally conductive fillers, they tend to form continuous conductive paths in the matrix, which seriously impairs the electrical insulation of the composites. Coating the surface of the filler can improve the surface properties of the filler, thus hindering the movement of electrons between the fillers. Grafting coupling agent on the filler surface can enhance the interfacial force between the filler and the matrix, effectively reducing the voids at the phase interface and the defects generated by filler agglomeration at high filling, thus inhibiting the migration of carriers [63, 64]. These two methods improve the interfacial compatibility while blocking the formation of conductive pathways, which is conducive to improving the electrical insulation of composites. This simultaneously solves the problem of poor compatibility at the filler–matrix interface and the problem of thermal conductivity without insulation.

Wang et al. [65] proposed a simple method to modify the surface of metal fillers to improve the thermal conductivity, insulating and mechanical properties of PI composites. A partially fluorinated alumina layer (PFAL) is constructed on the surface of metallic aluminum powders (APs) by direct fluorination of F_2 and N_2 mixtures. The surface of the shell–core structured aluminum fluoride powders (FAPs) forms an electron-rich layer with strong interactions, which enables effective interactions with the electron-absorbing imine ring of PI. The strong interaction between FAPs and PI reduces the interfacial thermal resistance between the filler and the matrix, while maintaining good insulation and mechanical properties. The in-plane and out-of-plane thermal conductivities of PI/FAPs 50 wt% are enhanced to 16.83 W/(m·K) and 1.42 W/(m·K), respectively, with a mechanical strength of 87.2 MPa. The PI composite films show low electrical conductivity over the entire frequency range. At a frequency of 67 kHz, the conductivity is less than 1×10^{-8} S/cm, which can be attributed to the surface-modified FAPs hindering the formation of conductive pathways. This innovative approach opens up new avenues for metal-filled PI thermally conductive composites.

The insulating properties of thermally conductive composites are critical to the safe operation of electronic devices. In complex environments such as high-voltage and high current, insulating materials need to be able to withstand higher electric field strengths to prevent failures such as current leakage and short circuits. Conventional silicon materials are difficult to meet the needs of modern electronic technology due to the limitations of their thermal conductivity and insulating properties. Carbon materials show great potential as thermally conductive fillers, but how to fully utilize the inherent thermal conductivity of carbon materials while lowering their conductive threshold is a technical problem that needs to be overcome. The structural design according to specific needs is more conducive to improving performance. In the future, there is a need to focus on the design and development of multifunctional thermally conductive and electrically insulating composites.

6.4.3 Thermal Conductivity and Low Permittivity

In terms of fast signal transmission and high energy density, electronic devices have increasingly stringent requirements for packaging materials. Permittivity (or dielectric constant), a key parameter that measures the response of a material to an electric field, has a significant impact on the signal propagation time in electronic components. The composites with low dielectric constant and low dielectric loss can significantly reduce signal delay and parasitic capacitance, thus ensuring efficient and accurate signal transmission. Therefore, the composites not only need to have good thermal conductivity to ensure that the heat can be quickly dispersed, but also need to have a low dielectric constant to ensure the stability and efficiency of the signal transmission process [66, 67]. Designing a composite material with both high thermal conductivity and low dielectric constant becomes the key to improve the signal transmission quality of intelligent terminals and effective heat dissipation of electronic devices.

Song et al. [68] prepared polyimide/aramid fiber/boron nitride @ carbon nitride (PI/ANFs/SBN@CN) composites with high thermal conductivity and low permittivity by in situ polymerization and hot pressing. The ANFs and SBN@CN form a foliage-like network structure, which makes the filler arranged in a directional and orderly manner, thus improving the thermal conductivity of the composites. When the filling amount of SBN@CN is 30 wt%, the thermal conductivity is elevated to 1.162 W/(m·K), which is 8 times higher than that of the pure PI film. The dielectric constant and dielectric loss of the PI composite were reduced to 3.81 and 0.0034 ($f = 10^8$ Hz), respectively. The PI/ANFs/SBN@CN composites exhibit good overall performance and provide a new perspective on the application of PI in electronic packaging.

The "crystal-like phase" strategy proposed by Dong et al. [69] broke through the bottleneck of high thermal conductivity and low dielectric constant in PI composites, which is difficult to be synergistically regulated. IL-CaF_2 /PI films are prepared by using ionic liquid (IL) as pore-causing agent and calcium fluoride (CaF_2) as filler. The thermal conductivity of 10 vol% IL-CaF_2/PI film can reach 7.22 W/(m·K), and the dielectric constant and dielectric loss are as low as 2.14 and 0.01 ($f = 10^6$ Hz), respectively. IL induces the formation of a three-dimensional porous thermal conductivity network, which helps to reduce the dielectric constant of the PI film. The intermolecular charge transfer between CaF_2 and PI contributes to the "birefringent" behavior and the formation of an ordered structure similar to that of liquid crystals, which creates favorable conditions for improving the thermal conductivity of PI films. Based on the characteristics of high thermal conductivity and low permittivity, IL-10 vol% CaF_2/PI film is designed as ultra-thin flexible ultrasonic generator, which provides a new material for wearable ultrasonic detection.

The influence of interface should also be considered in regulating the thermal conductivity and low dielectric properties of PI composites. When the two-phase components have different polarities or different conductivities, the space charge is difficult to migrate between the different phases, resulting in interfacial polarization at the interface. Doping inorganic fillers into the PI matrix will introduce a large number of organic–inorganic interfaces, resulting in more charge migration to the interface to form charge accumulation [70]. The interfacial compatibility can be improved by surface modification of the inorganic filler and optimization of the preparation process, so as to achieve effective regulation of the thermal conductivity and dielectric properties of PI composites.

6.5 Conclusions and Outlook

This chapter reviews the thermal conductivity of PIs and presents the research progress of intrinsically thermally conductive PIs and filled thermally conductive PIs. These functional materials have widespread applications in electronic packaging, battery thermal management, flexible electronic devices, and the aerospace industry. The development of filled thermally conductive PIs focuses on lowering the interfacial thermal resistance to improve their thermal conductivity and enhancing their performance in specific applications. The selection and optimization of fillers, the improvement of the preparation process, and the study of structure and properties remain key points for future research.

The key to filled thermally conductive PI lies in the selection of appropriate fillers. Factors such as thermal conductivity of the filler, compatibility and dispersion of the PI matrix, and cost need to be considered. In recent years, researchers have further enhanced the thermal conductivity of PIs by optimizing the shape, size, and surface properties of the fillers and developing new fillers. The study of the structure–property relationship of thermally conductive PI provides theoretical support for the design of multifunctionalized thermally conductive PI composites. The directional structure design primarily focuses on reducing the interfacial thermal resistance caused by the disordered overlap of fillers through improvements in the preparation process, thus building thermally conductive pathways with ordered arrangements. The non-oriented structure design is based on the interaction between fillers or between fillers and matrix, which is based on the surface of fillers with reactive groups. The multilayer structural design provides a new research idea for obtaining thermally conductive composites with excellent mechanical properties and electromagnetic shielding properties.

At present, the thermally conductive PI composite has made important breakthroughs and is generally in a high-speed development stage, but it still lacks deeper theoretical research. The key parameters and laws of action of interface structure on thermal conductivity need to be clarified to further improve the heat transfer mechanism. Based on the theoretical research of intrinsically thermally conductive polymers, copolymer of PI is developed according to specific needs, and new molecular chain structure is constructed to promote the structural design of PI. In the future, the relationship between interface structure and performance should continue to be studied, and various performance improvement mechanisms should be integrated with thermal conductivity mechanisms to promote the rapid development of new intelligent thermally conductive PI composite materials. In the future, further research should be conducted on the relationship between interface structure and performance. Various performance enhancement mechanisms should be integrated with thermal conductivity mechanisms to promote the rapid development of new intelligent thermally conductive PI composite materials.

References

1 Lin, Y., Li, P.L., Liu, W.J. et al. (2024). Application driven high thermal conductivity polymer nanocomposites. *ACS Nano* 18 (5): 3851–3870.
2 Guo, Y.Q., Ruan, K.P., Shi, X.T. et al. (2020). Factors affecting thermal conductivities of the polymers and polymer composites: a review. *Composites Science and Technology* 193: 108134.
3 Chen, Q.G., Yang, K.L., Feng, Y. et al. (2023). Recent advances in thermal-conductive insulating polymer composites with various fillers. *Composites Part A: Applied Science and Manufacturing* 178: 107998.
4 Xiao, M. and Du, B.X. (2016). Review of high thermal conductivity polymer dielectrics for electrical insulation. *High Voltage* 1 (1): 34–42.
5 Chen, Y., Zhang, H.G., Chen, J. et al. (2022). Thermally conductive but electrically insulating polybenzazole nanofiber/boron nitride nanosheets nanocomposite paper for heat dissipation of 5G base stations and transformers. *ACS Nano* 16 (9): 14323–14333.
6 Chen, R., Li, X., Ma, J.R. et al. (2023). Covalently modified graphene and 3D thermally conductive network for PEEK composites with electromagnetic shielding performance. *Composites Part A: Applied Science and Manufacturing* 173: 107633.
7 Chu, Q.D., Lin, H., Ma, M. et al. (2022). Cellulose nanofiber/graphene nanoplatelet/MXene nanocomposites for enhanced electromagnetic shielding and high in-plane thermal conductivity. *ACS Applied Nano Materials* 5 (5): 7217–7227.
8 Fu, C.J., Li, Q., Lu, J.B. et al. (2018). Improving thermal conductivity of polymer composites by reducing interfacial thermal resistance between boron nitride nanotubes. *Composites Science and Technology* 165: 322–330.
9 Zhou, Z.H., Zhou, N., Jia, X.L. et al. (2023). Research progress of filled-type high-thermal-conductivity flexible polyimide composites: a review. *Journal of Materials Science* 58 (41): 15973–16001.

10 Wan, B.Q., Zheng, M.S., Yang, X. et al. (2023). Recyclability and self-healing of dynamic cross-linked polyimide with mechanical/electrical damage. *Energy & Environmental Materials* 6 (1): e12427.

11 Burger, N., Laachachi, A., Ferriol, M. et al. (2016). Review of thermal conductivity in composites: mechanisms, parameters and theory. *Progress in Polymer Science* 61: 1–28.

12 Liu, J. and Yang, R.G. (2012). Length-dependent thermal conductivity of single extended polymer chains. *Physical Review B* 86: 104307.

13 Feng, C.P., Bai, L., Bao, R.Y. et al. (2019). Superior thermal interface materials for thermal management. *Composites Communications* 12: 80–85.

14 Wang, Z.H., Carter, J.A., Lagutchev, A. et al. (2007). Ultrafast flash thermal conductance of molecular chains. *Science* 317: 787–790.

15 Tao, K.K., Sun, G.H., Feng, C.C. et al. (2023). Structural design and research progress of thermally conductive polyimide film–a review. *Macromolecular Rapid Communications* 44 (13): 2300060.

16 Guo, Z., Lee, D., Liu, Y. et al. (2014). Tuning the thermal conductivity of solar cell polymers through side chain engineering. *Physical Chemistry Chemical Physics* 16 (17): 7764–7771.

17 Ma, H. and Tian, Z.T. (2019). Chain rotation significantly reduces thermal conductivity of single-chain polymers. *Journal of Materials Research* 34 (1): 126–133.

18 Kaya, İ. and Kamaci, M. (2018). Synthesis, optical, and thermal properties of polyimides containing flexible ether linkage. *Journal of Applied Polymer Science* 135 (31): 46573.

19 Liu, X.Y., Ai, Q., Xu, J.Z. et al. (2023). Improving inherent thermal conductivity of epoxy resins based on contribution components of thermal conductivity: a molecular dynamics study. *European Polymer Journal* 198: 112407.

20 Zhang, X., Wang, Y., Xia, R. et al. (2020). Effect of chain configuration on thermal conductivity of polyethylene—a molecular dynamic simulation study. *Chinese Journal of Polymer Science* 38: 1418–1425.

21 Xu, X., Chen, J., Zhou, J. et al. (2018). Thermal conductivity of polymers and their nanocomposites. *Advanced Materials* 30 (17): 1705544.

22 Huang, C.L., Qian, X., and Yang, R.G. (2018). Thermal conductivity of polymers and polymer nanocomposites. *Materials Science and Engineering: R: Reports* 132: 1–22.

23 Liu, B.H., Zhou, Y., Dong, L. et al. (2022). Enhanced thermal conductivity in copolymerized polyimide. *Iscience* 25 (11): 105451.

24 Ruan, K.P., Guo, Y.Q., and Gu, J.W. (2021). Liquid crystalline polyimide films with high intrinsic thermal conductivities and robust toughness. *Macromolecules* 54 (10): 4934–4944.

25 Luo, D.C., Huang, C.L., and Huang, Z. (2018). Decreased thermal conductivity of polyethylene chain influenced by short chain branching. *Journal of Heat Transfer-Transactions of the ASME* 140 (3): 31302.

26 Liu, J., Ju, S.H., Ding, Y.F. et al. (2014). Size effect on the thermal conductivity of ultrathin polystyrene films. *Applied Physics Letters* 104 (15): 153110.

27 Shen, S., Herry, A., Tong, J. et al. (2010). Polyethylene nanofibers with very high thermal conductivities. *Nature Nanotechnology* 5 (4): 251–255.

28 Ruan, K.P., Shi, X.T., Zhang, Y.L. et al. (2023). Electric-field-induced alignment of functionalized carbon nanotubes inside thermally conductive liquid crystalline polyimide composite films. *Angewandte Chemie* 135 (38): e202309010.

29 Shoji, Y., Ishige, R., Higashihara, T. et al. (2013). Cross-linked liquid crystalline polyimides with siloxane units: their morphology and thermal diffusivity. *Macromolecules* 46 (3): 747–755.

30 Shoji, Y., Ishige, R., Higashihara, T. et al. (2010). Synthesis and liquid crystalline behavior of laterally substituted polyimides with siloxane linkages. *Macromolecules* 43 (21): 8950–8956.

31 Ruan, K.P., Zhong, X., Shi, X.T. et al. (2021). Liquid crystal epoxy resins with high intrinsic thermal conductivities and their composites: a mini-review. *Materials Today Physics* 20: 100456.

32 Zhang, L., Ruesch, M., Zhang, X.L. et al. (2015). Tuning thermal conductivity of crystalline polymer nanofibers by interchain hydrogen bonding. *RSC Advances* 5 (107): 87981–87986.

33 Ran, X.N., Wang, Y.X., Lu, J.B. et al. (2022). Effect of hydrogen bonds on the thermal transport in a precisely branched polyethylene with ordered and amorphous structures. *Computational Materials Science* 205: 11191.

34 Nicholls, R.A., Pellister, M., Perez, Y. et al. (2019). Thermomechanical characterization of thermoplastic polyimide–polyurea to improve the chain interaction via internal hydrogen bonds. *Polymer Engineering & Science* 59 (9): 1948–1959.

35 Kim, G., Lee, D., Shanker, A. et al. (2015). High thermal conductivity in amorphous polymer blends by engineered interchain interactions. *Nature Materials* 14 (3): 295–300.

36 Fuller, J.J. and Marotta, E.E. (2001). Thermal contact conductance of metal/polymer joints: an analytical and experimental investigation. *Journal of Thermophysics and Heat Transfer* 15 (2): 228–238.

37 Guo, Y.Q., Xu, G.J., Yang, X.T. et al. (2018). Significantly enhanced and precisely modeled thermal conductivity in polyimide nanocomposites with chemically modified graphene via in situ polymerization and electrospinning-hot press technology. *Journal of Materials Chemistry C* 6 (12): 3004–3015.

38 Zhang, X.D., Dong, J.W., Pan, D. et al. (2021). Constructing dual thermal conductive networks in electrospun polyimide membranes with highly thermally conductivity but electrical insulation properties. *Advanced Composites and Hybrid Materials* 4: 1102–1112.

39 Liu, D.X., Chi, H.T., Ma, C.G. et al. (2022). Improving in-plane and out-of-plane thermal conductivity of polyimide/boron nitride film with reduced graphene oxide by a moving magnetic field induction. *Composites Science and Technology* 220: 109292.

40 Ding, D.L., Shang, Z.H., Zhang, X. et al. (2020). Greatly enhanced thermal conductivity of polyimide composites by polydopamine modification and the 2D-aligned structure. *Ceramics International* 46 (18): 28363–28372.

41 Choy, C.L., Luk, W.H., and Chen, F.C. (1978). Thermal conductivity of highly oriented polyethylene. *Polymer* 19 (2): 155–162.

42 Losego, M.D., Grady, M.E., Sottos, N.R. et al. (2012). Effects of chemical bonding on heat transport across interfaces. *Nature Materials* 11 (6): 502–506.

43 Majumdar, S., Sierra-Suarez, J.A., Schiffres, S.N. et al. (2015). Vibrational mismatch of metal leads controls thermal conductance of self-assembled monolayer junctions. *Nano Letters* 15 (5): 2985–2991.

44 Meier, T., Menges, F., Nirmalraj, P. et al. (2014). Length-dependent thermal transport along molecular chains. *Physical Review Letters* 113 (6): 060801.

45 Tian, Z., Marconnet, A., and Chen, G. (2015). Enhancing solid-liquid interface thermal transport using self-assembled monolayers. *Applied Physics Letters* 106 (21): 211602.

46 Wang, Y., Zhang, X., Ding, X. et al. (2020). Imidization-induced carbon nitride nanosheets orientation towards highly thermally conductive polyimide film with superior flexibility and electrical insulation. *Composites Part B: Engineering* 199: 108267.

47 Wan, B.Q., Li, X., Zeng, X.L. et al. (2024). Covalent-assisted construction of "scale-like" boron nitride/polyimide thermal interface materials with high thermal conductivity. *Composites Communications* 45: 101803.

48 Lan, H.Y., Wu, B., Yan, Y.Y. et al. (2023). Enhanced in-plane thermal conductivity of polyimide-based composites via in situ interfacial modification of graphene. *Nanoscale* 15 (8): 4114–4122.

49 Tan, S.H., Tang, L.M., and Chen, K.Q. (2014). Phonon scattering and thermal conductance properties in two coupled graphene nanoribbons modulated with bridge atoms. *Physics Letters A* 378: 1952–1955.

50 Xie, Z.L., Wu, K., Liu, D.Y. et al. (2021). One-step alkyl-modification on boron nitride nanosheets for polypropylene nanocomposites with enhanced thermal conductivity and ultra-low dielectric loss. *Composites Science and Technology* 208: 108756.

51 Huxtable, S.T., Cahill, D.G., Shenogin, S. et al. (2003). Interfacial heat flow in carbon nanotube suspensions. *Nature Materials* 2 (11): 731–734.

52 Cao, L., Wang, J.J., Dong, J. et al. (2020). Preparation of highly thermally conductive and electrically insulating PI/BNNSs nanocomposites by hot-pressing self-assembled PI/BNNSs microspheres. *Composites Part B: Engineering* 188: 107882.

53 Xu, L.B., Chen, G.F., Wang, W. et al. (2016). A facile assembly of polyimide/graphene core–shell structured nanocomposites with both high electrical and thermal conductivities. *Composites Part A: Applied Science and Manufacturing* 84: 472–481.

54 Guo, Y.Q., Qiu, H., Ruan, K.P. et al. (2022). Hierarchically multifunctional polyimide composite films with strongly enhanced thermal conductivity. *Nano-Micro Letters* 14: 1–13.

55 Wang, Y., Wang, H.T., Liu, F. et al. (2020). Flexible printed circuit board based on graphene/polyimide composites with excellent thermal conductivity and

sandwich structure. *Composites Part A: Applied Science and Manufacturing* 138: 106075.

56 Chen, X., Wang, Y.Y., Zheng, K. et al. (2024). High thermally conductive polyimide film designed with heterostructural Cring-C_3N_4 filler and multilayer strategy. *Journal of Alloys and Compounds* 983: 173907.

57 Wu, Z.Q., Dong, J., Li, X.T. et al. (2022). Interlayer decoration of expanded graphite by polyimide resins for preparing highly thermally conductive composites with superior electromagnetic shielding performance. *Carbon* 198: 1–10.

58 Huang, X., Tuersun, Y., Luo, P.J. et al. (2023). In-situ reduction of AgNPs on MXene surfaces for synthesis of efficient thermally conductive composites with powerful electromagnetic shielding capabilities. *Colloids and Surfaces A: Physicochemical and Engineering Aspects* 677: 132444.

59 Shang, Y., Ji, Y.X., Dong, J.W. et al. (2021). Sandwiched cellulose nanofiber/boron nitride nanosheet/$Ti_3C_2T_x$ MXene composite film with high electromagnetic shielding and thermal conductivity yet insulation performance. *Composites Science and Technology* 214: 108974.

60 Zhang, Y.L., Ruan, K.P., Zhou, K. et al. (2023). Controlled distributed $Ti_3C_2T_x$ hollow microspheres on thermally conductive polyimide composite films for excellent electromagnetic interference shielding. *Advanced Materials* 35: 2211642.

61 Guo, F.J., Wang, Y.Y., Xue, K.L. et al. (2024). Epoxy composites with satisfactory thermal conductivity and electromagnetic shielding yet electrical insulation enabled by Al_2O_3 platelet-isolated MXene porous microsphere networks. *Composites Science and Technology* 248: 110425.

62 Bian, F.P., Huang, R., Li, X.B. et al. (2024). Facile construction of chestnut-like structural fireproof PDMS/Mxene@BN for advanced thermal management and electromagnetic shielding applications. *Advanced Science* 11: 2307482.

63 Ma, M., Xu, L., Qiao, L.L. et al. (2020). Nanofibrillated cellulose/MgO@rGO composite films with highly anisotropic thermal conductivity and electrical insulation. *Chemical Engineering Journal* 392: 123714.

64 Lin, H., Ma, M., Chu, Q.D. et al. (2022). MgAl layered double oxide/reduced graphene oxide composites with reduced interfacial thermal resistance for thermal conductivity and electrical insulation applications. *ACS Applied Nano Materials* 5 (12): 18955–18964.

65 Wang, X., Li, L.M., Liu, S.Y. et al. (2023). Partially fluorinated metallic surface constructing robust, thermally conductive and electrically insulating polymer composite. *Composites Science and Technology* 243: 110225.

66 Xiao, C., Tang, Y.L., Chen, L. et al. (2019). Preparation of highly thermally conductive epoxy resin composites via hollow boron nitride microbeads with segregated structure. *Composites Part A: Applied Science and Manufacturing* 121: 330–340.

67 Li, R., Yang, X., Li, J. et al. (2022). Review on polymer composites with high thermal conductivity and low dielectric properties for electronic packaging. *Materials Today Physics* 22: 100594.

68 Song, Q.Y., Wang, M.Q., Lai, Y.M. et al. (2023). Preparation and properties of branch-leaf-like polyimide composite films with high thermal conductivity and low dielectric loss. *Journal of Applied Polymer Science* 140 (41): e54520.

69 Dong, X.D., Wan, B.Q., Zheng, M.S. et al. (2024). Dual-effect coupling for superior dielectric and thermal conductivity of polyimide composite films featuring "crystal-like phase" structure. *Advanced Materials* 36: 2307804.

70 Xie, S.H., Zhu, B.K., Li, J.B. et al. (2004). Preparation and properties of polyimide/aluminum nitride composites. *Polymer Testing* 23: 797–801.

7

Polyimide Nanocomposites for Electromagnetic Interference Shielding

Yuchao Li

Liaocheng University, Shandong Province, School of Materials Science and Engineering, No.1 Hunan Road, Liaocheng City, Shandong Province 252059, China

7.1 Introduction

With the skyrocketing growth of the electronics industry and information technology, various electronic devices and communication technologies have mushroomed in the modern life. The technological revolution brings great convenience and at the same time inevitably produces a large number of byproducts, including electromagnetic wave (EMW) pollution, the fourth largest pollution preceded by water pollution, air pollution, and noise pollution, leading to increasingly severe electromagnetic pollution issues [1, 2]. Moreover, the advent of 5G technology and the Internet of Things (IoT) has also introduced a significant concern regarding electromagnetic interference (EMI). Excessive EMWs not only cause interference to electronic instruments and communication facilities, but also incur serious harm to human health. Hence, the development of highly efficient EMI-shielding materials has been paid more and more attention in the past few decades [3]. In addition, advanced technologies such as artificial intelligence (AI) and new energy vehicles drive toward high integration, high power density, and high frequency of semiconductor devices (CPU, communication IC, and high-power device IGBT). Moreover, radar, infrared, and other detection technologies have been widely used in the field of aerospace and aviation with the development of the electronic information technology. Therefore, there is also an urgent need to develop ultra "thin, light, wide, and strong" high-performance EM wave absorption materials, other than EMI-shielding materials.

Metals remain a popular choice for EMI shielding due to their high conductivity and effectiveness in blocking electromagnetic interference. However, in certain applications where weight, cost, corrosion, flexibility, or transparency is an important factor, alternative materials, such as conductive polymers or polymer-based composite materials, may be widely considered for EMI-shielding purposes because of their advantages of low density, corrosion resistance, low cost, and good machining performances. As is well known, most commercial polymer

Functional Polyimide Dielectrics: Structure, Properties, and Applications, First Edition.
Edited by Jun-Wei Zha.

substrates are inherently insulating and cannot endure harsh environments with high temperature, humidity, and salt spray, which severely limits their applications in EMI shielding and causes serious service reliability and service lifetime problems for EMI-shielding materials and equipment in such environments. Therefore, researchers have paid more attention to solving the problem of poor EMI-shielding performance and low reliability by properly choosing polymer matrixes and hierarchically combining functional conductive nanofillers, including various metals, conductive ceramics, and carbonaceous materials.

Nowadays, polyimide (PI) is regarded as a high-performance polymer due to its excellent comprehensive properties, such as stable chemical resistance, excellent thermal stability, outstanding insulation, favorable mechanical strength, good flexibility, and versatility. Polyimides (PIs) are thus greatly desirable for applications in cutting-edge aviation, medical, electronic, and aerospace technologies, e.g., liquid crystal, EMI shielding, electronic packaging, electrical diaphragm, thermal insulation, and moisture resistance [4]. Unfortunately, most bare PIs are transparent to EMI radiation due to their nonmagnetic and insulative nature. To address this limitation, manufacturers often incorporate additives or coatings into PIs to enhance their EMI-shielding performance, making it more suitable for applications where EMI protection is required [5].

In this chapter, we reviewed and discussed the recent developments of PI-based materials for EMI-shielding and -absorbing applications, from the principle of EMI shielding to the structural design of PI composites. These improvements enabled PI-based materials to become a viable option for EMI-shielding applications in harsh environment. For example, as detection technology rapidly advances, the design and exploration of radar stealth materials with excellent microwave absorption capabilities for use in special environments have become of paramount significance.

7.2 Theoretical

When microwaves interact with matter, the electrons and magnetic dipoles will oscillate and generate an oscillating electric current (and consequentially an electric field) and an oscillating magnetic field within the material, which are in phase with the incident microwaves [1, 6]. The magnitude of these fields depends on the conductivity and susceptibility of the material. Therefore, the interaction of the material with microwaves is related to its electrical (complex permittivity) and magnetic (complex permeability) properties. When an EMW interferes with matter, the following different phenomena will occur, as shown in Figure 7.1: reflection (and secondary reflection), absorption, internal multiple reflection, and transmission (and secondary transmission).

Reflection arises when there is a change in the impedance between two materials, where some of the microwaves are reflected back. The amount of reflection depends on the magnitude of impedance mismatch, the angle of incident wave, and its s or p polarization. Absorption occurs when a wave passes through matter: the interaction

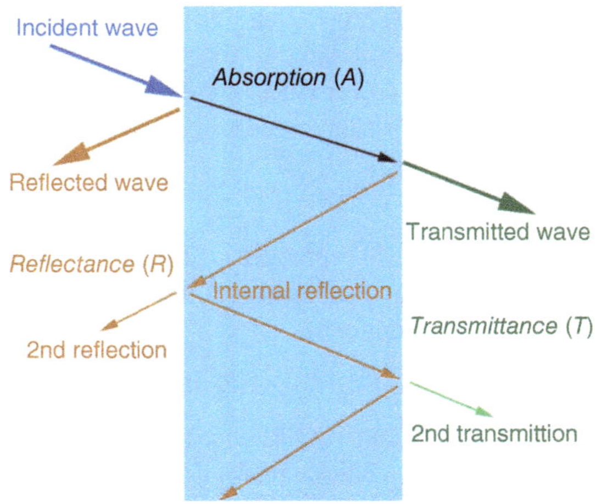

Figure 7.1 Schematic illustration of microwave interaction with matter.

between wave and matter constituents causes the transfer of part of its energy to the material, which is usually converted in heat generation. Absorption is a complex phenomenon made of several dissipation mechanisms, including dielectric losses, magnetic losses, and conductive losses. Therefore, absorption mechanisms are related to constitutive parameters of the material, i.e., dielectric permittivity, magnetic permeability, and conductivity. Usually, one or more mechanisms simultaneously occur, dispersing the energy of radiation. Internal multiple reflection is the reflection that occurs between the two walls of the shield. It is favored by the presence of several internal interfaces, such as nanostructures and inhomogeneities within the material. It is worth noting that certain structures, such as foams and porous aerogels, promote multiple reflection. All these contributions cause energy loss of the incident waves, and only a reduced portion of the waves passes through the thickness of the material. Therefore, electromagnetic shields are aimed to maximize these phenomena, i.e., absorbing or reflecting the unwanted signals and reducing their intensity, in order to insulate a portion of space from electromagnetic pollution. Therefore, there are generally two kinds of EMI-shielding materials according to the shielding mechanisms, the reflection- and absorption-dominant types. The reflection type is based on the electric vector interaction of incident EMWs with free charge carriers (electrons and holes) of the shielding material. Increasing the electrical conductivity can improve the direct reflection as well as the EMI-shielding performance. The absorption type contains electric and magnetic dipoles that can interact with the electric and magnetic fields, converting the incident EMWs to heat dissipation. The EMI-shielding mechanisms of traditional EMI-shielding materials are mostly dominated by reflection, which makes it easy to cause secondary electromagnetic pollution after reflection. With the booming of telecommunication and microelectronics, it is essential to develop EMI-shielding materials with low reflectivity characteristics to avoid the secondary pollution from EMWs [7].

The ability of a material to shield EMWs can be expressed as the electromagnetic interference shielding effectiveness (EMI SE) or simply the total shielding effectiveness (SE_T). It evaluates the ability of a material to attenuate the incident EM radiation, expressing as the logarithm of the ratio between transmitted (t) and incident (i) power P. SE_T can be also expressed as the ratio between electrical (E)/magnetic (H) transmitted and incident field (Eq. 7.1) [8].

$$SE_T(\text{dB}) = 10\log_{10}\frac{P_t}{P_i} = 20\log_{10}\frac{E_t}{E_i} = 20\log_{10}\frac{H_t}{H_i} \tag{7.1}$$

The SE_T can be also expressed as the sum of the shielding contributions (absorption A, reflection R, and internal multiple reflection M) as shown in Eq. 7.2, where dB (decibel) is the unit of SE.

$$SE_T(\text{dB}) = SE_A(\text{dB}) + SE_R(\text{dB}) + SE_M(\text{dB}) \tag{7.2}$$

The decomposition of SE_T into the individual contributions can be expressed in Eqs. (7.3)–(7.5) as follows [6, 8]:

$$SE_A(\text{dB}) = 20\log e^{t/\delta} = 8.68t\sqrt{\frac{\omega\sigma_{ac}\mu_r}{2}} \tag{7.3}$$

$$SE_R(\text{dB}) = 20\log\frac{\eta}{4\eta_0} = 10\log\frac{\sigma_{ac}}{16\omega\varepsilon_0\mu_r} \tag{7.4}$$

$$SE_M(\text{dB}) = 20\log_{10}\left(1 - e^{-\frac{2t}{\delta}}\right) = 20\log_{10}\left(1 - 10^{-\frac{SE_A}{10}}\right) \tag{7.5}$$

where t is the thickness of the shield, $\omega = 2\pi f$ is the angular frequency, $\sigma_{ac} = \omega\varepsilon_0\varepsilon''$ is the electrical conductivity (where ε_0 is vacuum permittivity and ε'' is the imaginary part of the permittivity of the shield), μ_r is the relative magnetic permeability, δ is known as "skin depth" (the distance required to attenuate a wave to 37% of its power) and equals to $8.68t/SE_A$. The impedance mismatch is represented by the characteristic impedance of the medium (η_0) and the impedance of the shielding material (η).

As shown from Eqs. (7.3)–(7.5), the σ_{ac} has a positive effect on both SE_A and SE_R, while the μ_r has opposite effect. Consequently, magneto-conductive materials exhibit good absorption properties. The greater the μ_r of the shield, the greater the contribution of SE_A with respect to SE_R. In addition, the thickness of the shield plays a key role since the amplitude of electromagnetic radiation decreases exponentially as it passes through the shield. Therefore, the contribution of internal multiple reflection is normally negligible if the δ is lower than t of the shield.

The performance of EMI SE was usually measured by a vector network analyzer at room temperature in the radio frequency range of 8.2–12.4 GHz (X-band). The scattering parameters S_{11} (or S_{22}) and S_{12} (or S_{21}) obtained from the experiments were used to calculate the SE values by the following Eqs. (7.6)–(7.9):

$$SE_T = -10\log T = -10\log\left(|S_{12}|^2\right) = -10\log\left(|S_{21}|^2\right) \tag{7.6}$$

$$SE_R = -10\log(1-R) = -10\log\left(1 - |S_{11}|^2\right) = -10\log\left(1 - |S_{22}|^2\right) \tag{7.7}$$

$$SE_A = -10\log\left(\frac{T}{1-R}\right) = -10\log\left(\frac{|S_{12}|^2}{1 - |S_{11}|^2}\right) \tag{7.8}$$

$$SE_M = SE_T - SE_R - SE_A \tag{7.9}$$

The EMI-shielding efficiency (%) can more clearly express the percentage of EMWs blocked. The relationship between the EMI-shielding efficiency and SE_T is as follows (Eq. 7.10) [9]:

$$\text{Shielding efficiency}(\%) = 100 - \left(\frac{1}{10^{SE_T}/10}\right) \times 100 \tag{7.10}$$

To incorporate the effects of thickness t and density D, three related quantities, namely shielding effectiveness per unit thickness (SE_t), specific shielding effectiveness (SSE), and absolute shielding efficiency (SSE_t), were also calculated using the following Eqs. (7.11)–(7.13) [10]:

$$SE_t = \frac{SE_T}{t}\left(\text{dB}\cdot\text{cm}^{-1}\right) \tag{7.11}$$

$$SSE = \frac{SE_T}{D}\left(\text{dB}\cdot\text{cm}^{3}\cdot\text{g}^{-1}\right) \tag{7.12}$$

$$SSE_t = \frac{SSE}{t}\left(\text{dB}\cdot\text{cm}^{2}\cdot\text{g}^{-1}\right) \tag{7.13}$$

In addition, the EMW absorption ability of materials can be evaluated by calculating the reflection loss (RL) and effective absorption bandwidth (EAB). According to the transmission line theory, the RL is calculated according to the following formula (Eqs. (7.14)–(7.15)) [4]:

$$RL = 20\log\left|\frac{Z_{in}-1}{Z_{in}+1}\right| \tag{7.14}$$

$$Z_{in} = Z_0\sqrt{\frac{\mu_r}{\varepsilon_r}}\tanh\left[j\left(\frac{2\pi}{c}\right)ft\sqrt{\varepsilon_r\mu_r}\right] \tag{7.15}$$

where Z_{in} is the normalized input impedance, c is the free-space light velocity, and f represents the frequency. $RL < -10$ dB means that the material has absorbed 90% of the electromagnetic energy, and the corresponding frequency range is referred to as EAB.

Excellent microwave-absorbing coatings should meet the following requirements: (i) exhibit strong microwave absorption properties over a wide frequency range; (ii) be thin and lightweight; (iii) have a simple coating layer structure and take less time to work during urgent processes. For the EMW-absorbing materials, beneficial impedance matching and powerful EMW attenuation capability are necessary to achieve high EMW absorption performance. Generally, improving conductivity and constructing architectures with multiple reflection interfaces are regarded as two effective pathways to enhance EMI-shielding performances [11]. According to their EMW loss characteristics, the EMW-absorbing materials are typically classified as dielectric loss materials and magnetic loss materials. The magnetic loss materials include magnetic metallic materials, ferrite (Fe_3O_4, Co_3O_4, NiCo, etc.). However, traditional magnetic absorption materials do not meet the requirements of lightweight owing to the high density. Furthermore, the μ_r usually decreases significantly with increasing frequency due to the Snoek limit, weakening microwave attenuation capabilities. Therefore, dielectric absorbers without magnetic component have attracted wide attention as lightweight EMA

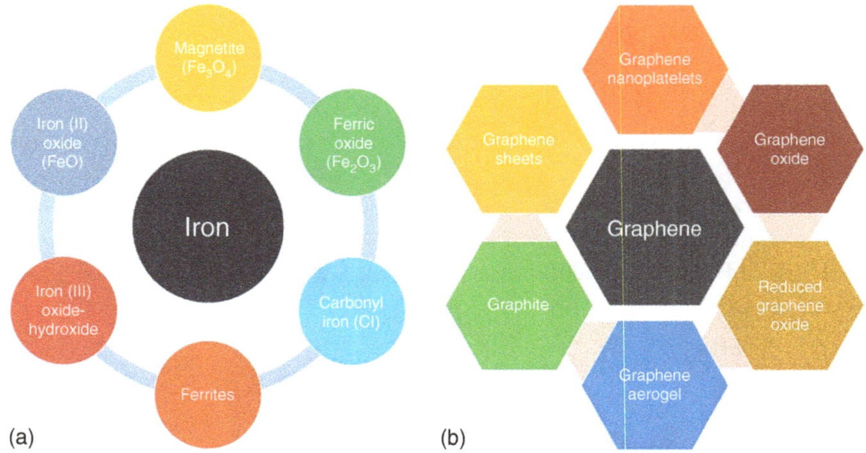

Figure 7.2 (a) Iron-based magnetic components and (b) dielectric absorber of graphene family for EMI shielding. Source: Ref. [12]/MDPI/CC BY 4.0.

materials, meeting the development demands of "thin, light, wide, and strong" microwave absorbers [4]. The commonly used dielectric loss materials are carbon materials (graphene, carbon nanotubes, carbon fibers, etc.), conductive polymers. Figure 7.2 displays basic iron-based magnetic components and graphene-based dielectric absorbers for EMI-shielding applications. [12]

7.3 Nanofillers Used for PI Electromagnetic Shielding

Researchers have effectively solved the problem of poor EMI-shielding performance of PIs by combining different nanofillers within the polymer matrix. They vary from metals, magnetic particles, to carbonaceous nanofillers in different dimensions. They can be zero-dimensional (0D) silver nanoparticles (NPs), one-dimensional (1D) carbon nanotubes, two-dimensional (2D) graphene or MXene, etc. characterized by a large conductivity and surface area that enhance the interaction probability with EMWs [13].

7.3.1 Magnetic Fillers

According to the theory, one way to improve the current electromagnetic PIs is to reinforce them with ferromagnetic NPs [14, 15]. These microwave absorbers contribute mainly to the magnetic losses in comparison with those of dielectric losses, thus achieving broader bandwidth and higher absorption at a smaller layer thickness. Regardless of weight, many magnetic particles [16–20], such as ferroferric oxide (Fe_3O_4), iron(III) oxide (Fe_2O_3), cobalt oxide (Co_3O_4), cobalt ferrite ($CoFe_2O_4$), nickel ferrite ($NiFe_2O_4$), and carbonyl iron, are good adsorbers to enhance the magnetic loss of shielding materials.

Jiang et al. designed a type of dual-functional magnetic microcapsules containing a phase change material core (magnetic microcapsules containing Fe_3O_4 NPs) and an organo-silica shell for the electromagnetic shielding and thermal regulating PI films [19]. The hybrid films were endowed with both temperature retention and microwave-absorbing capabilities due to the introduction of magnetic microcapsules. Results showed that more than 99.9% EMWs could be blocked by the hybrid films.

Carbonyl iron (CI) particles have long attracted researchers' attention as the main magnetic components of EWA absorbers for the advantages of high saturation magnetization, high Curie temperature, and large value of magnetic permeability that can be obtained in GHz above Sneok's limit and have been applied in many EMI materials. A common drawback of the carbonyl iron is the easy oxidation to Fe_3O_4 at high temperature, which leads to the decrease in magnetic property. Wang et al. introduced different shape of CI absorbers into PI. Results showed that both the complex permittivity and complex permeability increased with increasing the CI volume fraction. The absorbing properties of the spherical CI composite were better than flaky CI system due to the higher impedance matching degree [17].

Thin electromagnetic absorbers are required to have high saturation magnetization, high electric resistivity, and high magnetic loss in the high-frequency range. Co, Ni, Fe are well-known magnetic absorbers and have positive effect on the EMW-absorbing properties [21, 22]. Ferromagnetic metal alloy thin films have also been studied as a candidate material because of their excellent soft magnetic properties. Kim et al. prepared soft magnetic metal (CoFe/NiFe)/PI hybrid films with high electrical resistivity by thermal imidization and selective oxidation between the metal alloy layer and poly(amic acid) (PAA) layer. The saturation magnetization of the CoFe/NiFe hybrid film was three times higher than that of the NiFe hybrid film. The Co–Fe/Ni–Fe hybrid resistive film showed an excellent electromagnetic absorption characteristic due to high soft magnetic properties and can be applicable as a thin electromagnetic-absorbing material for high-frequency devices operated over 3 GHz [16].

For significant absorption of the radiation, the shield must require electric and/or magnetic dipoles which interact with the electromagnetic fields in the incident radiation. Materials with high values of dielectric constant (SiO_2, ZnO, TiO_2, and $BaTiO_3$) and magnetic permeability (Ni, Co, and Fe_3O_4) are commonly used together for the development of EMI absorption materials. Wang et al. prepared a microwave-absorbing material by introducing SiO_2-coated CI into PI matrix via compression molding method. The SiO_2@CI shows good absorbing properties with RL_{min} decreased from −25 dB to −33 dB with the thickness increases from 1.5 to 2.1 mm [23].

7.3.2 Metals

Metals themselves are highly effective for electromagnetic shielding and are widely used in various industries to protect electronic equipment from interference. Metals, such as silver, copper, gold, nickel, and aluminum in diverse forms, can also be combined with PIs to create EMI-shielding materials, because they offer a good balance of conductivity, flexibility, and lightweight properties, making them a popular choice in industries such as aerospace, automotive, and electronics. Among the metal-based nanofillers, silver, particularly silver nanowire, has been extensively studied as shielding materials because of the large aspect-ratio and ultrahigh conductivity.

One common approach is to coat PI substrate with a thin layer of metal NPs using techniques like sputtering, electroplating, spray-coating, and self-metallization [24, 25]. The metal coating provides necessary conductivity for EMI shielding, while the polymer substrate provides the flexibility and durability. Zhang et al. realized a self-metallization of PI by silver ion/PAA precursor ion exchange and thermal reduction method. The surface-metallized PI composite exhibited a maximum EMI SE of 81 dB even above 300 °C [24]. Zhao et al. successfully prepared a nickel-metallized PI nonwoven fabric. The effective construction of metallized surface on PI fibers facilitated the formation of multiple scattering for incident EMWs, leading to an excellent broadband EMI performance. More importantly, the strong binding force between Ni plating and PI fiber substantially improved the high/low-temperature resistance, endowing the composite with excellent stability and durability in EMI performance [26].

Another approach is to incorporate metal particles or fibers into PI matrix to create a metal–polymer composite material. These materials can be molded or extruded into various shapes and forms, making them suitable for a wide range of applications. Mehmood et al. introduced high-quality 2D silver nanoplates (AgNPls) into PI matrix. The resulting 12 wt% AgNPls/PI composite possessed appreciable conductivity of 5.5 S/m^1 at only 130 μm thickness, effectively blocks electromagnetic radiation in X-band, resulting in substantially high specific shielding effectiveness and absolute shielding effectiveness of 22.48 and 1729.23 $dB{\cdot}cm^3/g$, respectively. Additionally, the nanocomposites remained thermal stability up to 500 °C in oxidative environment and possessed an appreciably high storage modulus of 3.1 GPa at 50 °C [27].

The incorporation of metallic fillers into PI does not seem to get much attention as evidenced from the published literatures, which may be ascribed to the high density, lack of flexibility, and poor corrosion resistance. For another, the mechanism is mainly based on reflection, which in turn results in secondary EMW pollution.

7.3.3 MXene

Among the reported dielectric absorbers, the newly emerged 2D transition metal carbonitride MXene and MXene-derived hybrid carbon aerogels displayed outstanding EMI-shielding performance due to their hierarchical architecture, metal-like conductivity, excellent mechanical property, and rich active surfaces. For example, MXene can offer an EMI SE of 116 dB at a thickness of 40 μm (8.2 GHz) [28], and the MXene-derived TiC/carbon hybrid aerogels own an EMI SE of 84.3 dB with an ultralow density of 47.1 mg/cm^3 [29]. The construction of conductive networks of MXenes in PI matrix is regarded as an effective method to obtain highly efficient EMI-shielding films. Chu et al. fabricated the 2D MXene/PI films by in situ polymerization and a simple scratch method. The obtained 30 wt% MXene/PI film with a thickness of 0.11 mm exhibited an EMI SE of 34.73 dB in the frequency range of 8.2–12.4 GHz (X-band). In addition, the films had a tensile strength of 67.58 MPa, and a high decomposition temperature of 510 °C, enormously promoting the application in the field of lightweight and flexible EMI-shielding materials [9]. It should be mentioned that MXene is easily oxidized to TiO_2 under high temperature and high humidity conditions, resulting in reduced conductivity and EMI-shielding performance. Therefore, appropriate modification of MXene is usually needed to get a better dispersion and adhesion. Yan et al. developed MXene-POSS/PI nanocomposite films with excellent EMI shielding, water resistance, and flame retardance. The nanocomposite film with 20 wt% MXene-POSS content achieved a desirable EMI SE of 12.7 dB and a water uptake of only 1.4% under 70% relative humidity [30].

The unique structure and properties of MXene make it a very good candidate for developing effective EMI-absorbing material [31–34]. Pu et al. provided a unique 3D "lamella-pillar" microporous architecture of PINF/MXene composite aerogel using short PI nanofibers (PINFs) as cross-linking and supporting building blocks. More interestingly, the parallelly aligned leaf vein-like lamellae empower the PINF/MXene composite with prominent wave absorption performance of RL_{min} −40.45 dB at 15.19 GHz and an EAB of 5.66 GHz (12.34–18 GHz), making it a promising candidate for wearable strain sensors and microwave absorbers [31].

7.3.4 Graphene

Graphene is another 2D material composed of a single layer of carbon atoms arranged in a hexagonal lattice. It has a range of unique properties, including high electrical conductivity, good thermal conductivity, excellent mechanical strength, which make it a highly sought-after material for various applications. Graphene is thus utilized for a wide range of electromagnetic applications, including EMI shielding, antennas, sensors, terahertz technology, energy harvesting, and stealth technology [10]. Ongoing research and development in this field is expected to lead to further innovations and commercialization of graphene-based electromagnetic materials.

Xu et al. prepared a reduced graphene oxide (rGO)/PI composite film by optimized thermal annealing process for EMI-shielding application. The final product showed good conductivity of 500 S/cm. The thin sample (thickness < 0.1 mm) containing 10% PI showed an enhanced EMI-shielding performance of 55–65 dB [35]. The functional groups of hydroxyl and carboxyl can participate in esterification reaction, thereby promoting filler dispersion and developing strong bonding with PI matrix. Mehmood et al. reported the rGO/PI composites through esterification reaction between —OH groups on the rGO and ester (—COOR) terminals of PI. Consequently, the 20 wt% rGO/PI composites possessed the highest conductivity of 15.27 S/cm. Ultimately, the composite performed an outstanding SEM SE, SSE, and SSE_t of 46 dB, 138 dB·cm^3/g, and 8358 dB·cm^2/g in the X-band (8–12 GHz), respectively. Additionally, the composite material was highly stiff (3.945 GPa) yet flexible (easily bent through 180°), lightweight (~0.34 g/cm^3), and capable of forming thin films (162 mm) [10].

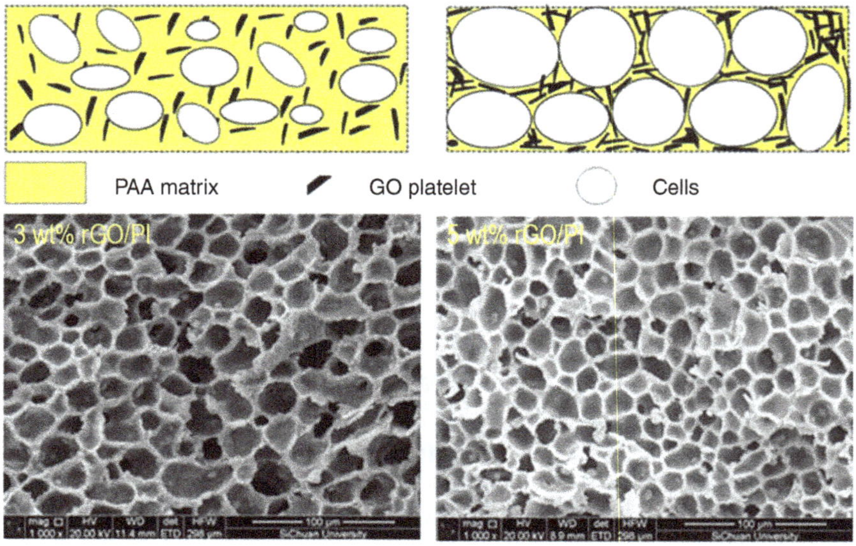

Figure 7.3 Porous rGO/PI composite foams with closed cell structure.
Source: Ref. [37]/John Wiley & Sons.

rGO/PI composites with porous structure can further enhance EMI-shielding and -absorbing performance [36–38]. Yang et al. reported a porous in situ thermally reduced rGO/PI composite film (Figure 7.3). The resultant 8 wt% rGO/PI composite film (500 μm thickness) exhibited effective electrical conductivity of 0.015 S/m and excellent SSE value of 693 $dB/cm^2/g$. In addition, the thermal stability and mechanical strength of the rGO/PI composite films was also dramatically enhanced [37]. Zhang et al. prepared an ultralight graphene/PI foam with ordered lamellar structure. More lamellar interfaces formed inside the foam per unit volume effectively facilitating the layer-by-layer dissipation for the vertical incident EMWs, thereby endowing the foam with efficient broadband electromagnetic absorption performance. The optimized foam with an ultralow density of 9.10 mg/cm^3 presents an RL_{min} value of −61.29 dB at 9.25 GHz and an EAB of 5.51 GHz (7.06–12.57 GHz, covering the whole X band) when the thickness is 4.75 mm [38].

7.3.5 Carbon Nanotube

Carbon nanotubes (CNTs) exhibit exceptional electrical conductivity, surpassing that of copper and other traditional conductive materials, making them suitable for various electromagnetic applications [39–42]. The 1D nanoscale structure also facilitates the formation of conductive network and hierarchical structures, showing potential value in obtaining lightweight, flexible, and tunable EMI SE materials.

Liu et al. designed a regular radial needle-like microporous structure through radial freeze-drying methods. The fabricated CNT/PI aerogels showed excellent thermal stability, compressive mechanical properties, and EMI-shielding performance of −53.1 dB with a thickness of 2 mm in the axial direction. The SE_A/SE_T value is as high as 94.5%, indicating an excellent wave absorption performance [39]. Wang et al. introduced CNT into PI with the aid of PVP as surfactant. A lightweight and robust CNT/PI composite foam was developed with a density of only 32.1 mg/cm^3. The EMI SE and absorption coefficient are up to −41.1 dB and 82.3%, respectively, which represents one of the highest EMI SE values compared to previously reported polymer-based foams [40].

Jiang et al. manufactured the CNT/PI composite films with excellent EMW absorption ability. In particular, the RL peak of the 6.0 wt% CNTs/PI film is −18.62 dB at 10.34 GHz and its EAB is 2.72 GHz with the matching thickness of only 2.0 mm [43]. Ni et al. synthesized carboxyl functionalized multi-walled carbon nanotubes/PI (MWCNTs-COOH/PI) composite aerogels with aligned slit-like channels structure by unidirectional freezing method. The PI/MWCNTs-COOH composite aerogel showed anisotropic EMW absorption performance with RL_{min} of −52 dB at 12.8 GHz and EAB of 6.7 GHz (10.6 ~ 17.3 GHz) with a thickness of only 2.5 mm in the perpendicular direction of slit-like channels, while only −11 dB at 15.9 GHz in the orientation direction [44].

7.3.6 Other Fillers

Carbon nanofiber (CNF) is a lightweight and strong 1D material that is commonly used in the aerospace and automotive industries. It is also a good conductor of electricity, making it suitable for use in electromagnetic applications. Nayak et al. prepared a conductive CNF/PI composites at very low percolation threshold by in situ polymerization of PI in the presence of highly graphitized CNFs. The composite of 0.07 mm thickness showed EMI SE of above 12 dB at 5 wt% of CNF loading in a frequency range of 8.2–12.4 GHz (X-band) [45].

Cabon Black (CB) is a cost-effective and efficient 0D filler in strengthening and antistatic polymer materials. CB-reinforced conducting polymer composites are used in a wide range of industries, including electronics, automotive, aerospace, and energy storage [46]. Kim et al. fabricated a highly flexible EMI-shielding CB/PI composite using a very simple hot-pressing method. The study found that the PI sponge with 50 wt% CB composite showed highly brittle behavior which made it impossible to use as an EMI-shielding film. However, after applying the hot-pressing method, all the films recovered the flexible property on par with neat PI film. The final product of hot-pressed EMI-SE film showed increase in both EMI SE and excellent mechanical properties [47].

Qing et al. for the first time studied the dielectric and microwave absorption properties of Ti_4O_7 NPs/PI composite. Interestingly, due to the formation of Ti^{3+} and oxygen vacancies during the annealing process, the special crystal structure will form electron channels that promote dark Ti_4O_7 NPs to possess a higher electrical conductivity and show outstanding dielectric loss compared to pristine TiO_2 powders, endowing the 60 wt% Ti_4O_7/PI hybrid with an exceptional RL reaching −49.3 dB of 13.7 GHz at 1.25 mm thickness [48].

Unlike the above inorganic nanofillers, conductive polymers, which are intrinsically lightweight, strong, and exceptionally durable, have emerged as promising candidates for multifunctional materials. Furthermore, the distinctive conjugated π-electronic structure of conducting polymers imparts extraordinary electrical properties to the absorber, including a good electron mobility, excellent electrical conductivity, and high electron affinity. These characteristics have consequently motivated numerous researchers to explore the use of conducting polymers as multifunctional EMW-absorbing materials. Currently, conductive polymers widely used in various research fields as well as industrial applications, encompass PI, polyaniline (PANI), polypyrrole (PPy), and polythiophene (PT). He et al. prepared an electrically conductive PANI/PI composite fabric by in situ polymerized aniline on PI fabric in the acid solution using ammonium persulfate as the oxidant. The conductivity of PANI/PI composite fabric was 0.53 S/m. Remarkably, the prepared composite fabric exhibited excellent EMW absorption features (RL of −57.99 dB) [49].

Table 7.1 summarizes PI-based composites with different nanofillers for EMI shielding, while Table 7.2 lists absorption-dominant PI based binary composites. PI composites with high aspect ratio fillers and hierarchical structure have a wide range of benefits and can be customized to meet the specific requirements of EMI-shielding materials.

7.4 Ternary Composites

With the rapid development of 5G technology and the increasing demand for high-performance electromagnetic materials, it is true that traditional binary electromagnetic materials may not always meet all the requirements for advanced applications. The unique challenges posed by 5G technology, such as higher frequencies, increased data rates, and more complex electromagnetic environments, necessitate the development of more sophisticated and versatile materials. Ternary or multicomponent materials are increasingly being explored and developed to address the specific requirements of these advanced technologies. In the case of ternary electromagnetic composites, different types of fillers are combined to create a synergistic effect that improves the overall performance of the composite material in terms of electromagnetic shielding. By incorporating additional components such as conductive fillers, magnetic materials, or specialized polymers, ternary or hybrid composites can provide enhanced electrical conductivity, improved magnetic shielding, and optimized electromagnetic absorption properties. The synergistic combination of multiple components allows for the optimization of EM-shielding performance while maintaining other desirable properties. PI-based ternary composites for EMI related applications have also gained significant attention in recent years. They are basically classified into the following two aspects.

7.4.1 PI/Conductive Components/Magnetic NPs

A suitable combination of magnetic and dielectric fillers provides better attenuation of EM waves via taking full use of the conductivity of conducting components and the magnetic properties of the magnetic NPs. For example, Liu et al. have constructed a 3d/4d-bimetallic based crystalline precursor of dielectric/magnetic synergistic system by embedding polyoxometalates (POMs) within the Co-MOFs. The proper pyrolysis temperature induced the homogeneously distributed metallic

Table 7.1 PI-based binary composites for EMI shielding.

EMI-shielding Materials	Contents (%)	EMI SE (dB)	SSE/t [SSE] (dB·cm^2/g [dB·cm^3/g])	Density (mg/cm^3)	Thickness (mm)	Conductivity (S/m)	Structure	References
MXene/PI	30 wt%	−34.73	315.73 dB/mm	—	0.11	27.8	Film	[9]
rGO/PI	20 wt%	−46	8358	340	162	1527	Film	[10]
$Ti_3C_2T_x$/PI	2 vol%	−54.5	15,527	390	0.09	16	Multilayer film	[11]
Ni@PI-HP	—	−68.6	13372	900	0.057	953	Fabric	[26]
Ag/PI	12 wt%	−9.89	22.48	450	0.13	550	Film	[27]
MXene-POSS/PI	20 wt%	−12.7	—	—	—	—	Film	[30]
GO/PI	70 wt%	−65	—	—	0.082	50000	Film	[35]
rGO/PI	16 wt%	−21	[75]	0.28	0.8	0.8	Foam	[36]
rGO/PI	8 wt%	−14.9	693	0.43	0.5	0.015	Porous film	[37]
CNT/PI	—	−53.1	[1285.1]	48.4	2	54.6	Aerogel	[39]
CNT/PI	67 vol%	−41.1	—	32.1	2	17	Foam	[40]
CNF/PI	5 wt%	−12.6	—	—	0.07	—	Film	[45]
Ag@PI	—	−89.4	[812.7]	110	0.3	278.0	Aerogel	[50]
Ni/PIC	65 wt%	−44	—	—	—	22300	Fabric	[51]
Graphene/PI	13 wt%	−28.8	1518	76	2.5	0.015	Aerogel	[52]
VGSs/PI	20 wt%	−31.37	—	—	0.151	220	Film	[53]
MXene@PI	—	−62.5	21,317	48.7	1.5	—	Foam	[54]
MXene/PI	—	−63	31651.2	—	0.25	661.3	Film	[55]
AgMS/PI	28 vol%	−90.6	14,613	110	0.02	106	Film	[56]
GO/PI	24 wt%	−48.96	19789	—	—	0.1	Foam	[57]
MXene/PI	20 wt%	−9.4	—	—	—	178	Film	[58]
MXene/PIF	49.1 wt%	−49.9	—	—	0.256	3787.9	Film	[59]
rGO/PI	5 wt%	−49	7485.5	—	—	33.0	Foam	[60]

Table 7.2 PI-based binary composites for EMI absorbing.

EMW-absorbing materials	Contents	RL_{min} (dB)	Frequency (GHz)	Thickness (mm)	EAB (GHz)	Conductivity (S/m)	Structure	References
MXene/PI	30%	−49.36	16.28	2.35	6.9	—	Aerogel	[4]
Carbonyl iron/PI	60 wt%	−11	12.4	1.5	0.6 (9.7–10.3 GHz)	—	Composites	[17]
Ti_3SiC_2/PI	30 wt%	−48.6	8.5	2.9	3.8	—	Composites	[32]
MXene/PI	31.6 wt%	−41.8	6.0	1.91	6.5 (11.5–18 GHz)	2.4	Aerogel	[33]
MXene/PI	0.084 vol%	−45.4	9.59	3.0	5.1	4.0	Aerogel	[34]
Graphene/PI	—	−61.29	9.25	4.75	5.51 (7.06–12.57 GHz)	—	Lamellar Foam	[38]
MWCNTs/PI	2 wt%	−29.1	10.70	1.7	2.04	—	Composites	[41]
CNT/PI	6 wt%	−18.62	10.34	2.0	2.72 (9.12–11.84 GHz)	—	Composites	[43]
MWCNTs-COOH/PI	16 phr	−52	12.8	2.5	6.7 (10.6–17.3 GHz)	0.1	Aerogel	[44]
Ti_4O_7/PI	60 wt%	−49.3	13.7	1.25	3.6 GHz (12.1–15.7 GHz)	—	Composites	[48]
PANI/PI	—	−57.99	8.32	3.5	3.0 (8.2–12.4 GHz)	0.53	Fabric	[49]
rGO/PI	—	−32.87	10.39	4.0	6.22 (8.26–14.48 GHz)	—	Foam	[61]
Fe_3O_4@PPy/PI	20 wt%	−31.0	9.0	6.0	4.4 (8.2–12.6 GHz)	2.7×10^3	Foam	[62]
Fe_3O_4/PI HPC	22.5%	−20.10	11.76	5.5	1.5	—	Composite	[63]
MXene/PINF	23.5%	−37.9	6.57	6.00	3.3	—	Aerogel	[64]

Co and MoCx hetero-units into carbon matrix with modified porous defect engineering to enhance EMW. Owing to the brilliant synergistic effect of polarization, magnetic loss, and impedance matching, the superior RL_{min} of −47.72 dB at 11.76 GHz at the thickness of 2.0 mm and a wide EAB of 4.58 GHz (7.44–12.02 GHz) covered the whole X-band at the thickness of 2.5 mm for η-MoC/Co@NC-800 are obtained. [15] Jiang et al. introduced CNT and nano-Fe_3O_4 into PI by in situ polymerization. CNT and nano-Fe_3O_4 contributed to the attenuation of EMW through microwave absorption and reflection loss, and the porous structure was able to provide multiple reflection paths for incident EMW and achieved higher attenuation. Particularly, RL_{min} value of the PI/CNTs/nano-Fe_3O_4 composite fibers was demonstrated to be −54.61 dB with the matching thickness of 2.10 mm [65].

Li et al. reported a cotton-derived carbon fabric wrapped with CoFe alloy NPs (CCF@CoFe) by pyrolysis and subsequently encapsulation of PI resin to obtain the CCF@CoFe/PI films. It was found that the EMI-shielding performance of the composite films can be adjusted by changing the CoFe loading content and pyrolysis temperature, and an EMI-shielding efficiency of ~62 dB was achieved by stacking 6 layers of nanocomposite films, shielding 99.999937% of incident EMW with thickness of only 0.98 mm [66]. In the meanwhile, Li et al. introduced PI into $NiCo_2O_4$ nanowire arrays wrapped carbon fibers (CF@$NiCo_2O_4$) by pyrolysis and followed vacuum-assisted filtration approach. The obtained flexible CF@NiCo/PI composite film displays superior EMI SE of 87 dB with a thickness of only 1.08 mm. Particularly, the composite film exhibits extremely low SE_R of ~6 dB, which surpasses those of most previously reported composites materials with similar thickness. Figure 7.4 shows the schematic illustration of the synthesis of CF@NiCo and the fabrication of CF@NiCo/PI films [67].

7.4.2 PI/Conductive Components with Diverse Dimensions

PI composites combined with diverse conductive nanofillers of different dimensions, including 0D NPs, 1D nanotubes, and 2D nanosheets, are widely used for EMI-shielding materials. These advanced composite materials offer enhanced

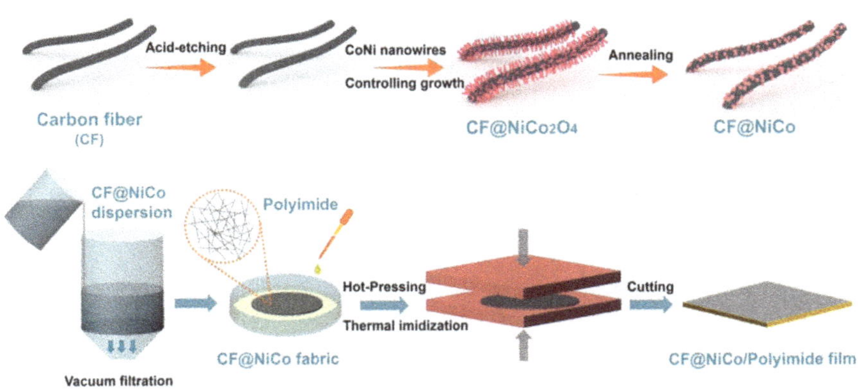

Figure 7.4 Schematic illustration of the synthesis of CF@NiCo and the fabrication of CF@NiCo/PI films. Source: Ref. [67]/with permission of Elsevier.

EMI-shielding performance due to the synergistic effects of PI matrix and the hierarchical nanostructures. The tailored design of PI composites with NPs, nanotubes, and nanosheets allows for the development of lightweight, flexible, and high-performance materials that can effectively shield electronic devices and equipment from electromagnetic interference.

PI with diverse conductive components can be designed for various structures by different techniques. For example, Zheng et al. developed an innovative ultrathin CNF/rGO/PI composite with excellent strength, shape memory properties, thermal stability, and wide temperature range tolerance. Gradient-thick Ag films are then deposited on both sides of the composite by magnetron sputtering to form sandwich structure composites. The obtained material exhibited an extremely high electrical conductivity and excellent EMI SE of up to 112 dB in the 8.2–12.4 GHz band [68]. Zhang et al. reported a highly conductive MXene@rGO/PI composite aerogel via coating 2D MXene onto the rGO/PI skeleton using vacuum impregnation method. Due to the dual electrically conductive network formed by rGO and MXene, the MXene@rGO/PI composite aerogels exhibit exceptional EMI SE of 58.7 dB in the X-band at a density of 0.063 g/cm^3 [69]. Xue et al. prepared MXene/CNT/PI aerogel frame with integrated gradient-conductive structure via 3D printing technology, rendering the composite with excellent EMI-shielding performance (68.2 dB) and ultralow reflection (R = 0.23). The obtained aerogel frame can effectively block wireless power transmission as shown in Figure 7.5 [70].

Literatures show there are also hybrid PI composites incorporated with more than two nanofillers. The synergistic combination of PI with two or more different types of nanofillers provides a versatile and customizable approach to designing high-performance materials for electromagnetic shielding, absorption, and other related applications. For example, Zhu et al. reported a 3D ordered hierarchical porous Fe_3O_4-decorated CNT (Fe_3O_4@CNT)/MXene/cross-linked aramid nanofiber

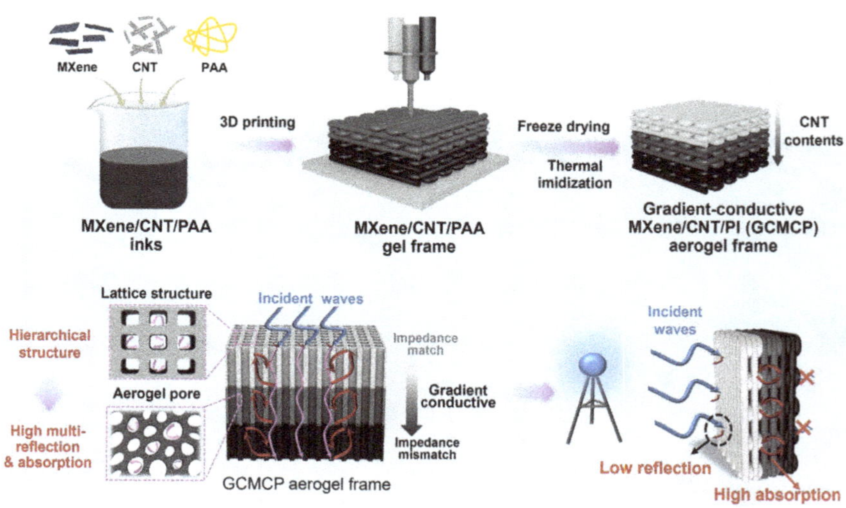

Figure 7.5 Preparation of gradient-conductive PI aerogel frame with integrated 2D MXene and 1D CNT via 3D printing technology. Source: Ref. [70]/Springer Nature/CC BY 4.0.

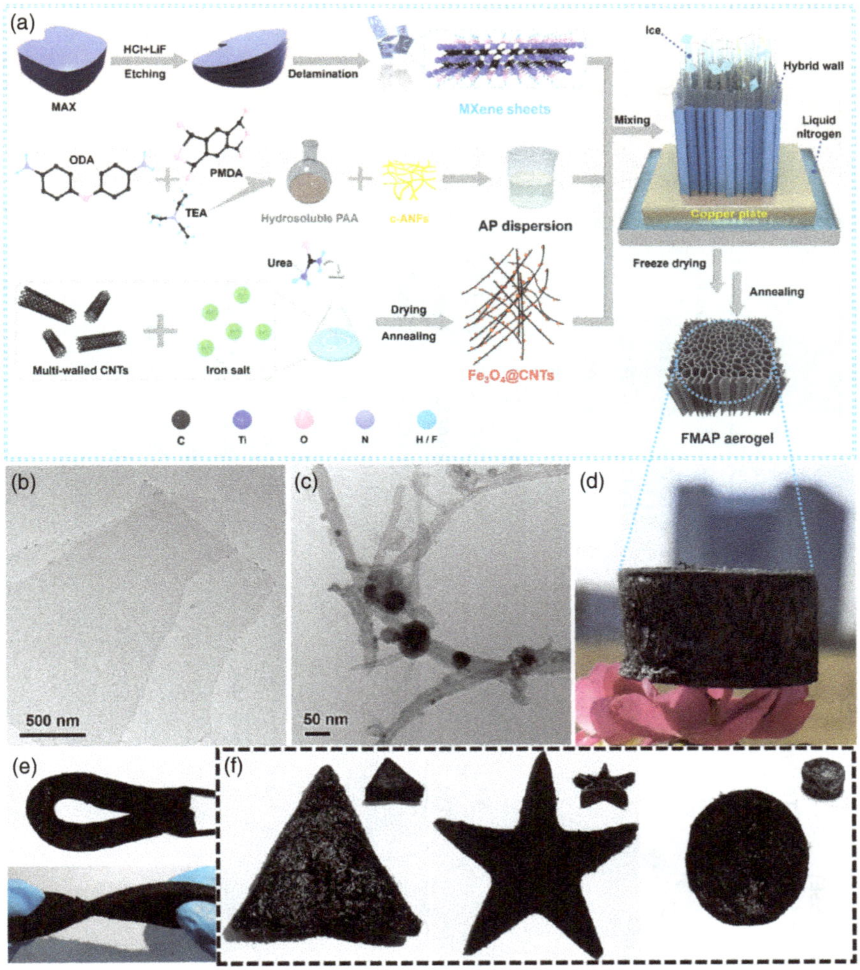

Figure 7.6 (a) Schematic illustration of the synthetic strategy for the FMAP aerogel, TEM images of (b) $Ti_3C_2T_X$ nanosheets, and (c) Fe_3O_4@ CNTs, digital photograph of (d) the FMAP aerogel, (e) twisted FMAP aerogels, and (f) different shapes. Source: Ref. [71]/American Chemical Society.

(c-ANF)/PI aerogels via a unidirectional freezing strategy (Figure 7.6). Benefiting from the magnetic loss effect of Fe_3O_4 magnetic NPs, the conductive and dielectric loss effects of CNTs, and the multiple reflections induced by the 3D ordered hierarchical porous structure, the aerogels with the same contents of 8 wt% of Fe_3O_4@CNTs and MXene exhibit a high EMI SE of up to 67.42 dB and a microwave reflection (SE_R) of 0.60 dB [71].

Tables 7.3 and 7.4 summarize the EMW-shielding and -absorbing properties of PI-based hybrid materials. These materials offer a combination of properties that make them effective in mitigating EMI and absorbing EMW energy. The ability to tailor the microstructure and composition of these materials allows for the optimization of their EMW-shielding and -absorbing properties for specific applications.

Table 7.3 PI-based hybrid materials for EMI shielding.

Materials	EMI SE (dB)	SE_A	SSE/t [SSE] ($dB \cdot cm^2/g$ [$dB \cdot cm^3/g$])	Thickness (mm)	Conductivity (S/m)	Fabrication	Structure	References
CCF@CoFe/PI	−62	99.99%	—	0.98	—	Pyrolysis & encapsulation	Multilayer	[66]
CF@$NiCo_2O_4$/PI	−87	80.9	[1549.2]	1.08	$13.2 \times 10^{-3} \Omega \cdot m$	Pyrolysis and filtration	Film	[67]
Fe_3O_4@CNTs-MXene-ANFs-PI	−67.42	66.8	62,654	3		Unidirectional freezing	Multilayer	[71]
Ni-W-P/PANI/PI	−103	—	—	0.487	0.08 Ω/sq	Electroless plating	PI fabric	[72]
(Fe_3O_4/PI)–Ti_3C_2Tx–(Fe_3O_4/PI)	−85	62	—	0.075	—	Sacrificial template	Multilayer	[73]
rPET/FeCo@C/AgNWs/PI	−49.35	36.527	[4112.5 dB/cm]	0.12	7710	Filtering & taping	Multilayer	[74]
PI@Ni,Cu,Ni/rGO/Fe_3O_4	−53.9	—	—	4	23750	Electro-plating&hot-pressing	Multilayer	[75]
Ag/$BaFe1_2O_{19}$ microfiber/PI	−98.9	97.6	3297	0.1	2.6×10^6	Blow spinning	Film	[76]
Fe_3O_4@rGO/WPU-Ni@PI	−32.4	26.5	—	3	—	Hot compression molding	Film	[77]
MXene/Fe_3O_4/PI	−66	58	—	0.075	—	Electrospinning, polymerization	Film	[78]
Ni-Co-Fe-P/PANI/PI	−69.4	62.5	—	0.2	6170	Electroless plating	Fabric	[79]
carbonyl iron/MWCNT/PI	−24	19	—	1.4	—	Molding pressing	Composite	[80]
PPy/Ni-CAT/PI	−41.0	—	—	—	841	Gas-phase polymerization	Film	[81]
PI/PPy/CF	−74.59	70	—	0.62	3.02	Solution	Film	[7]
MXene@rGO/PI	−58.7	51	[932.3]	5	39.35	Freeze-drying and annealing	Aerogel	[69]
MXene/CNT/PI	−68.2	67.1	[448.7]	5	25.5	3D printing	Aerogel	[70]
CNF/rGO/PI	−112	86	—	0.146	19180	Solution	Film	[68]

MXene@ANFs/PI	−48.9	45.1	1645.5	4	210.13	Thermal imidization	Foam	[82]
MXene/Ag/PI	−79.54	65	3500	0.15	1500	Electrospinning, hot pressing	Film	[83]
MWCNTs/PEEK/TPI	−12.41	10.27	349.5	2	2.404	Sc-CO_2-Foaming	Foam	[84]
PI-MXene-AgNPs	−52.97	36.42	8755.37	0.110	—	Self-reduction; imidization	Film	[85]
CF/CB/PI	−23.9	14.85	—	0.35	0.398	Solution and curing	Film	[86]
Graphene/MXene/PI	−45	35	—	0.08	—	Thermal annealing	Sandwich	[87]
PI/rGO/MWCNTs	−18.2	—	823	0.5	1.87	Evaporation-induced phase Separation	Film	[88]
PI/NCF/GO	−45	30	—	—	7000	Bidirectional freezing Casting and hot-pressing	Film	[89]
CNT/graphene/PI	−28.2	—	7050	0.02	—	Freeze-drying and Annealing	Foam	[90]
Graphene/CF/PI	−73.01	62.46	3331	2	2000	Solution method	Composites	[91]
PANI-g-GO/MWCNT	−28.2	—	—	2	160	Polymerization, non-Solvent-induced phase Separation	Porous film	[92]
AgNW-PI/MXene/AgNW-PI	−40.73	—	3885.6 dB/cm	0.097	1055	Electrospinning	Sandwich	[93]
Gr-PANI@PI	−21.3	—	4096.2	0.04	2.1	Polymerization and casting	Film	[94]

Table 7.4 PI-based hybrid materials for EMI absorption.

Ternary composites	RL_{min} (dB)	Frequency (GHz)	Thickness (mm)	EAB (GHz)	Fabrication	Structure	References
CNT/Fe_3O_4/PI	−54.61	8.65	2.1	2.02 GHz (8.20–10.22 GHz)	Wet spinning process	Composite	[65]
CNTs@$CoFe_2O_4$/PI	−54.4	15.05	2.6	7.15 GHz (8.95–16.1 GHz)	Directional freezing	Aerogel	[95]
$NiFe_2O_4$@PANI/PI fabric	−19.2	16.1	1.5	5.1 GHz	Coating	Fabric	[96]
Ni@CNT/PI	−56.56	10.08	2.6	4.2	3D printing	Aerogel	[97]
Nb_2O_5/CNTs/PI	−41.66	12.40	2.1	2.63	Freeze drying	Aerogel	[98]
Graphene/Fe_3O_4/PI	−38.5	13.5	2.8	9.5 GHz (7.0–16.5 GHz)	Freeze-drying, imidization	Aerogel	[99]
$NiFe_2O_4$/PANI/PI	−42.5	12.5	0.041	3.4 GHz (12.4–15.8 GHz)	Coating	Fabric	[100]
MXene/CNTs/PI	−50.03	13.7	1.8	5.6 (11.3–16.9 GHz)	Freeze drying & annealing	Aerogel	[101]
MXene/ANF/PI	−39.62	11.2	2.4	3.61	Freeze drying	Aerogel	[102]
CB/MWCNT/PI	−19.8	9.9	2.2	3.04	Thermal imidization	Composite	[103]

7.5 Structural Design

A well-designed structure can ensure that EMI shielding is effective and reliable, providing a high level of protection for sensitive electronic components. It is important to consider factors such as the material used, the thickness and design of the shielding, as well as the overall construction of the structure in order to achieve optimal EMI-shielding performance. PI-based EMI-shielding composites encompass uniform structure, fabric structure, porous structure, layered structure, gradient structure, or complex of the above [104, 105]. The uniform structure refers to the dispersion of conductive filler in the matrix in a uniform manner. Common methods for preparing EMI-shielding materials with such a structure include solution mixing, melt blending, as well as in situ polymerization. These methods are preferred due to their low cost and simplicity. However, a significant drawback is that it is necessary to increase the filler content in the matrix to enhance the EMI SE. It, in turn, leads to a high percolation value of the material, ultimately impacting its mechanical properties.

7.5.1 Metallized PI Fibric

Metallized PI fabric is a type of electrically conductive fabric that is commonly used for electromagnetic shielding applications. It consists of a PI base material that has been coated with a thin layer of metals or conductive polymers, typically silver, aluminum, nickel, copper, or PANI etc., to provide conductivity and EMI-shielding properties [26, 50, 72]. Metallized PI fabric offers a combination of the unique properties of polyimide, such as high temperature resistance, chemical resistance, and mechanical strength, with the electrical conductivity of the metal coating, making it an ideal material for applications that require both EMI shielding and heat resistance.

Xue et al. designed a polydopamine-assisted Ag deposited on PI (Ag@PI) aerogel fabrics for EMI shielding (Figure 7.7). The Ag@PI aerogel fabric integrates highly conductive Ag and leverages the three-dimensional porous structure of aerogel to facilitate multiple reflections and interface polarization of EMWs. The Ag@PI aerogel fabrics exhibit ultrahigh EMI SE of 89.4 dB owing to their unique porous structure, while the Ag@PI fabric without porous structure is only 65.1 dB [50]. Zhou et al. decorated a conductive nickel layer on the surface of PI fibers by electroless metal deposition (ELD). The resultant Ni/PI fiber containing 65 wt% Ni exhibited a conductivity of 223 S/cm, and the corresponding fabric exhibited an EMI SE of 44 dB in the X-band [51].

7.5.2 Aerogel

Aerogel is a new type of porous solid material with high surface area, high porosity, low density, good flexibility, low thermal conductivity, etc. Because of the presence of the porous structure, random electromagnetic reflection/scattering would repeatedly occur at the interfaces of the cell wall, resulting in the transfer of electromagnetic energy to be dissipated as heat in the form of microcurrent. The

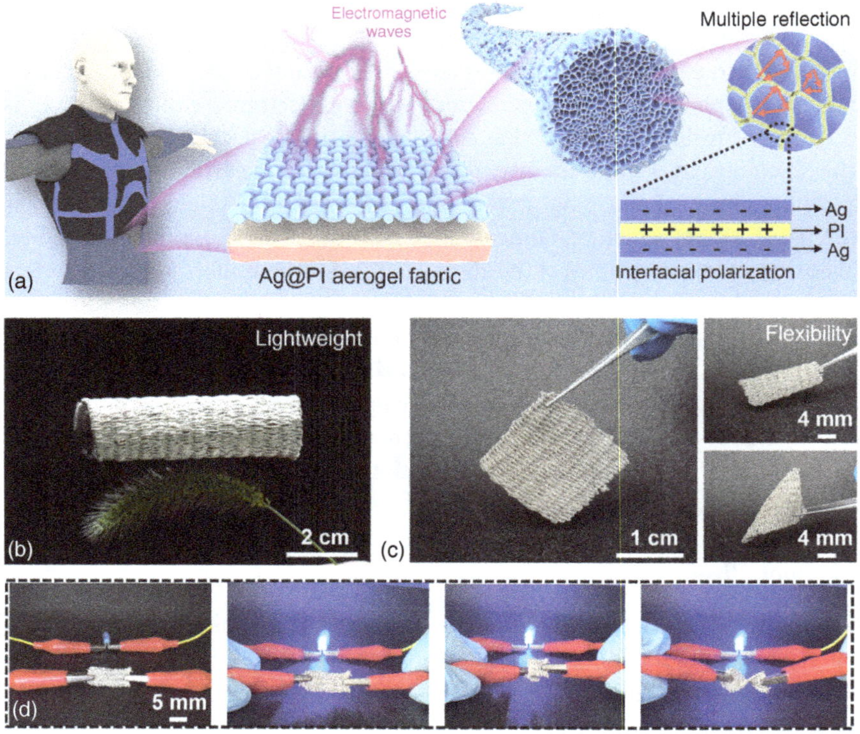

Figure 7.7 (a) Schematic of the EMI of Ag@PI aerogel fabric. (b–c) digital photograph of Ag@PI aerogel fabric. (d) The brightness of LED is constant under different bending of Ag@PI aerogel fabric. Source: Ref. [50]/with permission of Elsevier.

aerogel prepared with PI as matrix combines the excellent properties of PI and the porous structure of aerogels, which is the ideal material for aerospace, electronics, EMI-shielding or EMW absorption applications [33]. Freeze-drying method is an effective method that has been extensively used to prepare porous PI aerogels owing to the advantages of simple process, more controllable, friendly to the environment, etc. Figure 7.8 displays a representative PI/SiO_2 aerogel fabricated by freeze-dying method, exhibiting porous morphology.

PI aerogels can be printed into various structures. Wu et al. presented a 3D printed "aerogel-in-aerogel" structure nanocomposite by embedding nanoparticulate silica aerogel grains in a nanofibrillar PI aerogel matrix. The printed PI-silica aerogels are non-brittle ($E = 6.7$ MPa) with a super-insulating thermal conductivity (20.3 m·W/m/K) and high thermal stability ($T_{5\,wt\%}$ = 447 °C). In addition, they display excellent low-loss dielectric properties and microwave transmission over all relevant communication bands and can be functionalized for electromagnetic interference (EMI) shielding [107].

7.5.3 Foam

Similar to aerogel material, PI foam is used to create a 3D structure with unique properties, such as being lightweight, flexible, heat resistant, and insulating. The 3D

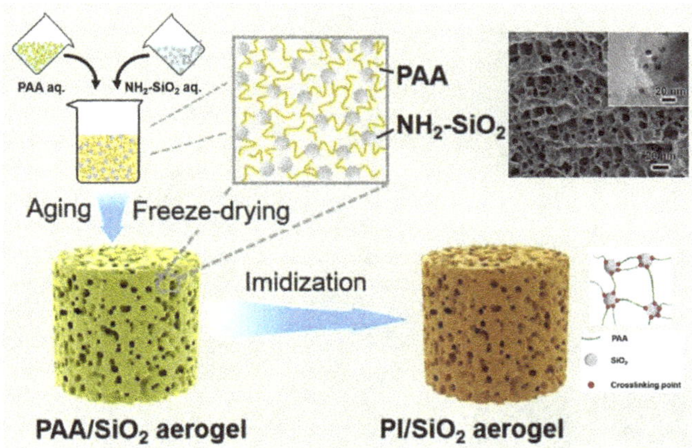

Figure 7.8 PI/SiO_2 aerogel fabricated by freeze-drying method. Source: Ref. [106]/with permission of Elsevier.

foam structure can enhance the multiple reflections of incident EMWs and improves the microwave absorption ability [108, 109].

Ma et al. presented three ultralightweight PI composite foams filled with silver nanospheres (AgNSs), silver nanowires (AgNWs), and silver nanowires–silver nanoplatelets (AgNWPs) by a facile and effective one-pot liquid foaming process. It was found that the AgNWPs/PI composite foam exhibited the highest EMI SE owing to the denser 3D conductive network of AgNWPs compared to AgNWs and AgNSs, in which the seamlessly interconnected AgNWPs network provided fast electron transport channels inside foams. Maximum specific EMI SE value of 216–249 dB/g cm^3 at 8–12 GHz was achieved at 4.5 wt% AgNWPs loading, which far surpassed the best values of other composite materials [110, 111].

Zhuo et al. designed PI-based PI@MXene@ANFs/PI hybrid foams assembled with MXene and PI bilayer coatings via a facile freeze-drying plus thermal imidization process followed by two-step dip-coating method. MXene coating enabled the PI-based hybrid foams to form a complete conductive network, and PI coating protected MXene coating from oxidation, thus endowing the PI-based hybrid foams excellent EMI-shielding and anti-oxidation performance. The PI@MXene@ANFs/PI hybrid foam exhibits an absorption-dominant EMI-shielding mechanism. After being heat treated at 250 °C for 100 h, the average EMI SE_T of PI@MXene@ANFs/PI-35 foam was slightly decreased from 46.3 to 44.0 dB, presenting great prospects for the long-term application in extreme conditions [82].

7.5.4 Sandwich and Multilayer Film Structure

Some scholars have designed sandwich and multilayer film structures to obtain low-reflection EMI-shielding materials [31, 83, 112]. These structures consist of alternating layers of conductive and dielectric materials, which help to absorb and attenuate EMWs, thereby reducing the amount of electromagnetic interference that passes through the material. Additionally, these structures offer advantages such as flexibility, durability, and ease of manufacturing, making them a promising

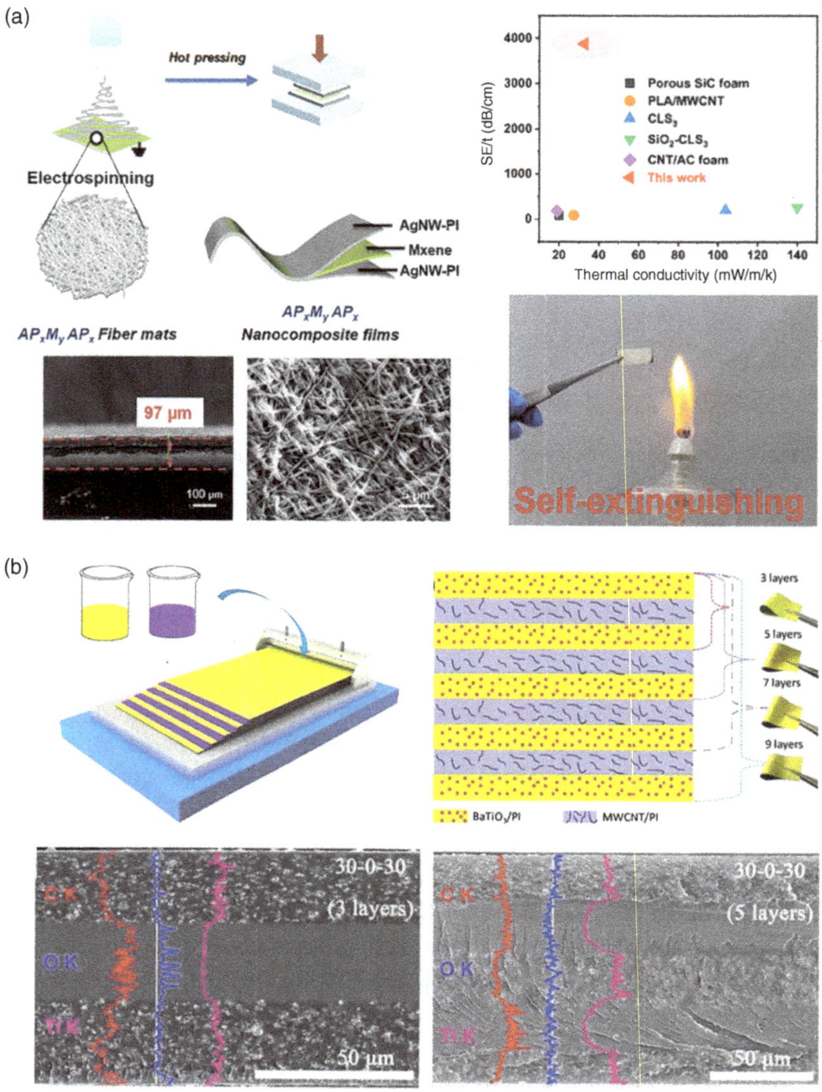

Figure 7.9 PI composites with (a) sandwich and (b) multilayer structure for EMI-shielding applications [93, 113].

solution for addressing EMI-shielding requirements in modern electronic devices and systems. Figure 7.9 illustrates PI-based composites with sandwich/multilayer structures.

There are many reports using conductive MXene as the mid-layer to build sandwich PI materials to shield EMWs [73, 83, 112]. Zhang et al. prepared a porous/sandwich (Fe_3O_4/PI)–$Ti_3C_2T_x$–(Fe_3O_4/PI) composite film with controllable pore sizes and distribution of $Ti_3C_2T_x$ hollow microspheres by sacrificial PMMA microsphere template method. The composite films exhibited excellent EMI SE of 85 dB when the pore size of the $Ti_3C_2T_x$ hollow microspheres is 10 μm and the mass ratio of PMMA/$Ti_3C_2T_x$ is 2:1 [73]. Sang et al. constructed a flexible, hydrophobic, mechanically robust, and EMI-shielding PTFE/MXene/PI layered material through a very simple and fast "cutting and sticking" method by taping PTFE and PI tapes onto the surface of MXene film. The commercially available PTFE and PI tapes endowed the sandwich PTFE/MXene/PI structure with ideal mechanically robustness and hydrophobic self-cleaning function; thus, it can work in different environments [114]. Zhang et al. fabricated a sandwich film consisting of two identical layers of AgNW-PI nanofiber mats and a highly conductive layer of $Ti_3C_2T_x$ MXene nanosheets (Figure 7.9a). The AgNW-PI/MXene/AgNW-PI films obtained after hot pressing have excellent mechanical properties and EMI-shielding performances (3885.6 dB/cm), and long-term serviceability in harsh environments. [93].

Magnetic nanofillers are also introduced into the mid-layer of PI composite films to enhance the microwave absorbing. Piao et al. developed a multifunctional rPET/FeCo@C/AgNWs/PI sandwich membrane with innovative properties, including Joule heating, EMI-shielding, antibacterial, and self-cleaning functions. The EM-shielding property with 49.35 dB at 120 μm thickness was obtained in the X-band (8.2–12.4 GHz) [74]. Feng et al. constructed a heterogeneous alternating multilayer structure with conductive gradient material for EMI shielding. Combining the synergetic electromagnetic loss and interface polarization loss formed by the selective distribution of metallized chopped PI fiber (PI@Ni/Cu/Ni), rGO, and Fe_3O_4 fillers, as well as multilevel EMW propagation paths, the penetrating microwaves would undergo a particular "absorption–reflection–reabsorption" process in the composite. The asymmetric multilayer composite with an absorption-dominated shielding feature exhibits a high EMI SE of 53.90 dB [75].

7.5.5 Anisotropic Structure

Anisotropic structure materials exhibit different electrical and magnetic properties in different directions, which allows these materials to selectively block EMI in specific directions, providing enhanced shielding efficiency compared to isotropic materials with uniform properties in all directions [33]. Anisotropic materials can be designed and engineered using various techniques, such as incorporating aligned conductive fillers or creating the above sandwich/layered structures with different conductive properties along different axes. These structures can be tailored to target specific EMI frequencies or directions of incoming EMWs, offering superior shielding performance in critical areas.

The above-discussed foams and aerogels with oriented porous structure thus have anisotropic EMI-shielding performances. Zhuo et al. developed a novel anisotropic MXene@PI hybrid foam for efficient EMI shielding using a synergistic approach involving directional freezing, freeze-drying, thermal imidization, and a subsequent dip-coating step. The foam exhibited a tubular structure in a vertical frozen direction while adopting a honeycomb-like structure in a parallel freezing direction. The anisotropic MXene@PI-5 hybrid foam demonstrated remarkable EMI SE of up to 62.7 dB in the vertical freezing direction, 158.02 % higher than that observed in the parallel direction, and 109.70 % higher than that in the isotropic foam with the same loading of MXene [115]. Luo et al. prepared the anisotropic CNTs@$CoFe_2O_4$/PI aerogels by directional freeze-drying and thermal imidization strategy. Benefiting from the synergistic effect between adjustable electromagnetic components and anisotropic structure, the aerogel displays interesting attributes in different directions. The optimum reflection loss always emerges in radial direction and the maximum value is −54.4 dB with the thickness of 2.6 mm, whereas maximum EAB appears in axial direction and which is 7.15 GHz. In addition, the composite aerogel also exhibits obvious anisotropic thermal insulation behavior with lowest thermal conductivity of 29.6 mW/m/K in the radial direction. The anisotropy of the heat conduction promotes realization of high thermal insulation in the radial direction, while facilitates heat diffusion in the axial plane to avoid localization overheating [95]. Yu et al. obtained an anisotropic graphene/PI aerogel by the same unidirectional freezing method. The one-way growing ice crystals endowed the composite aerogels with highly arranged tubular pores. The graphene/PI composite aerogels possessed anisotropic conductivity, electromagnetic interference (EMI) shielding, heat transfer, and compression performance. The EMI SE and its specific EMI SE value of 13 wt% graphene/PI are 26.1–28.8 dB and 1373–1518 dB·cm^2/g, respectively [52].

7.5.6 Gradient Conductive Structure

A gradient conductive structure for EMI shielding is a material or composite engineered with a gradient in the distribution of conductive elements. By strategically designing the gradient in conductivity, the material can effectively attenuate EMWs across a broad frequency range. The gradual change in conductivity helps to mitigate reflection and absorption of electromagnetic radiation, thus improving overall shielding performance [70].

Zou et al. designed a gradient-conductive Ag/$BaFe_{12}O_{19}$ microfiber/PI composite film with exceptional EMI SE by blow spinning. The gradient-conductive structure effectively minimized EMW reflection. The incorporation of $BaFe_{12}O_{19}$ microfiber further enhanced the EMI SE of the film. At a thickness of 100 μm, the EMI SE of the composite film can reach 98.9 dB with an absorption coefficient of 0.75, confirming an absorbing EMI-shielding material [76]. After packaging and drying, PI/PPy/CF composite films with a gradient structure are obtained. With the increase of the PPy layers, the EMI-shielding absorption loss of composite films increases gradually. It is believed that the PI/ PPy/CF composite films with high EMI SE, excellent mechanical properties, high flexibility, and temperature resistance hold great promise and are expected to be widely used in aerospace, communication engineering, electronic equipment, and smart wearable fields [7]. Pu et al. designed a PI/graphene composite foam with a multi-hierarchical impedance gradient and controllably constructed by two-step vacuum impregnation, effectively facilitates further attenuation of EMWs entering the graphene skeleton due to the generation of interfacial polarization relaxation and internal multiple reflection/scattering. the optimized PI–GP–rGO composite foam exhibits a minimum reflection loss of −32.87 dB with an effective bandwidth of 6.22 GHz (8.26–14.48 GHz) at 10.39 GHz when the thickness is 4.0 mm [61].

7.5.7 Other Structure

Core–shell structure

Wang et al. coated the PI fabric with core–shell $NiFe_2O_4$@PANI NPs to obtain an excellent microwave absorption performance (Figure 7.10a). The resultant fabric possessed a high microwave attenuation property with an RL_{mini} value of 19.2 dB (>90% attenuation) at 16.1 GHz and the EAB was 5.1 GHz. The high performance was attributed to the uniform dispersion of core–shell $NiFe_2O_4$@PANI NPs, better impedance matching, and intensive synergistic effect between dielectric loss caused by the PANI shell and magnetic loss from the $NiFe_2O_4$ core [96].

Needle-like porous structure

Liu et al. designed a regular radial needle-like microporous PI/CNT aerogel through radial freeze-drying methods (Figure 7.10b). The fabricated aerogels show excellent thermal stability, compressive mechanical properties, and electromagnetic interference shielding performance (~53.1 dB with a thickness of 2 mm in the axial direction). The SE_A/SE_T value of the PI/CNT aerogels is as high as 94.5% [39].

Urchin-like nanostructure

Li et al. designed an isotropic urchin-like nanostructure of vertical graphene nanosheet (VGSs)/PI composites by composing VGSs with PI via in situ polymerization (Figure 7.10c). With the urchin-like nanostructure and agglomeration-free dispersion of VGSs, rich electrons conduction paths, enclosure nanostructure, and countless interfaces were constructed in PI matrix. The optimal composite film showed superior EMI SE of 31.37 dB in the frequency range of 8.2–12.4 GHz with a thickness of 151 μm [53].

Spiral-like structure

Feng et al. designed a multilayer composite with unique spiral structure consisting of PI fabric and waterborne polyurethane (WPU) created by hot compression molding method for lightweight and flexible electromagnetic interference (EMI)-shielding materials (Figure 7.10d) [77]. The Fe_3O_4@rGO/WPU-Ni@PI composite film consisted of Ni@PI and Fe_3O_4@rGO/WPU film; the two components were rolled together to form a cylinder and then underwent hot compressive molding with the molding temperature of 100 °C and pressure of 1 MPa for 5 min, respectively. Then, a lightweight Fe_3O_4@rGO/WPU-Ni@PI composite with a density of 1.098 g/cm^3 was prepared.

Honeycomb structure

Wang et al. fabricated three kinds of EMW-absorbing honeycombs which were composed of graphene/PI, MWCNT/PI, and Ni-coated MWCNT/PI (Ni@CNT/PI) composites by a simple, green method combining hydrogel-ink-based 3D printing and two-stage curing process (Figure 7.10e). The impedance matching and multiple scattering of honeycomb was enhanced by its macrostructure and waviness in the cell walls at a low filling level. The obtained Ni@CNT/PI honeycomb demonstrated an optimal RL_{min} of −56.56 dB at 10.08 GHz with a thickness of 2.6 mm and a filler loading of 5 wt. %. Furthermore, the EAB of the graphene/PI honeycomb covered the entire X-band [97]. Chu et al. developed a

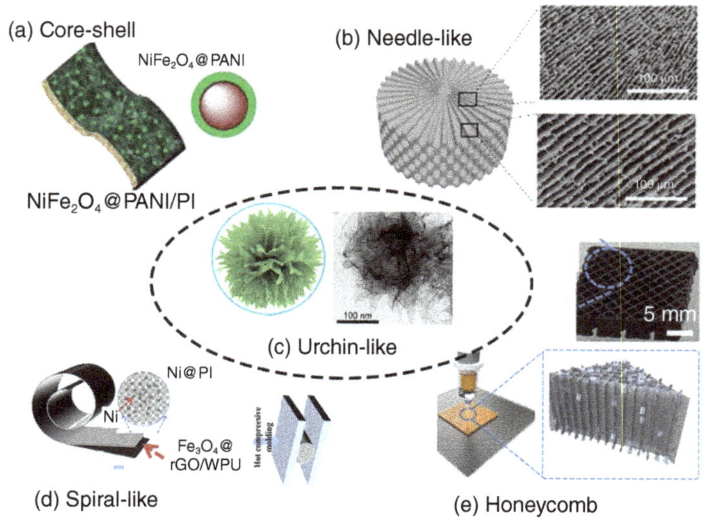

Figure 7.10 PI-based composites with inner structures.

highly absorption-dominated composite EMI-shielding foam with "honeycomb" structure through chemical foaming and subsequent in situ vapor deposition of PPy within PI foams. The optimal reflection loss (RL) of the composite PI/Fe_3O_4-20@PPy foam reached up to −31 dB at 6.0 mm and ≤ -10 dB in the whole X band with a thickness of ≥4 mm [62].

7.6 Future Development and Prospects

7.6.1 Ultralight, Ultrathin, and Ultraflexible

Lightweight, portability, and miniaturization are important characteristics for the next generation of high-performance EMI-shielding equipment. This is particularly important as the demand for reliable electromagnetic interference protection continues to grow in industries such as telecommunications, automotive, aerospace, and medical devices.

Cui et al. constructed an ultralight 3D MXene/CNTs/PI aerogel material aided by controlling the growth of ice peaks. The highly porous structure reduced the density of MCP-1, giving it the advantage of being ultralight. Due to the excellent impedance matching and special microstructure, MXene/CNTs/PI has an RL_{min} of −50.03 dB@13.7 GHz, and an EAB of 5.6 GHz. Meanwhile, MXene/CNTs/PI also exhibits significant photothermal conversion performance [101].

Zeng et al. manufactured a lightweight, ultraflexible, and robust crosslinked MXene-coated PI (C-MXene@PI) porous composites via a scalable dip-coating followed by facile chemical crosslinking approach. Consequently, the 1.5 mm-thick C-MXene@PI composite foam possessed a density of 48.7 mg/cm^3 and a satisfactory X-band EMI SE of 62.5 dB. Because of the excellent flexible characteristics, the obtained C-MXene@PI composites can be effortlessly attached to human body as wearable sensors for detecting the human motions [54].

7.6.2 High Durability

Feng et al. designed a multilayer Fe_3O_4@rGO/WPU-Ni@PI composite with unique spiral structure. The addition of Fe_3O_4 NPs improves the EMI SE of Fe_3O_4@rGO/WPU composites, especially the absorption efficiency. The continuous conductive network included in the Fe_3O_4@rGO/WPU-Ni@PI composite firmly locks EMW as they enter the interior and provide more interfaces to increase multiple reflection losses. Therefore, a maximum SE value of 36.8 dB with only 6.2 wt% Ni and 8.53 wt% Fe_3O_4@rGO loading is obtained, breaking through the bottleneck of high filler loading and low SE of conductive polymeric composites. Furthermore, the composite with spiral structure retains significant EMI SE after 1000 bending cycles [77].

7.6.3 Multifunctional EMI PI

With advancing aviation technologies, hypersonic aircraft is constantly developing in military applications. The components of aircraft suffer from high temperatures during ultrahigh-speed flight. Stealth technology is essential to avoid detection by enemy radar and IR detectors. Therefore, it is of great importance for scientists and engineers to develop multifunctional materials that possess both EMW attenuation and thermal insulation capability, as well as low density and good mechanical properties for harsh environments. While in the fields of electromagnetic radiation protection such as wearable devices, the EMI-shielding composite films are expected to not only have excellent EMI-shielding performances, but also have good self-heating performances, mechanical properties and hydrophobicity, etc., to effectively improve the comfort in temperature, wearability, and safety of wearable devices. Wu et al. reported a MWCNTs/poly ether ether ketone/thermoplastic polyimide (MWCNTs/PEEK/TPI) composite foam. The density, electrical conductivity, SE_T could still retain more than 90% after immersing in different solvents for 30 days, implying good corrosion resistance. What's more, it can withstand high temperature of 300 °C for a long time, which was suitable for extreme environmental service [84].

Recently, multifunctional materials have attracted tremendous attention in intelligent equipment and human–computer interaction, requiring the integration of electrical conductivity, sensing property, and heating management. Sang et al. constructed a multifunctional PVDF/MXene/PI wearable electronic by depositing conductive MXene on the PVDF film surface and followed by integrating PI tape. The obtained PVDF/MXene/PI film exhibited outstanding EMI SE of 40 dB. Besides, it possessed excellent mechanosensation ability, which can sensitively monitor different human movements. Furthermore, it presented prominent Joule heating capability and achieved a temperature of over 80 °C at 6.5 V voltage. Insulated high-temperature resistant PVDF and PI tape as protective layers ensured the stability and safety of PVDF/MXene/PI heater, showing potential applications in human thermotherapy, human–machine interaction, and EMW protection [116].

Particularly, portable thermal-sensitive electronic devices have more stringent requirements for thermal insulation performances. In most cases, ultrathin EMI-shielding materials integrated with ultralow thermal conductivity are not easy to be achieved. Ding et al. fabricated a dual protective composite film based on $Ti_3C_2T_x$ MXene and PI are fabricated by sacrificing polymethyl methacrylate (PMMA) templates. By optimizing the contact thermal resistance and Kapitza resistance, the composite film presents superior thermal insulation performances with a thermal conductivity of 0.0136 W·m/K. Moreover, the hybrid porous film maintains superior EMI SE of 63.0 dB and high SSE/t of 31651.2 dB cm^2/g [55]. Wu et al. reported a strategy to boost the thermal conduction and EMI-shielding properties of PI-based composites by constructing large-size aligned graphite sheets inside PI matrix [117]. Zhang et al. provided a flexible multifunctional MXene/Fe_3O_4/PI EMI-shielding composite films by depositing $Ti_3C_2T_x$ nanosheets on the surface of the Fe_3O_4/PAA nanofiber films via vacuum-assisted filtration for important applications in the fields of 5G communication technology, wearable electronic devices, and artificial intelligence. Besides the good EMI-shielding performance, the composite films also had excellent electrothermal and photothermal conversion performances [78].

7.6.4 Extreme Conditions

Zhang et al. reported a highly conductive PI-MXene-AgNPs composite film by performing a self-reduction reaction on the surface of PAA nanofibers. The composite film has stable EMI-shielding performance even after high (300 °C) temperature, cryogenic (−196 °C), corrosive environment, multiple bending, and physical shaking. Furthermore, the films exhibit excellent flexibility and superior mechanical properties that can be maintained even after treatment under harsh conditions. In addition, the PI-MXene-AgNPs composite films show good versatility, such as photothermal conversion properties, and antibacterial properties [85]. Cheng et al. constructed hierarchically porous PI/$Ti_3C_2T_x$ films with consecutively conductive pathways via a unidirectional PI aerogel–assisted immersion and hot-pressing strategy. Contributed by special architectures and high conductivities, PI/$Ti_3C_2T_x$ films with 2.0 vol% $Ti_3C_2T_x$ have high absolute EMI SE up to 15,527 dB cm^2/g at the thickness of 90 μm. Superior EMI-shielding performance can be retained even after being subjected to hygrothermal or combustion environments, cryogenic (−196°C) or high (250°C) temperatures, and rapid thermal shock (ΔT = 446 °C), demonstrating high potential as high-performance EMI-shielding materials resisting harsh conditions [11].

7.6.5 Recoverable or Shape Memory EMI PI

Wan et al. obtained a smart PI film using a low-molecular-weight PI gene unit. The PI films combining hardness with softness are designed into three forms via a

"Mimosa-like" bionic strategy to adapt to different application scenarios. The films have good degradation efficiency, excellent recyclability, and can be self-healable, which makes them reusable. The films can be used in the preparation of ultrafast sensors with a response time ≈ 0.15 s and the application of corona-resistant films with 100% recovery [118].

Kong et al. for the first time reported an electromagnetic shielding shape memory PI (EMSMPI) material by incorporating 5% short carbon fiber and 4% carbon black into shape memory PI matrix. The glass transition temperature (308 °C) was higher than other EMI-shielding polymers, and it can hang a kettle 43,000 times heavier than itself. The 0.35 mm thick EMSMPI film exhibited an average EMI SE of 23.9 dB in X-band. Shape processability of EMSMPI endowed by shape memory effect was valuable for sophisticated devices as it can be processed into different shapes conveniently [86].

7.6.6 The Next 6G Technology

Information and communication technology has greatly improved human society, and the next 6G technology (using 0.1–10 THz EMWs) will fuse the physical and digital world seamlessly, promoting the development of human civilization largely. Therefore, it is highly important and urgent to seek high-efficiency EMI-shielding materials and fabricate smart EMW control devices for THz EMWs. Ji et al. reported a PI/MXene film that can be converted into porous LIG/TiCN materials in air, which can be used to fabricate high-performance THz EMI-shielding devices and flexible infrared response actuators. The laser-induced graphene (LIG) sample with a thickness of 80 μm has SE_T up to 15 dB in the frequency range of 0.1–3 THz, and the generation of TiCN NPs can improve the SE_T to 30 dB due to the increase of heteroatom interfaces in LIG. In addition, a multilayer sandwich structure is introduced into the EMI-shielding devices, which contributes to the impedance matching at resonance, hence, increasing the SE_T to 45 dB in the frequency range of 0.5–1.5 THz [87].

7.7 Summary

EMI-shielding PI-based composite materials have received immense attention from researchers owing to the rapid development in electronics and telecommunications. However, a shift toward rational materials' design rather than a simple production–testing approach can lead to more sustainable and efficient material development processes, driving innovation and pushing the boundaries of what is possible in various industries. Meanwhile, the demand of high EMI-shielding performances together with low density used in the high-tech aerospace and harsh environment is still a big challenge due to the high cost and difficulty to produce. Accordingly, cutting-edge materials such as MXene and graphene have been widely used in PI, while inexpensive inorganic materials are still far from groundbreaking and revolutionary advancements toward massive production of cheap and easily processable solutions. Nevertheless, we believe that the continuous research of innovative PI materials can lead the way toward a bright future in the field of EMI shielding.

References

1 Zecchi, S., Cristoforo, G., Bartoli, M. et al. (2024). A comprehensive review of electromagnetic interference shielding composite materials. *Micromachines* 15 (2): 187.

2 Orasugh, J. and Ray, S. (2023). Functional and structural facts of effective electromagnetic interference shielding materials: a review. *ACS Omega* 8 (9): 8134–8158.

3 Hwang, U., Kim, J., Seol, M. et al. (2022). Quantitative interpretation of electromagnetic interference shielding efficiency: is it really a wave absorber or a reflector? *ACS Omega* 7 (5): 4135–4139.

4 Dong, C., Wang, S., Zhang, P. et al. (2024). Insight into lightweight MXene/polyimide aerogel with high-efficient microwave absorption. *Materials Today Physics* 42: 101373.

5 Acevedo-Del-Castillo, A., Águila-Toledo, E., Maldonado-Magnere, S., and Aguilar-Bolados, H. (2021). A brief review on the high-energy electromagnetic radiation-shielding materials based on polymer nanocomposites. *International Journal of Molecular Sciences* 22 (16): 9079.

6 Wang, J., Wu, R., Feng, J. et al. (2022). Recent advances of electromagnetic interference shielding Mg matrix materials and their processings: a review. *Transactions of Nonferrous Metals Society of China* 32 (5): 1385–1404.

7 Li, H., Li, J., Chu, W. et al. (2024). Flexible hierarchical polyimide/polypyrrole/carbon nanofiber composite films for tunable electromagnetic interference shielding. *ACS Applied Nano Materials* 7: 7783–7793.

8 Omana, L., Chandran, A., John, R.E. et al. (2022). Recent advances in polymer nanocomposites for electromagnetic interference shielding: a review. *ACS Omega* 7 (30): 25921–25947.

9 Chu, N., Luo, C., Chen, X. et al. (2023). $Ti_3C_2T_x$ MXene/polyimide composites film with excellent mechanical properties and electromagnetic interference shielding properties. *Journal of Alloys and Compounds* 955: 170241.

10 Mehmood, Z., Shah, S., Omer, S. et al. (2024). Scalable synthesis of high-quality, reduced graphene oxide with a large C/O ratio and its dispersion in a chemically modified polyimide matrix for electromagnetic interference shielding applications. *RSC Advances* 14 (11): 7641–7654.

11 Cheng, Y., Li, X., Qin, Y. et al. (2021). Hierarchically porous polyimide/$Ti_3C_2T_x$ film with stable electromagnetic interference shielding after resisting harsh conditions. *Science Advances* 7 (39): eabj1663.

12 Ayub, S., Guan, B., Ahmad, F. et al. (2021). Graphene and iron reinforced polymer composite electromagnetic shielding applications: a review. *Polymers* 13 (15): 2580.

13 Wanasinghe, D., Aslani, F., Ma, G., and Habibi, D. (2020). Review of polymer composites with diverse nanofillers for electromagnetic interference shielding. *Nanomaterials (Basel, Switzerland)* 10 (3): 541.

14 Guo, Z., Park, S., Ht, H. et al. (2007). Magnetic and electromagnetic evaluation of the magnetic nanoparticle filled polyurethane nanocomposites. *Journal of Applied Physics* 101 (9): 09M511.

15 Liu, T., Zhang, Y., Wang, C. et al. (2024). Multifunctional M_0C_x hybrid polyimide aerogel with modified porous defect engineering for highly efficient electromagnetic wave absorption. *Small* 20 (31): e2308378.

16 Yoon, C.S. (2007). High-frequency electromagnetic properties of soft magnetic metal–polyimide hybrid thin films. *Journal of Magnetism and Magnetic Materials* 316 (2): e893–e895.

17 Wang, H., Zhu, D., Zhou, W., and Luo, F. (2014). Microwave electromagnetic properties of polyimide/carbonyl iron composites. *Journal of Polymer Research* 21 (6): 478.

18 Zhou, Y., Zhou, W., Li, R. et al. (2016). Electroless plating preparation and electromagnetic properties of Co-coated carbonyl iron particles/polyimide composite. *Journal of Magnetism and Magnetic Materials* 401: 251–258.

19 Wang, X. and Wu, D. (2016). Magnetic microencapsulated phase change materials with an organo-silica shell: design, synthesis and application for electromagnetic shielding and thermal regulating polyimide films. *Energy* 98: 225–239.

20 Li, X.-J., Wei, Q., and Wang, X. (2014). Preparation of magnetic polyimide/maghemite nanocomposite fibers by electrospinning. *High Performance Polymers* 26 (7): 810–816.

21 Yu, W., Lin, J., Zhao, Z. et al. (2024). Polyimide-based porous carbon and cobalt nanoparticle composites as high-performance electromagnetic wave absorbers. *RSC Adv.* 14 (14): 9716–9724.

22 Li, J., Chu, W., Gao, Q. et al. (2022). In situ fabrication of magnetic and hierarchically porous carbon films for efficient electromagnetic wave shielding and absorption. *ACS Applied Materials & Interfaces* 14: 33675–33685.

23 Wang, H., Zhu, D., Zhou, W., and Luo, F. (2015). Electromagnetic property of SiO_2-coated carbonyl iron/polyimide composites as heat resistant microwave absorbing materials. *Journal of Magnetism and Magnetic Materials* 375: 111–116.

24 Zhang, J., Ni, H., Gong, M. et al. (2021). Realization of electromagnetic shielding performance for polyimide-matrix composite by self-metallization. *Journal of Composite Materials* 55 (26): 3849–3860.

25 Liu, X.-Y., Zhan, M.-S., and Wang, K. (2013). Preparation and characterization of electromagnetic interference shielding polyimide foam. *Journal of Applied Polymer Science* 127 (5): 4129–4137.

26 Tang, X., Zhao, X., Lu, Y. et al. (2024). Flexible metalized polyimide nonwoven fabrics for efficient electromagnetic interference shielding. *Chemical Engineering Journal* 480: 148000.

27 Mehmood, Z., Shah, S., Omer, S. et al. (2024). Guar gum assisted synthesis of high-quality silver nanoplates and their polyimide matrix nanocomposites for electromagnetic interference shielding applications. *Journal of Applied Polymer Science* 141: e55389.

28 Iqbal, A., Shahzad, F., Hantanasirisakul, K. et al. (2020). Anomalous absorption of electromagnetic waves by 2D transition metal carbonitride $Ti_3C_nT_X$ (MXene). *Science* 369 (6502): 446–450.

29 Guo, Z., Ren, P., Wang, J. et al. (2023). Methylene blue adsorption derived thermal insulating N, S-co-doped TiC/carbon hybrid aerogel for high-efficient absorption-dominant electromagnetic interference shielding. *Chemical Engineering Journal* 451: 138667.

30 Chen, K., Moon, H., Park, S., and Ha, C.-S. (2024). Flexible and hydrophobic polyimide/MXene-POSS nanocomposite films for electromagnetic interference shielding. *European Polymer Journal* 202: 112655.

31 Lei, P., Liu, Y., Li, L. et al. (2021). Polyimide nanofiber-reinforced $Ti_3C_2T_x$ Aerogel with "Lamella-Pillar" microporosity for high-performance piezoresistive strain sensing and electromagnetic wave absorption. *ACS Applied Materials Interfaces* 13 (39): 47134–47146.

32 Wang, H., Zhu, D., Zhou, W., and Luo, F. (2015). Temperature dependence of the electromagnetic and microwave absorption properties of polyimide/Ti_3SiC_2 composites in the X band. *RSC Advances* 5 (105): 86656–86664.

33 Dai, Y., Wu, X., Liu, Z. et al. (2020). Highly sensitive, robust and anisotropic MXene aerogels for efficient broadband microwave absorption. *Composites Part B: Engineering* 200: 108263.

34 Liu, J., Zhang, H.-B., Xie, X. et al. (2018). Multifunctional, superelastic, and lightweight MXene/polyimide aerogels. *Small* 14 (45): e1802479.

35 Xu, L., Wang, L., Zhang, W. et al. (2022). The reinforced electromagnetic interference shielding performance of thermal reduced graphene oxide films via polyimide pyrolysis. *ACS Omega* 7 (13): 10955–10962.

36 Li, Y., Pei, X., Shen, B. et al. (2015). Polyimide/graphene composite foam sheets with ultrahigh thermostability for electromagnetic interference shielding. *RSC Advances* 5 (31): 24342–24351.

37 Yang, H., Li, Z., Zou, H., and Liu, P. (2017). Preparation of porous polyimide/*in-situ* reduced graphene oxide composite films for electromagnetic interference shielding. *Polymers for Advanced Technologies* 28 (2): 233–242.

38 Li, S., Tang, X., Fan, W. et al. (2022). Ultralight and ordered lamellar polyimide-based graphene foams with efficient broadband electromagnetic absorption. *Journal of Materials Science & Technology* 102: 97–104.

39 Liu, S., Xu, Q., Bai, Y. et al. (2022). Toward strong x-band-electromagnetic-wave-absorbing materials: polyimide/carbon nanotube composite aerogel with radial needle-like porous structure. *Journal of Materials Chemistry A* 10 (47): 25140–25147.

40 Wang, Y.-Y., Zhou, Z.-H., Zhou, C.-G. et al. (2020). Lightweight and robust carbon nanotube/polyimide foam for efficient and heat-resistant electromagnetic interference shielding and microwave absorption. *ACS Applied Materials & Interfaces* 12 (7): 8704–8712.

41 Wang, H., Zhu, D., Zhou, W., and Luo, F. (2015). High temperature electromagnetic and microwave absorbing properties of polyimide/multi-walled carbon nanotubes nancomposites. *Chemical Physics Letters* 633: 223–228.

42 Chen, J., Jia, X., Wang, G. et al. (2021). Humidification of high-performance and multifunctional polyimide/carbon nanotube composite foams for enhanced electromagnetic shielding. *Materials Today Physics* 21: 100521.

43 Jiang, M., Lin, D., Jia, W. et al. (2021). Preparation and properties of polyimide/carbon nanotube composite films with electromagnetic wave absorption performance. *Polymer Engineering & Science* 61 (10): 2691–2700.

44 Ni, L., Chen, S., Jiang, X. et al. (2022). Anisotropic electromagnetic wave absorption performance of polyimide/multi-walled carbon nanotubes composite aerogels with aligned slit-like channels structure. *Composites Part A: Applied Science and Manufacturing* 154: 106781.

45 Nayak, L., Chaki, T., and Khastgir, D. (2014). Electrical percolation behavior and electromagnetic shielding effectiveness of polyimide nanocomposites filled with carbon nanofibers. *Journal of Applied Polymer Science* 131 (24): 40914.

46 Wang, H., Zhu, D., Zhou, W., and Luo, F. (2014). Influence of heat treatment on electromagnetic properties of polyimide/carbon black composites. *Polymers for Advanced Technologies* 25 (12): 1616–1621.

47 Kim, G., Kim, S.-Y., Lee, S. et al. (2021). Fabrication of highly flexible electromagnetic interference shielding polyimide carbon black composite using hot-pressing method. *Composites Part B: Engineering* 221: 109010.

48 Qing, Y., Li, Y., Li, W., and Yao, H. (2021). Ti^{3+} self-doped dark TiO_2 nanoparticles with tunable and unique dielectric properties for electromagnetic applications. *Journal of Materials Chemistry C* 9 (4): 1205–1214.

49 He, M., Tang, J., Wang, Y. et al. (2022). Preparation and properties of PANI/PI composite fabrics with conductive nanofiber network structure. *Polymer* 240: 124498.

50 Xue, T., Fu, Z., Yu, D. et al. (2023). Lightweight and flexible Ag-wrapped polyimide aerogel fabrics for electromagnetic interference shielding. *Composites Communications* 43: 101732.

51 Zhou, S., Dong, J., Li, X. et al. (2023). Continuous surface metallization of polyimide fibers for textile-substrate electromagnetic shielding applications. *Advanced Fiber Materials* 5: 1892–1904.

52 Yu, Z., Dai, T., Yuan, S. et al. (2020). Electromagnetic interference shielding performance of anisotropic polyimide/graphene composite aerogels. *ACS Applied Materials & Interfaces* 12 (27): 30990–31001.

53 Li, Z., Lin, Z., Han, M. et al. (2021). Vertical graphene nanosheet/polyimide composite films for electromagnetic interference shielding. *ACS Applied Nano Materials* 4 (7): 7461–7470.

54 Zeng, Z.-H., Wu, N., Wei, J.-J. et al. (2022). Porous and ultra-flexible crosslinked MXene/polyimide composites for multifunctional electromagnetic interference shielding. *Nano-Micro Letters* 14 (1): 59.

55 Ding, R., Yan, Q., Xue, F. et al. (2023). Dual protective porous $Ti_3C_2T_x$ MXene/polyimide composite film for thermal insulation and electromagnetic interference shielding. *Small* 19 (50): e2304946.

56 Zou, K., Sun, H., Li, X. et al. (2023). Extreme environment-bearable polyimide film with a three-dimensional Ag microfiber conductive network for ultrahigh electromagnetic interference shielding. *Science China Materials* 66 (4): 1578–1586.

57 Li, C., He, X., Xiao, H. et al. (2021). Porous carbon foam based on coassembled graphene and adenine-polyimide for electromagnetic interference shielding. *Polymer* 236: 124328.

58 Moon, H., Yan, Y.-Z., Lee, J. et al. (2023). MXenes ($Ti_3C_2T_x$) incorporated polyimide nanohybrid composite films. *Molecular Crystals and Liquid Crystals* 765 (1): 37–53.

59 Sun, K., Wang, F., Yang, W. et al. (2021). Flexible conductive polyimide fiber/MXene composite film for electromagnetic interference shielding and joule heating with excellent harsh environment tolerance. *ACS Applied Materials Interfaces* 13 (42): 50368–50380.

60 Luo, Y., Wang, G., Yan, L. et al. (2023). Fabrication of flexible polyimide/rGO composite foams and their derived carbon foams for thermal insulation and EMI shielding applications. *Polymer* 284: 126296.

61 Pu, L., Li, S., Zhang, Y. et al. (2021). Polyimide-based graphene composite foams with hierarchical impedance gradient for efficient electromagnetic absorption. *Journal of Materials Chemistry C* 9 (6): 2086–2094.

62 Chu, W., Li, J., Lin, J. et al. (2024). Honeycomb-like polyimide/Fe_3O_4@PPy foam for electromagnetic wave shielding with excellent absorption characteristics. *Composites Science and Technology* 249: 110489.

63 Lv, Y., Fang, J., Zuo, X. et al. (2018). A new method for an efficient porous carbon/Fe_3O_4 composite based electromagnetic wave absorber derived from a specially designed polyimide. *Composites Part B: Engineering* 155: 148–155.

64 Guo, Z., Zhang, W., Ding, E. et al. (2024). A strategy to fabricate hierarchical microporous architecture of polyimide nanofibrous aerogels with efficient electromagnetic wave absorption and thermal insulation. *Composites Part A: Applied Science and Manufacturing* 177: 107940.

65 Jiang, M., Lin, D., Jia, W. et al. (2022). Preparation and characterization of porous polyimide fibers with electromagnetic wave absorption properties. *Polymer Engineering & Science* 62 (10): 3121–3131.

66 Li, J., Ding, Y., Gao, Q. et al. (2020). Ultrathin and flexible biomass-derived C@CoFe nanocomposite films for efficient electromagnetic interference shielding. *Composites Part B: Engineering* 190: 107935.

67 Li, J., Zhang, X., Ding, Y. et al. (2022). Multifunctional carbon fiber@NiCo/polyimide films with outstanding electromagnetic interference shielding performance. *Chemical Engineering Journal* 427: 131937.

68 Zheng, W., Xie, H., Li, J. et al. (2024). Depositing gradient thickness silver film on CNF/rGO/polyimide surface to form a sandwich structure for efficient electromagnetic interference shielding. *Surfaces and Interfaces* 47: 104159.

69 Li, H., Liang, H., Wang, H. et al. (2024). Polyimide aerogels with a dual electrically conductive network for electromagnetic interference shielding, piezoresistive sensing, and thermal management. *Materials Today Communications* 38: 108506.

70 Xue, T., Yang, Y., Yu, D. et al. (2023). 3D printed integrated gradient-conductive MXene/CNT/polyimide aerogel frames for electromagnetic interference shielding with ultra-low reflection. *Nano-Micro Letters* 15 (1): 45.

71 Zhu, L., Mo, R., Yin, C.-G. et al. (2022). Synergistically constructed electromagnetic network of magnetic particle-decorated carbon nanotubes and MXene for efficient electromagnetic shielding. *ACS Applied Materials & Interfaces* 14 (50): 56120–56131.

72 Ding, X., Wang, W., Wang, Y. et al. (2019). High-performance flexible electromagnetic shielding polyimide fabric prepared by nickel-tungsten-phosphorus electroless plating. *Journal of Alloys and Compounds* 777: 1265–1273.

73 Zhang, Y., Ruan, K., Zhou, K., and Gu, J. (2023). Controlled distributed $Ti_3C_2T_x$ hollow microspheres on thermally conductive polyimide composite films for excellent electromagnetic interference shielding. *Advanced Materials* 35 (16): e2211642.

74 Piao, S., Jiang, Z., Li, S. et al. (2024). Recycled polyethylene terephthalate/FeCo@C/Silver nanowire/polyimide sandwich membrane for electrothermal heating and electromagnetic interference shielding. *Journal of Alloys and Compounds* 987: 174256.

75 Feng, Y., Li, M., Yang, S., and Wang, J. (2023). Heterogeneous waterborne polyurethane composite with alternating electromagnetic layers for absorption-dominated electromagnetic interference shielding. *Langmuir* 39 (26): 9219–9229.

76 Zou, K., Li, N., Pan, X. et al. (2023). Flexible Ag/$BaFe_{12}O_{19}$ microfiber/polyimide composite film with high electromagnetic interference shielding and low reflection. *Science China Materials* 66 (12): 4825–4833.

77 Zhang, X., Zhong, Y., Yang, S., and Wang, J. (2023). Flexible spiral-like multilayer composite with Fe_3O_4@rGO/waterborne polyurethane-Ni@polyimide for enhancing electromagnetic shielding. *Colloids and Surfaces A: Physicochemical and Engineering Aspects* 662: 131006.

78 Zhang, Y., Ma, Z., Ruan, K., and Gu, J. (2022). Multifunctional $Ti_3C_2T_x$-(Fe_3O_4/polyimide) composite films with janus structure for outstanding electromagnetic interference shielding and superior visual thermal management. *Nano Research* 15 (6): 5601–5609.

79 Wang, W., Ding, X., and Yu, D. (2020). Multilayer-structured Ni-Co-Fe-P/polyaniline/polyimide composite fabric for robust electromagnetic shielding with low reflection characteristic. *Chemical Engineering Journal* 380: 122553.

80 Wang, H., Zhu, D., Zhou, W., and Luo, F. (2015). Effect of multiwalled carbon nanotubes on the electromagnetic interference shielding properties of polyimide/carbonyl iron composites. *Industrial & Engineering Chemistry Research* 54 (25): 6589–6595.

81 Zhang, R., Lv, W., Zhu, R. et al. (2022). Metal–organic framework-derived high-performance polypyrrole/Ni-CAT/PI fiber paper-based electromagnetic shielding composites for high-frequency electromagnetic wave absorption. *Polymer Composites* 43 (12): 8904–8913.

82 Cai, Y., Shen, D., Gou, P. et al. (2023). Anti-oxidation polyimide-based hybrid foams assembled with bilayer coatings for efficient electromagnetic interference shielding. *Chemical Engineering Journal* 451: 138808.

83 Zhang, Y., Gao, Q., Sheng, X. et al. (2023). Flexible, robust, sandwich structure polyimide composite film with alternative MXene and Ag NWs layers for electromagnetic interference shielding. *Journal of Materials Science & Technology* 159: 194–203.

84 Wu, T., Liang, L., Bai, Y. et al. (2023). Fabrication of corrosion-resistant and heat-resistant multi-walled carbon nanotubes/poly ether ether ketone/thermoplastic polyimide electromagnetic shielding foams based on double percolation structure. *Carbon* 215: 118423.

85 Zhang, S., Wu, J., Liang, W. et al. (2023). Flexible and multifunctional polyimide-based composite films by self-reducing reaction for electromagnetic interference shielding in extreme environments. *Carbon* 212: 118103.

86 Kong, D., Li, J., Guo, A., and Xiao, X. (2021). High temperature electromagnetic shielding shape memory polymer composite. *Chemical Engineering Journal* 408: 127365.

87 Ji, J., Wang, Y., Zhao, W., and Wang, G. (2023). Laser-induced graphene/TiCN on a polyimide/MXene film as interference shielding materials for terahertz electromagnetic waves. *ACS Applied Nano Materials* 6: 23401–23409.

88 Yu, Z., Wu, P., Zou, H., and Liu, P. (2018). Electromagnetic interference shielding effectiveness of microcellular polyimide/in situ thermally reduced graphene oxide/carbon nanotubes nanocomposites. *Applied Surface Science* 434: 318–325.

89 Dong, J., Li, X., Zhao, X. et al. (2023). Tough polyimide composites synergistically reinforced by carbon nanofiber-grafted carbon fiber and rGO for improved heat dissipation and electromagnetic interference shielding. *Journal of Materials Science & Technology* 149: 225–236.

90 Wang, Y.-Y., Sun, W.-J., Yan, D.-X. et al. (2022). Ultralight carbon nanotube/graphene/polyimide foam with heterogeneous interfaces for efficient electromagnetic interference shielding and electromagnetic wave absorption. *Carbon* 176: 118–125.

91 Li, X., Xu, T., Cao, W. et al. (2023). Graphene/carbon fiber network constructed by co-carbonization strategy for functional integrated polyimide composites with enhanced electromagnetic shielding and thermal conductive properties. *Chemical Engineering Journal* 464: 142595.

92 Lv, Q., Liu, Y., Wang, C. et al. (2021). Rational design and fabrication of lightweight porous polyimide composites containing polyaniline modified graphene oxide and multiwalled carbon nanotube hybrid fillers for heat-resistant electromagnetic interference shielding. *Polymer* 224: 123742.

93 Zhang, S., Wu, J., Liu, J. et al. (2021). $Ti_3C_2T_x$ MXene nanosheets sandwiched between Ag nanowire-polyimide fiber mats for electromagnetic interference shielding. *ACS Applied Nano Materials* 4 (12): 13976–13985.

94 Cheng, K., Li, H., Zhu, M. et al. (2020). In situ polymerization of graphene-polyaniline@polyimide composite films with high EMI shielding and electrical properties. *RSC Advances* 10 (4): 2368–2377.

95 Wang, Y., Qu, Z., Wang, W., and Yu, D. (2022). Anisotropic, multifunctional and lightweight CNTs@$CoFe_2O_4$/polyimide aerogels for high efficient electromagnetic wave absorption and thermal insulation. *Chemical Engineering Journal* 442: 136388.

96 Wang, Y., Wang, W., Zhu, M., and Yu, D. (2017). Electromagnetic wave absorption polyimide fabric prepared by coating with core–shell $NiFe_2O_4$@PANI nanoparticles. *RSC Advances* 7: 42891–42899.

97 Ma, S., Li, D., Zhao, J. et al. (2023). Direct ink writing of thermoresistant, lightweight composite polyimide honeycombs with tunable X-band electromagnetic wave absorption properties. *Additive Manufacturing* 70: 103554.

98 Zeng, Y., Long, L., Yu, J. et al. (2024). High-efficiency electromagnetic wave absorption of lightweight Nb_2O_5/CNTs/polyimide with excellent thermal insulation and compression resistance integration. *Composites Science and Technology* 250: 110531.

99 Shi, T., Zheng, Z., Liu, H. et al. (2021). Configuration of multifunctional polyimide/graphene/Fe_3O_4 hybrid aerogel-based phase-change composite films for electromagnetic and infrared bi-stealth. *Nanomaterials* 11 (11): 3038.

100 Wang, W. and Yu, D. (2017). Three-phase heterostructures F-$NiFe_2O_4$/PANI/PI EMI shielding fabric with high microwave absorption performance. *Applied Surface Science* 425: 518–525.

101 Cui, Y., Yang, K., Zhang, F. et al. (2022). Ultra-light MXene/CNTs/PI aerogel with neat arrangement for electromagnetic wave absorption and photothermal conversion. *Composites Part A-Applied Science and Manufacturing* 158: 106986.

102 Yao, J., Jin, L., Huyan, W. et al. (2022). Multifunctional $Ti_3C_2T_x$ MXene/aramid nanofiber/polyimide aerogels with efficient thermal insulation and tunable electromagnetic wave absorption performance under thermal environment. *Composites Part B: Engineering* 243: 110161.

103 Wang, H., Zhu, D., Zhou, W., and Luo, F. (2015). Electromagnetic and microwave absorbing properties of polyimide nanocomposites at elevated temperature. *Journal of Alloys and Compounds* 648: 313–319.

104 Wang, Y., Zhao, W., Tan, L. et al. (2023). Review of polymer-based composites for electromagnetic shielding application. *Molecules (Basel, Switzerland)* 28 (15): 5628.

105 Liang, C., Gu, Z., Zhang, Y. et al. (2021). Structural design strategies of polymer matrix composites for electromagnetic interference shielding: a review. *Nano-Micro Letters* 13 (1): 181.

106 Fan, W., Zhang, X., Zhang, Y. et al. (2019). Lightweight, strong, and super-thermal insulating polyimide composite aerogels under high temperature. *Composites Science and Technology* 173: 47–52.

107 Wu, T., Ganobjak, M., Siqueira, G. et al. (2023). 3D printed polyimide nanocomposite aerogels for electromagnetic interference shielding and thermal management. *Advanced Materials & Technologies* 8: 2202155.

108 Li, J., Ding, Y., Yu, N. et al. (2020). Lightweight and stiff carbon foams derived from rigid thermosetting polyimide foam with superior electromagnetic interference shielding performance. *Carbon* 158: 45–54.

109 Li, C., Zhou, C., Lv, J. et al. (2019). Bio-molecule adenine building block effectively enhances electromagnetic interference shielding performance of polyimide-derived carbon foam. *Carbon* 149: 190–202.

110 Ma, J., Wang, K., and Zhan, M. (2015). A comparative study of structure and electromagnetic interference shielding performance for silver nanostructure hybrid polyimide foams. *RSC Advances* 5 (80): 65283–65296.

111 Ma, J., Zhan, M., and Wang, K. (2015). Ultralightweight silver nanowires hybrid polyimide composite foams for high-performance electromagnetic interference shielding. *ACS Applied Materials & Interfaces* 7 (1): 563–576.

112 Liang, W., Wu, J., Zhang, S. et al. (2024). Porous $Ti_3C_2T_x$ MXene nanosheets sandwiched between polyimide fiber mats for electromagnetic interference shielding. *Nano Research* 17 (3): 2070–2078.

113 Zhang, C., Shi, Z., Mao, F. et al. (2018). Flexible polyimide nanocomposites with dc bias induced excellent dielectric tunability and unique nonpercolative negative-*k* toward intrinsic metamaterials. *ACS Applied Materials & Interfaces* 10 (31): 26713–26722.

114 Sang, M., Liu, G., Liu, S. et al. (2021). Flexible PTFE/Mxene/PI soft electro-thermal actuator with electromagnetic-interference shielding property. *Chemical Engineering Journal* 414: 128883.

115 Zhuo, L., Shen, D., Gou, P. et al. (2023). Anisotropic high-strength polyimide-based electromagnetic interference shielding foam based on cation-Π interaction. *Chemical Engineering Journal* 477: 146992.

116 Sang, M., Liu, S., Li, W. et al. (2022). Flexible polyvinylidene fluoride(PVDF)/MXene($Ti_3C_2T_x$)/polyimide(PI) wearable electronic for body monitoring, thermotherapy and electromagnetic interference shielding. *Composites Part A: Applied Science and Manufacturing* 153: 106727.

117 Wu, Z., Dong, J., Li, X. et al. (2022). Interlayer decoration of expanded graphite by polyimide resins for preparing highly thermally conductive composites with superior electromagnetic shielding performance. *Carbon* 198: 1–10.

118 Wan, B., Yang, X., Dong, X. et al. (2023). Dynamic sustainable polyimide film combining hardness with softness via a "Mimosa-like" bionic strategy. *Advanced Materials* 35 (2): e2207451.

8

Colorless Transparent Polyimides with Balanced Dielectric and Optical Properties

Xiaojie He[1] *and Qinghua Lu*[2]

[1] *School of Chemical Science and Engineering, Tongji University, Siping Road No. 1239, Shanghai, 200092, China*
[2] *Shanghai Key Lab of Electrical & Thermal Aging, School of Chemistry and Chemical Engineering, Shanghai Jiao Tong University, Dongchuan Road No. 800, Shanghai, 200240, China*

8.1 Introduction

In recent years, the rapid development of semiconductor devices and microelectronics has driven the advancement of fundamental materials. Traditional materials such as silicon dioxide and indium tin oxide (ITO) are widely used in electronics, optoelectronics, and optics due to their excellent electrical and optical properties [1]. However, these materials face limitations in flexible electronic applications, such as rigidity impacting device performance and suboptimal optical transparency. In contrast, polyimides (PIs) have successfully entered the semiconductor and integrated circuit fields due to their excellent electrical insulation, compatibility with semiconductor processes, chemical resistance, and highly patternable nature [2, 3]. Since Bogert and Renshaw first used 4-amino phthalic acid in 1908 at the University of California, various PIs have been developed, including thermosetting, thermoplastic, and siloxane-containing polyimides. In electronic materials, high-performance photosensitive PIs, transparent PIs, and low-dielectric PIs play crucial roles in the miniaturization, lightweight, and high integration of IC circuits [4, 5].

PIs in 5G and 6G applications exhibit high thermal stability, low dielectric constant and loss, excellent mechanical properties, chemical resistance, and high processability [6–8]. PIs offer high transparency, good flexibility, environmental resistance, and low thermal expansion coefficient in flexible display applications. Despite the advantages of traditional PIs, their typical golden appearance and low transmittance, along with the presence of polar imide rings in the main chain, limit their use in optoelectronic display materials.

In recent years, scientists have reviewed PIs from different perspectives (Table 8.1). For example, Ha et al. [9] focused on colorless PIs for optoelectronic applications, discussing recent advancements (2010–2018) in monomer design, synthesis, and the effects of various structures on thermal and optical properties. Wu et al. [10]

Functional Polyimide Dielectrics: Structure, Properties, and Applications, First Edition.
Edited by Jun-Wei Zha.

Table 8.1 Recent noteworthy review articles on PIs.

Authors	Main topics	Publication year	References
Ha et al.	Colorless PIs for optoelectronic applications	2019	[7]
Wu et al.	Design of high-transparency PI films	2022	[8]
Nagella et al.	Design of PIs with low dielectric constants and loss, low water absorption, and high optical transparency	2023	[9]
Zha et al.	Design strategies to improve the energy storage properties of PI dielectric materials	2023	[10]
Zhang et al.	Methods and strategies to reduce the dielectric properties of PI films	2023	[11]
He et al.	Application of PIs in lithium-ion batteries	2023	[12]

reviewed the design of high-transparency PI films, covering recent advances in PI molecular design, including synthesis strategies and performance improvements over the past five years. Nagella et al. [11] reviewed the design of PIs with low dielectric constants and loss, low water absorption, and high optical transparency for communication applications, describing the structure–property relationships and necessary structural modifications to achieve useful parameters. Zha et al. [12] discussed design strategies to improve the energy storage properties of PI dielectric materials from a multi-structure perspective, including introducing high-polarity groups, molecular segment structure control techniques, and organic PI blending methods. Zhang et al. [13] reviewed methods and strategies to reduce the dielectric properties of PI films, concluding that reducing molecular polarizability and increasing free volume fraction are key strategies, with the introduction of porous structures being the most effective method to achieve low dielectric constants, high thermal stability, and excellent mechanical properties. He et al. [14] discussed the application of PIs in lithium-ion batteries, including coatings, separators, adhesives, solid polymer electrolytes, and active storage materials, highlighting current technical challenges and proposing solutions.

High glass transition temperature and colorless transparent PI films have achieved significant academic and commercial breakthroughs [15, 16]. However, comprehensive reviews on the design, synthesis, and application of high-transparency, low-dielectric PIs are limited. This review focuses on recent developments in the design and synthesis of high-transparency, low-dielectric PIs, discussing their structural design and characterization methods, and emphasizing the relationship between PI structure, transparency, and dielectric properties.

8.2 Synthesis and Molecular Structure Design of PIs

PI is a high-performance polymer material characterized by the alternating connection of the imide ring and benzene ring in its chemical structure, which endows PI with excellent thermal stability, mechanical strength, and chemical resistance. The fundamental structural unit of PI is synthesized through a polycondensation reaction between diamine and tetracarboxylic dianhydride. During synthesis, tetracarboxylic dianhydride and diamine monomers are typically reacted in high-boiling-point solvents such as dimethyl sulfoxide (DMSO) or N, N-dimethylacetamide (DMAc) to produce polyamic acid [17]. The synthesis method for high-transparency, low-dielectric PI is similar to conventional PI. Following the polycondensation reaction, polyamic acid is converted into PI through imidization. Two primary methods for imidization are employed: thermal imidization and chemical imidization [18]. Thermal imidization involves heating the polyamic acid at high temperatures to induce dehydration and cyclization, forming imide groups. This method is straightforward and efficient but necessitates high temperatures. In contrast, chemical imidization involves the addition of dehydrating agents such as acetic anhydride and pyridine at lower temperatures to promote the dehydration and cyclization of polyamic acid. This method is suitable for temperature-sensitive systems, allowing for better control of reaction conditions and resulting in high-quality PI.

8.3 Characterization and Evaluation of Properties

Several common experimental methods are employed to evaluate the properties of colorless and transparent PIs. Optical properties, particularly colorlessness and transparency, are crucial for applications in optoelectronics and flexible displays. Key evaluation metrics include transmittance, refractive index measurements, and color parameters [19, 20]. A high transmittance value indicates superior transparency, which is vital for applications requiring clear and unobstructed light passage. UV–visible spectroscopy is commonly used to measure transmittance across a range of wavelengths, providing a comprehensive profile of the material's transparency. The refractive index determines how much light bends as it enters the PI film. Accurate refractive index control is essential for designing materials with specific optical properties. Ellipsometry is a widely used technique for measuring the refractive index, offering high precision and the capability to analyze thin films [21]. The color of PI films can be evaluated using three parameters: yellowness index, haze, and transmittance at 400 nm (T_{400}) [4, 22]. For films with a thickness of about 20 μm, a T_{400} of approximately 80% or higher generally ensures that the film is entirely colorless. Comprehending these optical properties allows materials to be designed and optimized for specific applications. For instance, adjusting monomer composition or incorporating certain additives can enhance transparency and control the refractive index to meet desired performance criteria. The dielectric properties of PIs are critical for their use in electronic applications. Key metrics include the dielectric constant (ε) and dielectric loss ($\tan \delta$). As shown in Figure 8.1, the dielectric mechanisms of PIs under an external alternating current electric field involve three types of polarization effects that contribute to the overall relative permittivity [23]. This permittivity's real and imaginary parts correspond

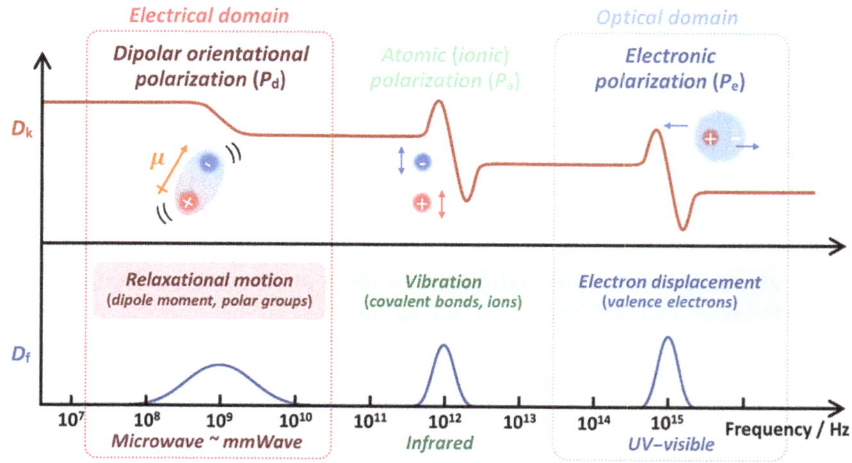

Figure 8.1 Frequency response of the dielectric mechanisms [23].

to the dielectric constant and dielectric loss. The dielectric constant measures the material's ability to store electrical energy. Lower dielectric constants are preferable for high-frequency applications as they reduce signal delay and energy loss [24]. The dielectric constant is typically measured using capacitance methods, where the polyimide sample is placed between two electrodes, and the capacitance is measured at various frequencies. Dielectric loss indicates the energy dissipated as heat within the material. Lower dielectric loss values are essential for minimizing energy loss in electronic circuits. Dielectric loss is measured alongside the dielectric constant using impedance analyzers, which provide data over various frequencies to evaluate the material's performance under different operating conditions [25]. Modifying the chemical structure or incorporating specific fillers can tailor the dielectric properties while maintaining or enhancing transparency. This holistic approach ensures that the developed materials meet the rigorous demands of advanced electronic and optoelectronic applications.

8.4 Advanced Design Strategies for Fabricating High-Transparency Low-Dielectric PIs

In recent years, significant progress has been made in synthesizing high-transparency, low-dielectric PIs. The charge transfer complexes (CTCs) formed by the electron donors and acceptors in the PI backbone, while endowing the material with excellent properties, also cause strong absorption in the visible range, giving it a yellow appearance [26, 27]. Additionally, the polar imide rings and CTC effects in the main chain exacerbate intermolecular polarization, increasing dielectric constant and dielectric loss, thereby limiting its dielectric performance. The main challenge in designing high-transparency, low-dielectric PIs is balancing their dielectric properties, optical transparency, and other properties (such as mechanical and thermal performance), as these properties are often contradictory [28, 29]. By introducing fluorinated groups [30, 31], alicyclic structures [32], bulky pendant groups (such as triphenyl, triphenylmethyl, and naphthyl) [33, 34], non-coplanar structures [35, 36], and silicon-based particles, researchers can significantly improve these materials' properties. This section will discuss these design strategies and performance analyses in detail, and explain the application of theoretical calculations and artificial intelligence in developing new materials.

8.4.1 Introduction of Fluorine Functional Groups

Incorporating fluorinated groups into polymer chains is an effective strategy for synthesizing highly transparent, low-dielectric PIs, with the trifluoromethyl group (CF_3) being the most common. The large volume of the CF_3 group hinders interactions between polymer chains and prevents the tight packing of molecular chains. Additionally, the low polarizability of the C-F bond helps to reduce the dielectric constant of PIs. The electron-withdrawing effect of the CF_3 group decreases the electron-donating ability of the diamine component, thereby reducing the formation of CTCs and enhancing the optical properties of the films [37, 38]. Hsiao et al.'s research focuses on improving the dielectric properties and optical transparency of PIs by introducing fluorinated groups, CF_3. They synthesized various CF_3-containing diamine monomers, including 1,3-, 2,6-, 1,1-, 2,7-, and 1,6-bis(4-amino-2-trifluoromethylphenoxy)naphthalene, and reacted them with different aromatic tetracarboxylic dianhydrides to create a series of novel fluorinated PIs [39–44]. These PIs exhibited significant advantages in solubility, mechanical properties, and thermal stability, with exceptional dielectric properties and optical transparency. The dielectric constant of the fluorinated PIs at 10 kHz reached as low as 2.49, mainly due to the bulky CF_3 groups that reduced the packing efficiency of polymer chains and increased free volume. Additionally, fluorine atoms' high electronegativity and low polarizability further contributed to lowering the dielectric constant. These PIs demonstrated excellent transparency, with UV–visible absorption cut-off wavelengths ranging from 364 to 436 nm. Yu et al. successfully synthesized a novel aromatic diamine, 2,2-bis[4-(2,4-diaminophenoxy) phenyl] hexafluoropropane (BDAPFP), and used it along with linear polyimide (LPAA0) from end-capped anhydrides to create a series of new star-shaped branched fluorinated PIs. The tetraamine BDAPFP, acting as a crosslinking point, reduced the formation of CTCs, resulting in robust, flexible, and transparent films with UV–visible absorption cut-off wavelengths between 366 and 398 nm. Furthermore, the CF_3 groups in the tetraamine structure hindered chain packing, increased free volume, and decreased polarity, thereby reducing dielectric loss. These star-shaped branched PI films exhibited dielectric constants ranging from 2.72 to 3.12 and dielectric losses between 0.0028 and 0.0040 at 1 MHz [45]. Jee et al. [46] prepared a series of PI films using 4,4′-(hexafluoroisopropylidene)diphthalic anhydride (6FDA) and bis(3-aminophenyl) sulfone (3DDS) as base materials, with varying amounts of terephthaloyl chloride (TPC) as a comonomer. All the prepared films exhibited high transparency with cut-off wavelengths ($\lambda_{cut\text{-}off}$) below 400 nm. The CF_3 groups in the 6FDA monomer and the sulfone (—SO_2—) groups in the 3DDS monomer effectively controlled the formation of CTCs in the PIs. As the TPC content increased, the transmittance improved, reaching around 90% at a TPC content of 10 mol%. Banerjee et al. [47] synthesized novel diamine monomers, 1,3-bis[3′-trifluoromethyl-4′-(4″-aminophenoxy)benzyl]benzene and 4,4′-bis[3′-trifluoromethyl-4′-(4-aminophenoxy)benzyl]biphenyl. The introduction of fluorine atoms reduced interactions between polymer chains, thereby decreasing the formation of CTCs. All PI films exhibited high optical transparency with cut-off wavelengths between 372 and 450 nm. Increased fluorine content enhanced the free volume between polymer chains, reducing the dielectric constant. These films demonstrated low water absorption (0.2%–0.7%) and low dielectric constants (2.74–3.2) at 1 MHz, along with high mechanical strength and modulus, making them suitable for applications in plastics, fibers, and membrane separations.

Huang et al. [48] synthesized a series of transparent aromatic PIs using 6FDA and biphenyl diamines with different substituents at the 2,2′ and 6,6′ positions (Figure 8.2). The results showed that introducing fluorine atoms or fluorinated groups effectively inhibited intermolecular and intramolecular charge transfer between PI chains, significantly enhancing the optical transparency of the PIs. Specifically, the film exhibited an average transmittance of nearly 90%. Additionally, larger substituents such as CH_3 and CF_3 increased the molar volume, reducing the dielectric constant.

Huang et al. [49] synthesized perfluorocyclobutyl biphenyl ether-based PI (PFCBBPPI) to develop a dielectric polymer with low viscosity, durability, and high thermal stability. The PFCBBPPI films exhibited a transmittance of 92% at 550 nm

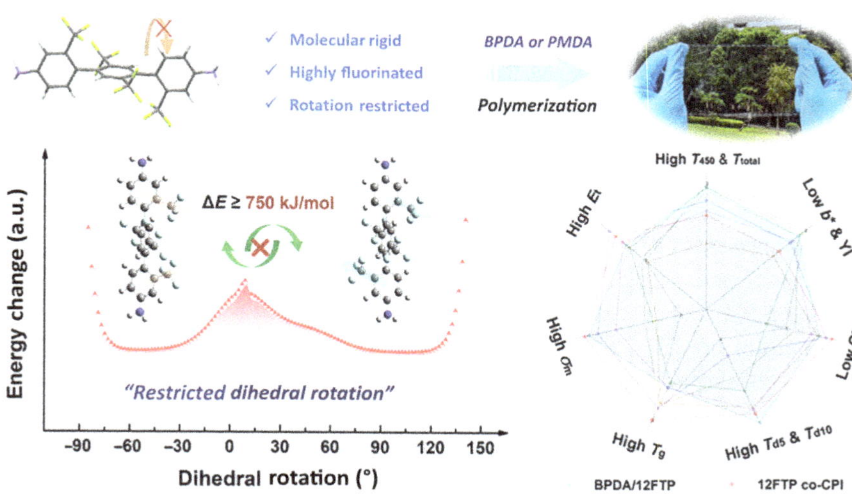

Figure 8.2 Molecular design of highly fluorinated aromatic diamines and unique geometric isomerism [48].

and a cut-off wavelength of 330 nm. The ε was 2.43 at 1 MHz, with an increase of less than 2% under varying humidity conditions. The highly hydrophobic PFCB aromatic ether group reduced water absorption, allowing PFCBBPPI to maintain a low dielectric constant.

Han et al. [50] synthesized a fluorinated diamine, 1,3-bis(4-amino-2-trifluoromethylphenoxy)benzene (mBATB), using resorcinol and 2-chloro-5-nitrobenzotrifluoride as starting materials, and prepared a series of soluble fluorinated aromatic PIs. The dielectric constants of the 6FDA-mBATB series ranged from 3.01 to 3.16 at 1 kHz. Compared to the corresponding rigid BTDA-based PIs, all fluorinated PIs with bulky CF_3 groups in the main chain had lower yellowness index (YI) values. The UV transmittance at 440 nm for ODPA and 6FDA was 82%–86%.

Liaw et al. [51] synthesized a novel diamine, 2,2-bis[4-(2-trifluoromethyl-4-aminophenoxy)-3,5-dimethylphenyl]propane (BTADP), which contains isopropyl, methyl-substituted aromatic ether, and trifluoromethyl groups. They then prepared a series of PIs via direct polycondensation with various aromatic tetracarboxylic dianhydrides. All PIs formed light-colored, optically transparent, and highly flexible films with low dielectric constants ranging from 2.84 to 3.09.

Wang et al. [52] designed and synthesized two novel diamines, 2,5-bis(2-trifluoromethyl-4-aminophenoxy)-1,4 : 3,6-dihydroxybenzene and 2,5-bis(2-methyl-4-aminophenoxy)-1,4 : 3,6-dihydroxybenzene, and reacted them with common dianhydrides to prepare PIs containing 1,4 : 3,6-dihydroxybenzene segments. These PIs exhibited transparency of up to 87% at 450 nm. The dielectric constants of these PIs ranged from 2.02 to 2.52 at 1 MHz. The introduction of bulky trifluoromethyl substituents effectively hindered the stacking of molecular chains and prevented intermolecular charge transfer. The presence of alicyclic segments in the PI backbone also weakened intermolecular and intramolecular charge transfer interactions due to the poor electron-donating ability of the isosorbide residues in the diamines.

Lu et al. [53] synthesized transparent PIs using fluorinated aromatic dianhydride 6FDA and three aromatic diamines with different amounts and positions of methyl side groups. These PIs exhibited moderate to high optical transparency and excellent solubility. The 2255TMB-PI had the highest transmittance of 82% at 400 nm due to the presence of four methyl groups at the 2,2′,5,5′ positions of the diamine. All these PIs demonstrated low dielectric constants ranging from 2.84 to 2.99.

Zhao et al. [54] successfully synthesized 2,2′-bis[4-(5-amino-2-pyridyloxy)phenyl] hexafluoropropane (6FBAPDP) and 2,2′-bis[4-(5-amino-2-pyridyloxy)phenyl] propane (BAPDP) and used them to prepare a series of new fluorinated PIs with various commercial aromatic dianhydrides. The resulting PI films exhibited excellent optical transparency, with UV–visible absorption cut-off wavelengths between 342 and 393 nm. The dielectric constants of these PI films ranged from 2.71 to 2.92 at 1 MHz. The fluorinated groups and pyridine units in the PI main chain increased the free volume between molecular chains and reduced the formation of CTCs, thereby enhancing the transparency of the polymers. The fluorinated groups also

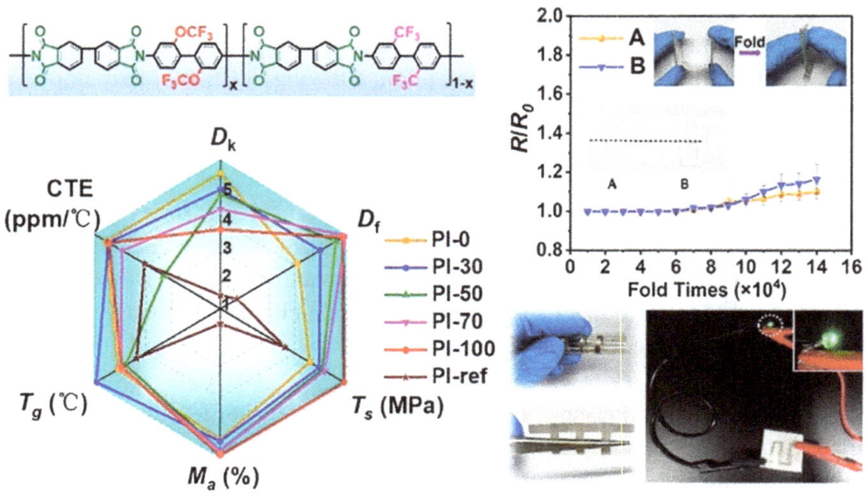

Figure 8.3 The derived PI films obtained by adjusting the molar ratio of TFMOB/TFMB and their excellent comprehensive performance [55].

had high electronegativity and low polarizability, which reduced the density and dielectric constant of the polymers.

Cheng et al. [55] synthesized a series of novel PI films containing trifluoromethoxy groups using 3,3′,4,4′-biphenyltetracarboxylic dianhydride (BPDA), 2,2′-bis(trifluoromethyl)biphenyl-4,4′-diamine (TFMB), and 2,2′-bis(trifluoromethoxy)biphenyl-4,4′-diamine (TFMOB) as raw materials (Figure 8.3). All the PI films exhibited excellent transparency, with transmittance exceeding 80% at 500 nm. The increased transparency was attributed to the introduction of fluorine atoms, which weakened the electron-donating effect of the diamine and disrupted the large π-π conjugation of the aromatic ring structure, thereby inhibiting the formation of CTCs in the molecular chains. As the TFMOB content increased, the dielectric constant of the PI films slightly increased, but the dielectric loss significantly decreased.

Chang et al. [56] synthesized polyamic acid (PAA) using 6FDA as the dianhydride, and 2,2-bis(3-aminophenyl)hexafluoropropane (6FAm) and 2,2-bis(3-amino-4-hydroxyphenyl)hexafluoropropane as diamines, and prepared PI composite films. All PI films had $\lambda_{cut\text{-}off}$ below 400 nm, indicating that they transmitted light before reaching the visible light region. The transmittance of the PI films ranged from 87% to 89%, with a very low YI (2–3), indicating that they were nearly colorless and transparent.

Ando et al. [57] synthesized two new diamine monomers, 1,4-bis(4-aminophenoxy)-2-[(3′,5′-ditrifluoromethyl)phenyl]benzene and 1,4-bis[2′-cyano-3′(4″-aminophenoxy)phenoxy]-2-[(3′,5′-ditrifluoromethyl)phenyl]benzene, using (3,5-ditrifluoromethyl)phenylhydroquinone as the starting material. All the polymers formed transparent, strong, and flexible films, with dielectric constants estimated from the average refractive index ranging from 2.69 to 2.89. The transparency of the polymers may be attributed to the flexible ether linkages in their molecular structures, which helped to reduce intermolecular interactions and aggregation, thereby enhancing the transparency of the polymers. The bis(trifluoromethyl) groups increased the free volume of the polymers, improved solubility and thermal stability, and reduced the dielectric constant.

8.4.2 Monomers Containing Cycloaliphatic Structures

Aliphatic structures, composed of single bonds, effectively reduce the electronic conjugation effects between molecular chains compared to rigid benzene ring structures. This suppression of charge transfer enhances the transparency of PIs. Table 8.2 shows the reported aliphatic/alicyclic-containing dianhydrides and aliphatic/alicyclic-containing diamines. Additionally, aliphatic or semi-aliphatic PIs exhibit lower packing density and molecular polarizability than aromatic PIs, demonstrating excellent dielectric properties and significant potential in optoelectronic display applications. Current research on aliphatic PIs focuses primarily on aliphatic dianhydrides. Chung et al. [58] studied the structure–property relationships of partially aliphatic copolymers based on aliphatic dianhydride (DAn) and aromatic dianhydrides. They found that incorporating a certain amount of flexible aromatic dianhydride into the PI backbone is essential for obtaining flexible PI films. Among these, PI-ODPA-10 films exhibited the highest transparency and colorlessness, as well as good flexibility. Fan et al. [59] synthesized a series of novel low-fluorine semi-aromatic PIs using 1,2,4,5-cyclohexanetetracarboxylic dianhydride (CHDA) and bicyclo[2.2.2]octane-7,2,3,5,6-tetracarboxylic dianhydride (BCDA) along with three aromatic diamines. These films demonstrated excellent optical transparency, with $\lambda_{cut\text{-}off}$ between 282 and 286 nm, transmittance above 89% at 500 nm, and a YI below 14.8. The improved optical transparency and color elimination are primarily due to the reduced electron-accepting capacity of the aliphatic dianhydrides, ortho-substitution, and the large steric hindrance caused by bulky pendant groups. The dielectric constants of these PIs ranged from 2.72 to 2.80. Tao et al. [79] synthesized a series of semi-aliphatic PIs using alicyclic dianhydrides and trifluoromethyl-containing aromatic diamines through high-temperature solution polymerization. Due to the low polarizability of the alicyclic groups and the large volume effect of the trifluoromethyl groups, the dielectric constant significantly decreased to 2.61–2.76. Furthermore, the semi-aliphatic PI films exhibited high transparency, with $\lambda_{cut\text{-}off}$ between 298 and 313 nm and transmittance above 85% at 500 nm. Chang et al. [80] synthesized six different types of polyamic acids by reacting 1,2,4,5-cyclohexane tetracarboxylic dianhydride with six diamine monomers, followed by various thermal treatments to prepare PI films. The results indicated that films containing -CF_3 substituents exhibited excellent optical transparency and ease of solubility, with a YI $\leq$ 4.1. Ando et al. [81] synthesized partially alicyclic PIs using Tröger base (TB) as a core structure (Figure 8.4). The resulting PIs were easily soluble in common organic solvents, and exhibited good mechanical properties, low dielectric constants (~2.66), and good thermal stability. The films were completely colorless and transparent, with transmittance above 77% at 400 nm. Hulubei et al. [82]

Table 8.2 Chemical structures of reported aliphatic/alicyclic-containing dianhydrides and aliphatic/alicyclic-containing diamines.

Aliphatic/alicyclic-containing dianhydrides.	References	Aliphatic/alicyclic-containing diamines.	References
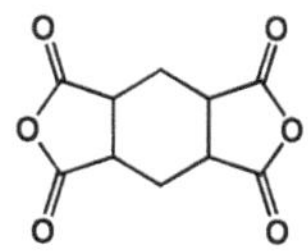	[60]	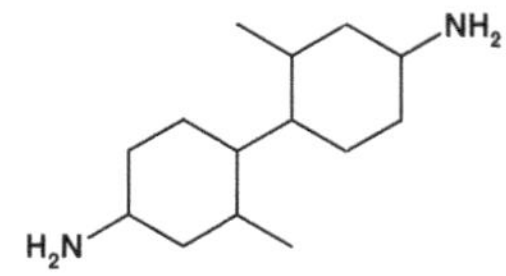	[61]
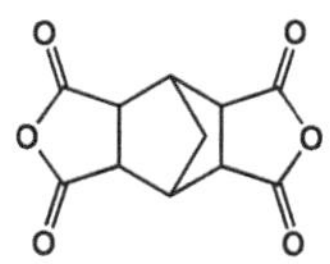	[62]	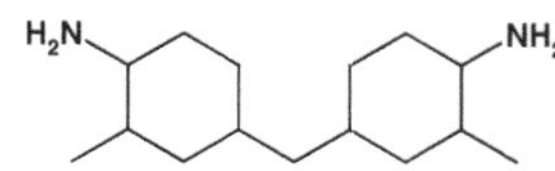	[63]
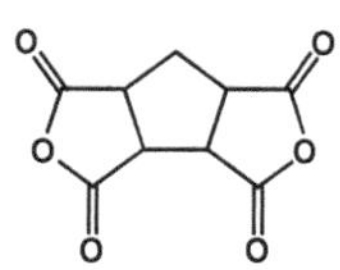	[64]	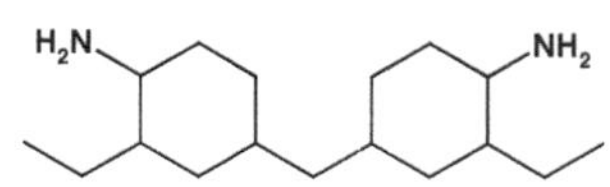	[65]
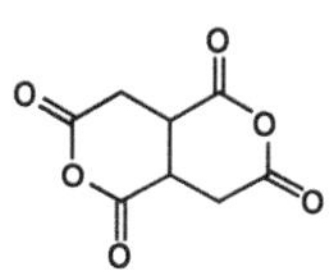	[66]	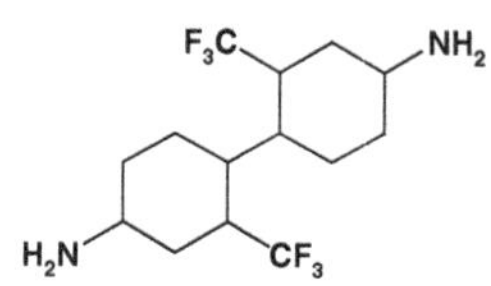	[61]
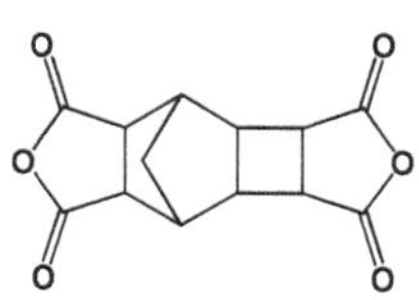	[67]	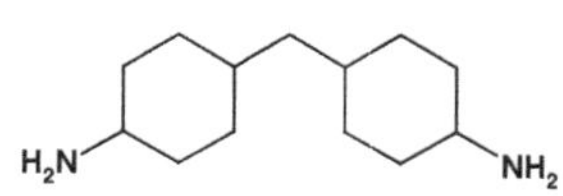	[68]
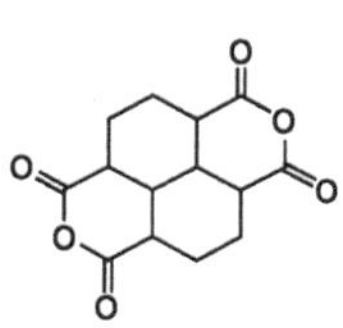	[69]	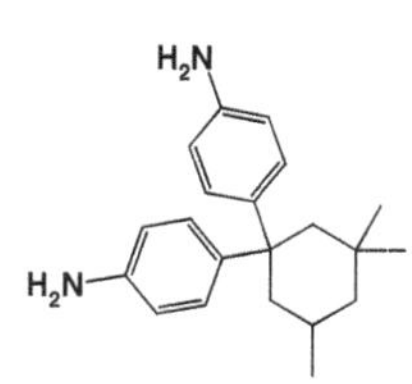	[70]
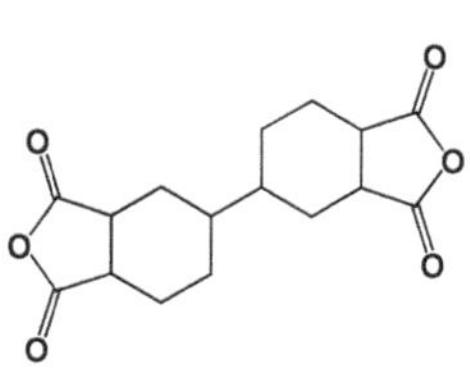	[71]	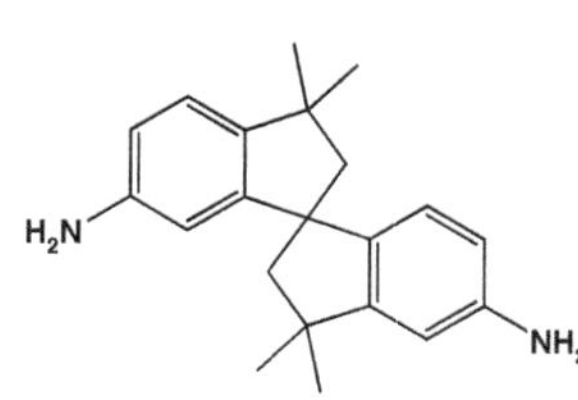	[72]

(Continued)

Table 8.2 (Continued)

Aliphatic/alicyclic-containing dianhydrides.	References	Aliphatic/alicyclic-containing diamines.	References
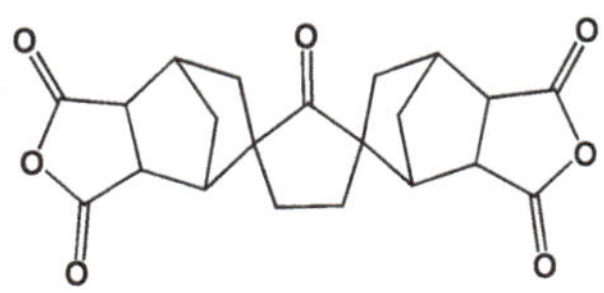	[73]	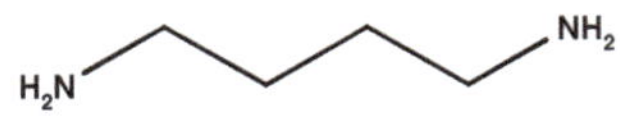	[74]
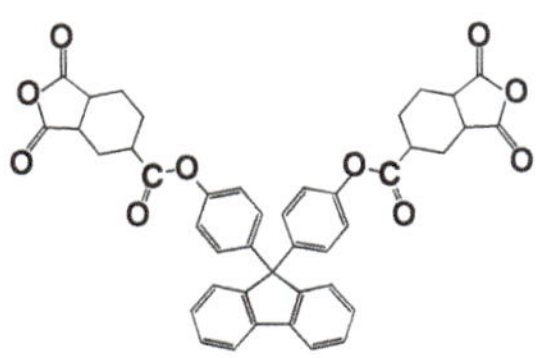	[75]	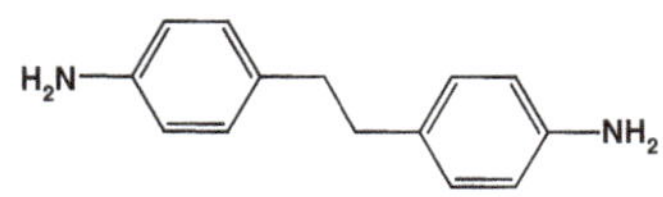	[76]
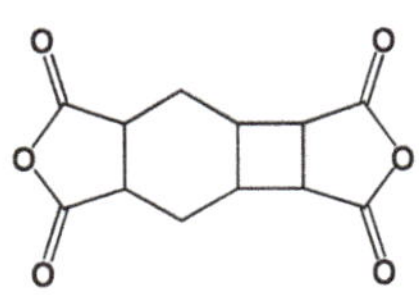	[77]	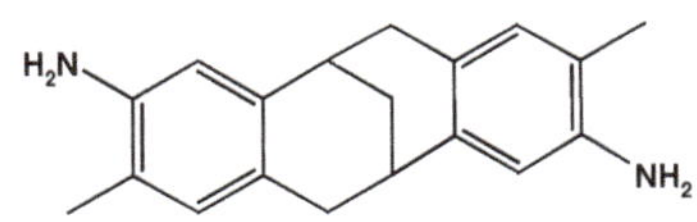	[78]

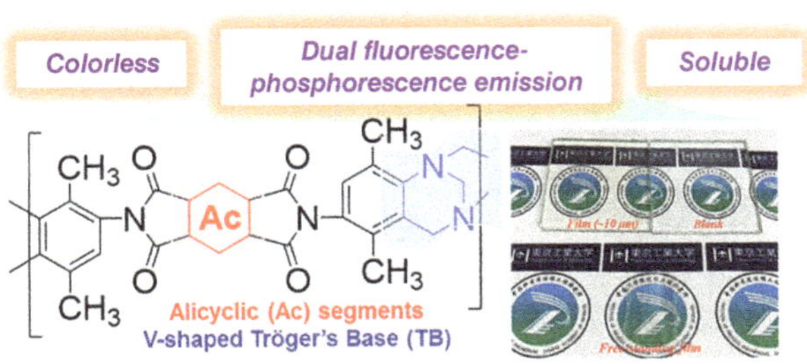

Figure 8.4 The synthesis of partially alicyclic PIs based on TB demonstrates a broad application prospect in the field of solar energy conversion [81].

designed certain aliphatic/aromatic PIs using rigid, dimensionally, and symmetrically different configurations of dihydride monomers. The use of semi-flexible and asymmetric dianhydride units reduced intermolecular polarization and CTC interactions, contributing to lower dielectric constants. Additionally, the larger intermolecular distances and asymmetric structures also helped lower the dielectric constants.

Besides aliphatic dianhydrides, aliphatic diamines and semi-aliphatic dianhydrides have also been developed. Hasegawa et al. [83] polymerized CHDA with pyromellitic dianhydride (PMDA) and 1,2,3,4-cyclobutanetetracarboxylic dianhydride (CBDA), followed by thermal imidization to obtain flexible and transparent PI films. The fully alicyclic CBDA/CHDA films showed lower CTE and a relatively low dielectric constant (2.66). Zhao et al. [84] successfully synthesized a semi-aromatic tetracarboxylic dianhydride, 1,2,3,4,5,6,7,8-octahydroanthracene-2,3,6,7-tetracarboxylic dianhydride (OHADA). They polymerized OHADA with 2,2′-bis(trifluoromethyl)biphenylamine (TFDB) and 9,9′-bis(4-aminophenyl) fluorene (BAFL) in m-cresol, obtaining two types of PIs. Both PIs exhibited good optical transparency, with $\lambda_{cut\text{-}off}$ of 315 nm and 305 nm, and transmittance above 74%/72% at 400 nm and 84%/80% at 500 nm, respectively. Wang et al. [85] synthesized three diamine monomers containing pyridine or methyl-substituted pyridine and cyclohexane. The prepared PI films exhibited transmittance between 94% and 97% at 500 nm and $\lambda_{cut\text{-}off}$ ranging from 345 to 359 nm. Introducing methyl, cyclohexane, and pyridine groups into the main chain improved the solubility and transparency of PI films. Additionally, they designed and synthesized two novel monomers containing cis-HBPA units: 4,4′-bis(4,4′-isopropylidenebiscyclohexyloxy)biphenyl dianhydride (HBPADA) and 4,4′-bis(4,4′-isopropylidenebiscyclohexyloxy)biphenylamine. PIs were obtained from HBPADA and five aromatic diamines, while PI-6 was obtained from HBPADA and a semi-aromatic diamine. All PIs formed flexible, tough, and transparent films with transparency up to 86% at 450 nm [86].

8.4.3 Monomers with Bulky and Pendant Substituents

Researchers have explored the mechanisms by which bulky side groups and pendant substituents affect the properties of PIs through molecular design and structural optimization. Bulky side groups typically refer to larger substituents that increase the molecular volume and spatial separation between molecules, facilitating the orderly arrangement of polymer chains. This reduces photon scattering between molecules, thereby enhancing transparency. In terms of dielectric properties, bulky side groups reduce intermolecular interactions and overall polarization, resulting in lower dielectric constants.

Zhang et al. [87] synthesized three PIs with identical bipyridyl cores but differing numbers of phenyl rings in the side chains. Their dielectric constants were 2.81 (PPy6F), 2.61 (mBPPy6F), and 2.44 (mTPPy6F), showing a decreasing trend with an increasing number of phenyl rings in the side chains (Figure 8.5). Cheng et al. [88] synthesized a novel PI using benzocyclobutene (BCB) and a trifluoromethyl-functionalized diamine monomer (BFDA), producing a series of copolyimides with varying BCB content. The PI-BFs exhibited excellent transparency, with $\lambda_{cut\text{-}off}$ ranging from 383 to 397 nm and transmittance at 700 nm between 85.7% and 88.3%. The bulky and low-polarizability trifluoromethyl groups increased the free volume and reduced the polarizability between molecules, resulting in dielectric constants ranging from 2.72 to 2.95 at 1 MHz. Wang et al. [89] used a one-step solution polymerization method to synthesize PIs with asymmetric pendant groups. These films exhibited low dielectric constants and high thermal stability, with dielectric constants ranging from 2.68 to 3.20 (at 1 MHz). Their UV $\lambda_{cut\text{-}off}$ were below 358 nm, and 80%

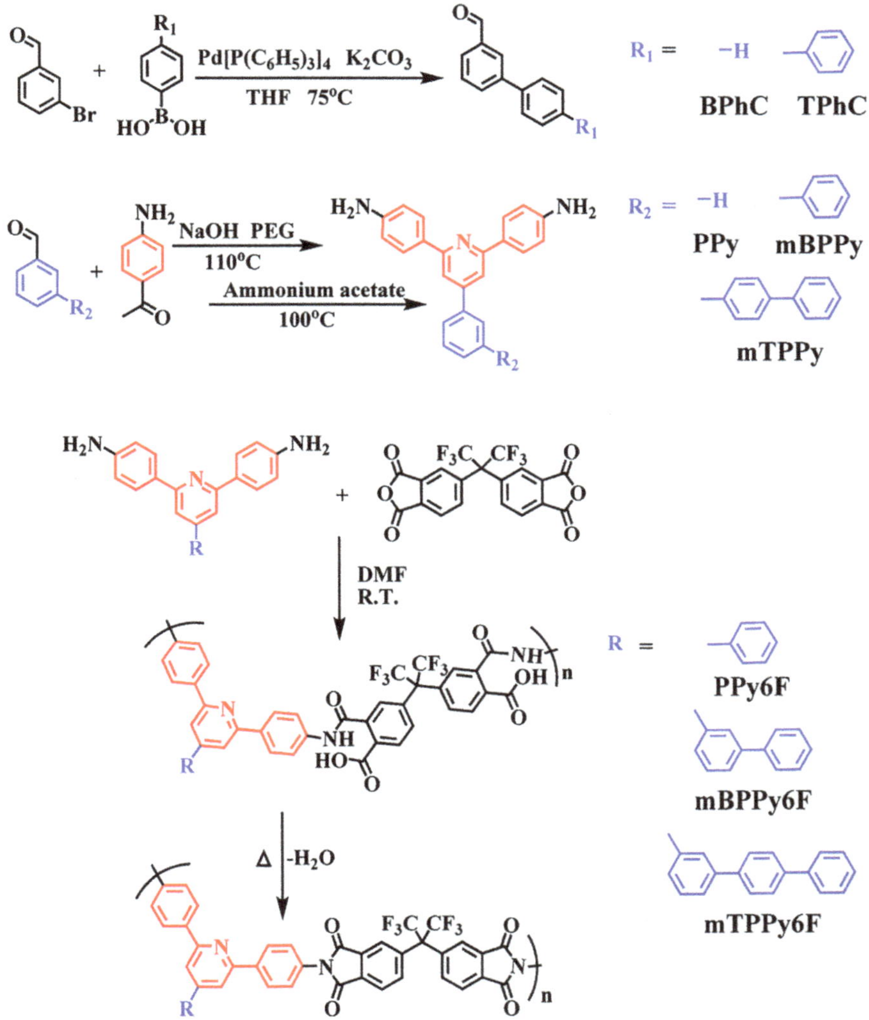

Figure 8.5 Three aromatic diamines (PPy, mBPPy, mTPPy) with a rigid nonplanar conjugated diphenylpyridine core and corresponding PIs [87].

transmittance wavelengths ranged from 431 to 468 nm. Li et al. [90] successfully synthesized a novel diamine monomer containing pyridine, 4-phenyl-2,6-bis [3-(4′-amino-2-trifluoromethylphenoxy)phenyl]pyridine (m-PAFP), and used various aromatic dianhydrides to synthesize a series of fluorine-containing pyridine-bridged polyetherimides. These PIs formed transparent flexible, and robust films, with dielectric constants ranging from 2.64 to 2.74 at 10 MHz. The synergistic effect of bulky side groups and pendant substituents significantly enhances the performance of PIs. This synergy can simultaneously improve transparency and reduce the dielectric constant. Hsiao et al. [91] used 1,4-bis(3,4-dicarboxyphenoxy)triphenyl dianhydride as a raw material and adopted a conventional two-step method to synthesize a series of new polyetherimides containing three-dimensional triphenyl structures. These triphenyl polyetherimides exhibited enhanced optical transparency, with $\lambda_{cut\text{-}off}$ below 400 nm and dielectric constants ranging from 2.22 to 3.40 at 1 MHz. Huang et al. [92] synthesized a series of new organic-soluble, optically transparent PIs using a novel diamine, 3,3′-dibutyl-4,4′-diaminodiphenyl-4′-morpholinophenylmethane (TAMPM), and various commercial aromatic dianhydrides via a one-step polymerization method. These PIs had $\lambda_{cut\text{-}off}$ ranging from 287 to 344 nm, excellent thermal stability, and high optical transparency. Additionally, all PIs exhibited low dielectric constants, ranging from 2.85 to 3.16 at 1 MHz.

8.4.4 Monomers with Non-coplanar Structures

Rigid non-coplanar structures, including fluorine units, spiro structures, adamantane structures, anthracene, and carbazole groups, can significantly enhance the transparency and dielectric properties of PIs. These structures disrupt the planar conjugation of polymer chains, limiting the formation of intermolecular and intramolecular CTCs. The reduction of CTCs effectively lowers the material's polarizability, thereby decreasing the dielectric constant. Additionally, non-coplanar structures prevent tight packing of polymer chains, increasing free volume, which helps reduce the dielectric constant and minimizes light scattering, enhancing the material's transparency. This enhanced transparency and dielectric properties are crucial for advanced electronic and optoelectronic applications. Sheng et al. [93] synthesized a new trifluoromethylated diamine monomer, 9,9-bis[4-(4-amino-2-trifluoromethylphenoxy)phenyl]xanthene (BATFPX), from 2-chloro-5-nitrobenzotrifluoride and 9,9-bis(4-amino-2-trifluoromethylphenoxy)benzene (Figure 8.6). Using a one-step polycondensation method in m-cresol, they prepared a series of novel fluorinated PIs. The resulting PI films were transparent and flexible, with low moisture absorption (0.28–0.51%) and a low dielectric constant (2.85–3.26 at 1 MHz). These films exhibited excellent optical transparency, with UV–visible cutoff wavelengths ranging from 352 nm to 410 nm. Spiro and spiral structures introduce significant twists in the polymer chains, which disrupt conjugation and enhance transparency and dielectric properties. Lu et al. [94] synthesized four new spirobisbenzoxazole-structured diamine isomers and polymerized them with 6FDA and CBDA to prepare PI films. These films showed excellent optical transparency with cutoff wavelengths between 346 nm and 380 nm and transmittance of 85% to 90% at 550 nm. The films also exhibited low dielectric constants (2.32–2.93). Li et al. [95] synthesized a novel spiral (fluorene-9, 9′-xanthene) backbone diamine monomer, 2, 7-bis (4-amino-2-trifluoromethylphenoxy)-spiral(fluorene-9,9′-xanthene), and polymerized it with various commercial aromatic tetracarboxylic dianhydrides through thermal imidization. The resulting PIs formed transparent, flexible, and robust films with moisture absorption of 0.35–0.64%, dielectric constants of 2.63–3.09 at 1 kHz, and UV–visible absorption cutoff wavelengths of 356–384 nm. The spiral backbone structure and trifluoromethyl groups reduced CTC formation,

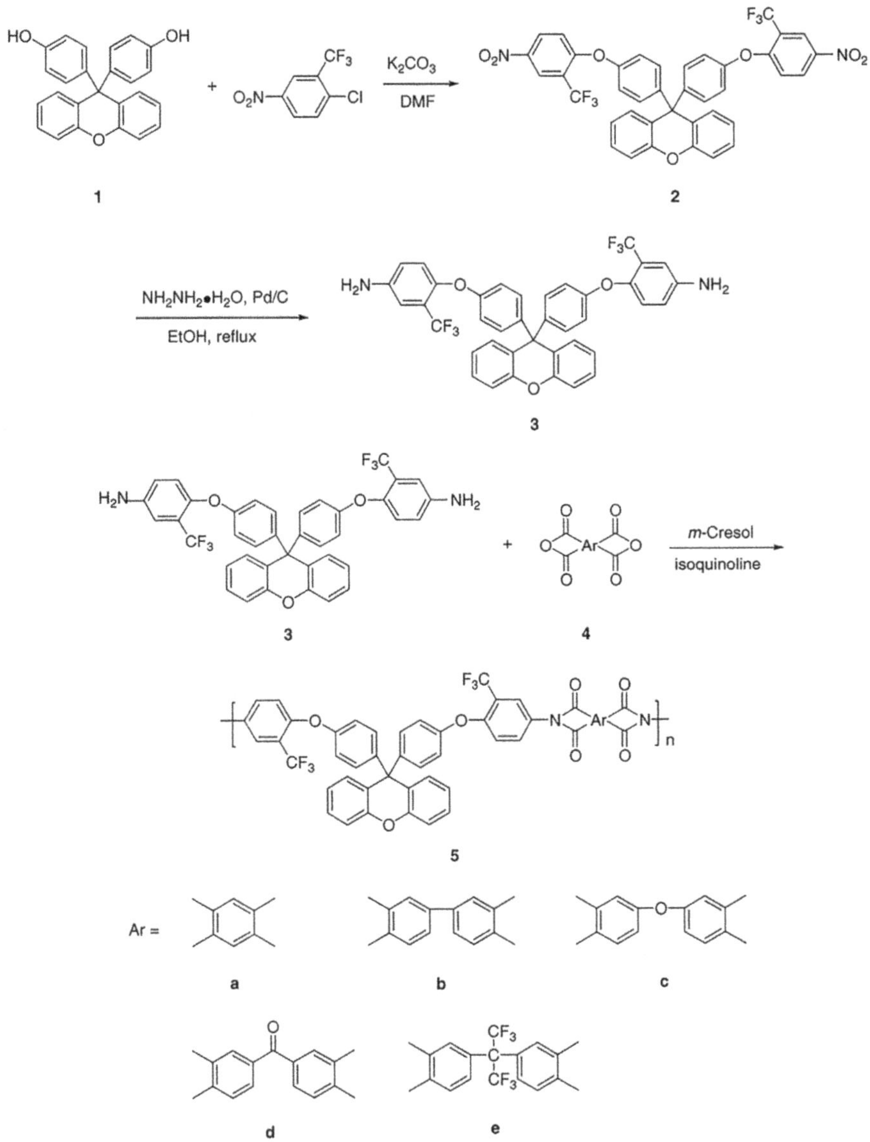

Figure 8.6 Synthetic route to BATFPX (3) and PIs 5a–5e [93].

enhancing transparency. Adamantine and diamantine structures are notable for their rigidity and non-coplanar nature, which prevents tight packing and reduces polarizability. Chern et al. [96] synthesized new adamantane-based PIs by reacting 1,3-bis[4-(4-aminophenoxy)phenyl]adamantane. These films exhibited low dielectric constants (2.77–2.91) and low moisture absorption (below 0.46%), forming tough, transparent films. The non-coplanar and rigid adamantane structure reduced polymer chain packing

Density, Enhancing Transparency And Dielectric Properties. Additionally, The Same Group Prepared Diamantane-Based Pis Using 1,6-Bis[4-(4-Aminophenoxy) Phenyl]Diamantane, Which Further Reduced The Dielectric Constant (2.56–2.78) Due To The Hydrophobicity And Low Polarity Of The Diamantane Units [97].

Anthracene and carbazole groups effectively increase rigidity and reduce CTC formation, leading to improved transparency and dielectric properties. Busireddy et al. [98] used ester-functionalized anthracene diamine monomers to synthesize a series of anthracene-based PIs with four commercial dianhydrides (Figure 8.7a). These films exhibited good optical transparency (Figure 8.7b), flexibility, thermal expansion coefficient (CTE), and mechanical properties. The non-planar and rigid anthracene structure resulted in a larger molar volume, increasing free volume and reducing the dielectric constant (Figure 8.7c). Ha et al. [99] synthesized new soluble PIs functionalized with carbazole groups on the side chains. These PIs showed excellent transparency in the visible range (below 400 nm) and dielectric constants below 3. The introduction of carbazole groups increased molecular rigidity and enhanced solubility and transparency through planar resonance structures. Additionally, the side chains introduced charge transport pathways, further reducing the dielectric constant. The use of hydroxyfluorene groups enhances transparency and dielectric properties by reducing CTC formation and increasing free volume. Zhao et al. [100] modified hydroxyfluorene-containing PIs through a simple esterification reaction, producing a series of hydroxyfluorene Cardo PIs. The modified PI films showed a significant increase in transmittance at 400 nm from 41.6–54.5% to 80.1–83.5%, and exhibited bright fluorescence. The dielectric constant at 1 MHz decreased to 2.58–2.90.

8.4.5 The Influence of Silicon-Based Particles

Incorporating silicon-based particles into PI matrices has become an effective strategy for enhancing dielectric properties and transparency. This modification is crucial for applications such as flexible electronics and high-frequency communication devices, where materials with low dielectric constants and high optical transparency are in high demand. The simultaneous improvement of transparency and dielectric properties of PI by silicon-based particles is based on the following elements. First, Si-O-Si bonds have lower polarizability than carbon-based bonds, reducing the overall dielectric constant. Second, the bulky nature of silicon particles introduces free volume within the polymer matrix, decreasing the density of polarizable groups and further lowering the dielectric constant [101]. Third, silicon groups, such as silanes, increase the hydrophobicity of PI films, reducing moisture absorption. Since moisture absorption significantly increases the dielectric constant (Figure 8.8), reducing it helps maintain low dielectric properties [102, 103]. Finally, silicon particles enhance the dispersion of the inorganic phase within the organic polymer matrix, reducing light scattering and improving transparency. For instance, PI films modified with 3,3′-(1,1,3,3-tetramethyldisiloxane-1,3-diyl)bis(propan-1-amine)

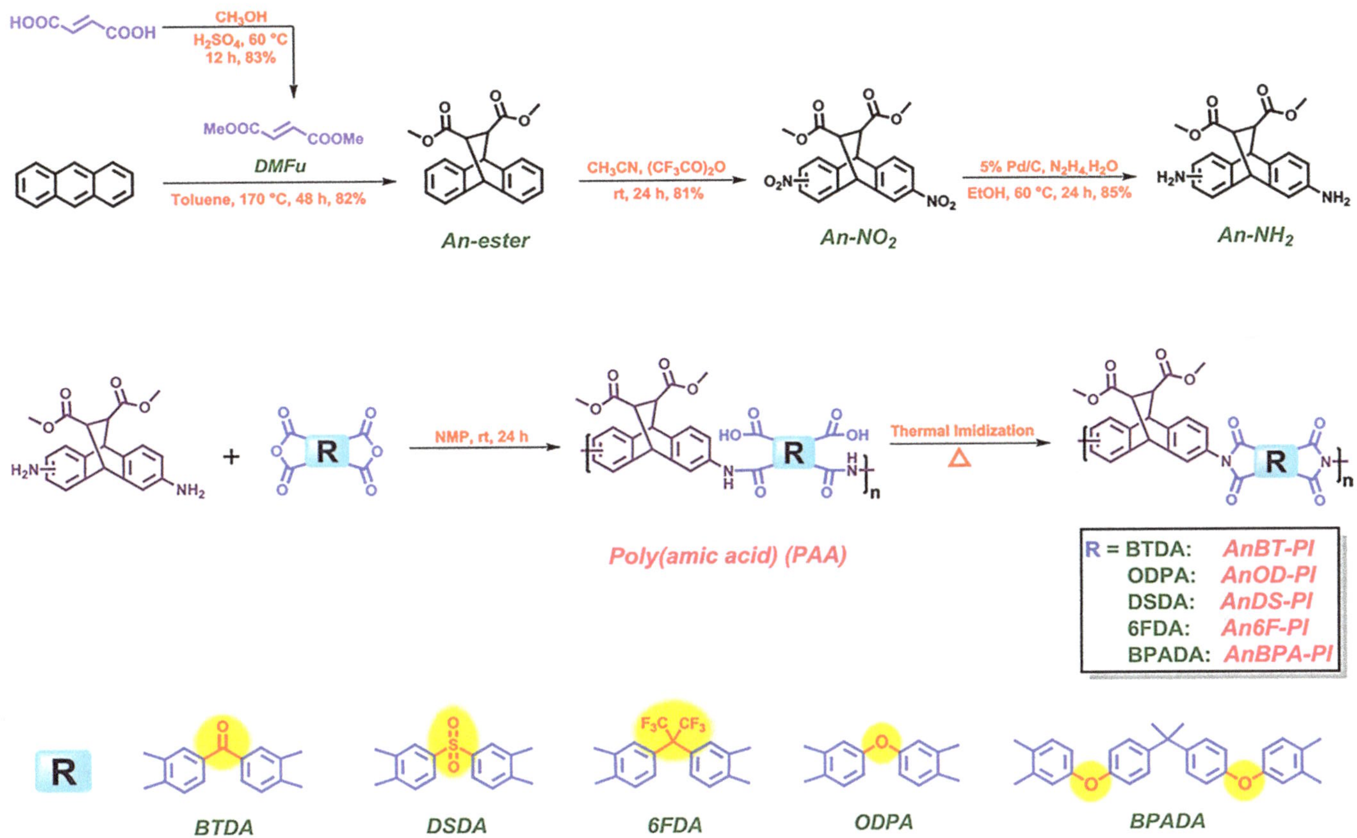

(a)

Figure 8.7 High thermally stable anthracene-based PIs with low dielectric constants and dissipation Factors [98]: (a) Synthetic routes of An-NH_2 Monomers and An-Based PIs. (b) UV–visible spectra of all An-based PI films. (c) Demonstration of the D_k and D_f values (measured at a frequency of 10 GHz) of An-based PI films.

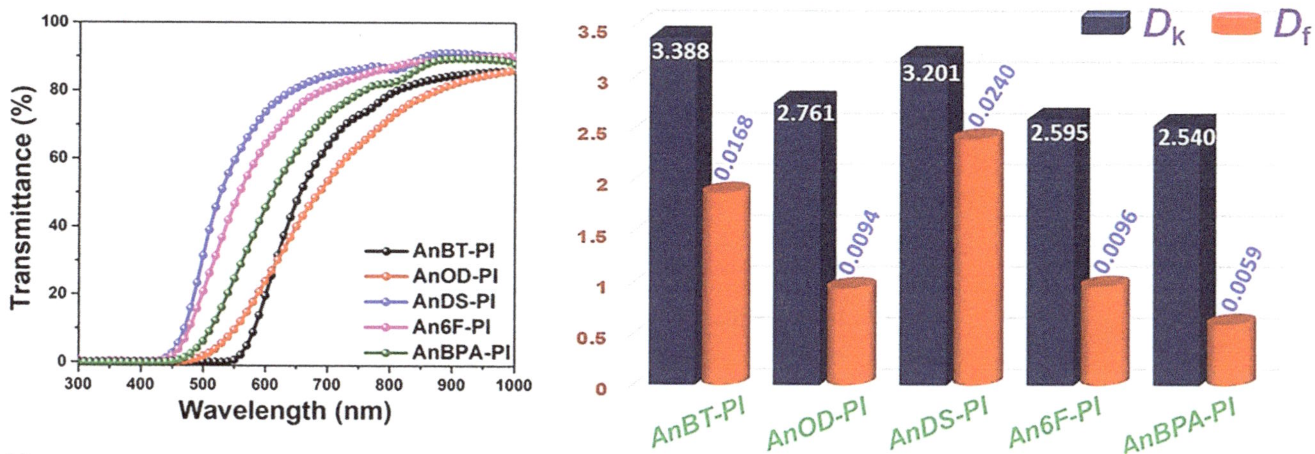

(b)

Figure 8.7 *(Continued)*

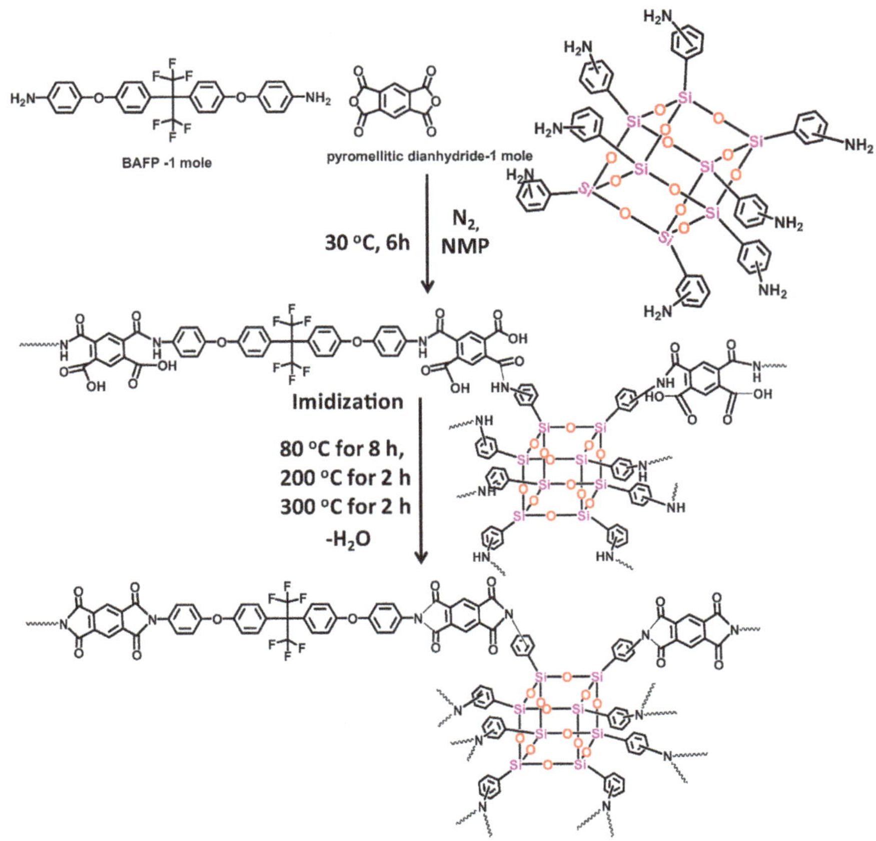

Figure 8.8 Preparation of PI nanocomposites with OAPS [103].

(APDS) demonstrated a dielectric constant as low as 2.48 and a dielectric loss (tan δ) of 0.00335 at 10 GHz, showcasing a significant improvement over unmodified PIs [104]. Introducing silane into the PI structure not only reduces the dielectric constant but also enhances the stability of dielectric properties across different frequencies. The dielectric constants of silane-containing PIs remain stable even at high frequencies, which is critical for high-speed communication technologies like 5G. The low polarizability of silicon atoms and their stable electronic structure contribute to these desirable properties. In addition to improving dielectric properties, silicon-based particles also enhance the transparency of PI films. The interaction between silicon compounds and the PI matrix reduces micro-phase separation, which often causes opacity in polymer films. For example, a hybrid PI composite containing 30% silane-modified hyperbranched PI and 20% SiO_2 exhibited a transmittance of 95% at 450 nm, a significant improvement compared to composites without silicon modification [105].

In conclusion, incorporating silicon-based particles into PI matrices is an effective strategy for developing materials with low dielectric constants and high transparency. The modification enhances the comprehensive performance of PIs and opens up new possibilities for their use in advanced optoelectronic devices.

8.4.6 Porous and Composite Structures in PIs

Another way to boost the transparency and dielectric properties of PI is by incorporating porous materials and composites into the PI matrix. The improvement mechanisms can be attributed to several factors:

Introduction of Porous Structures: The incorporation of porous structures increases the free volume within the PI matrix and introduces low-dielectric constant air, effectively reducing the overall dielectric constant of the material. For instance, when the porosity increases from 0 to 16.50%, the dielectric constant of porous PI films decreases from 3.07 to 2.13 [106]. On this basis, introducing low-dielectric constant materials such as polytetrafluoroethylene (PTFE) can further reduce the dielectric constant of PI. Studies have shown that when

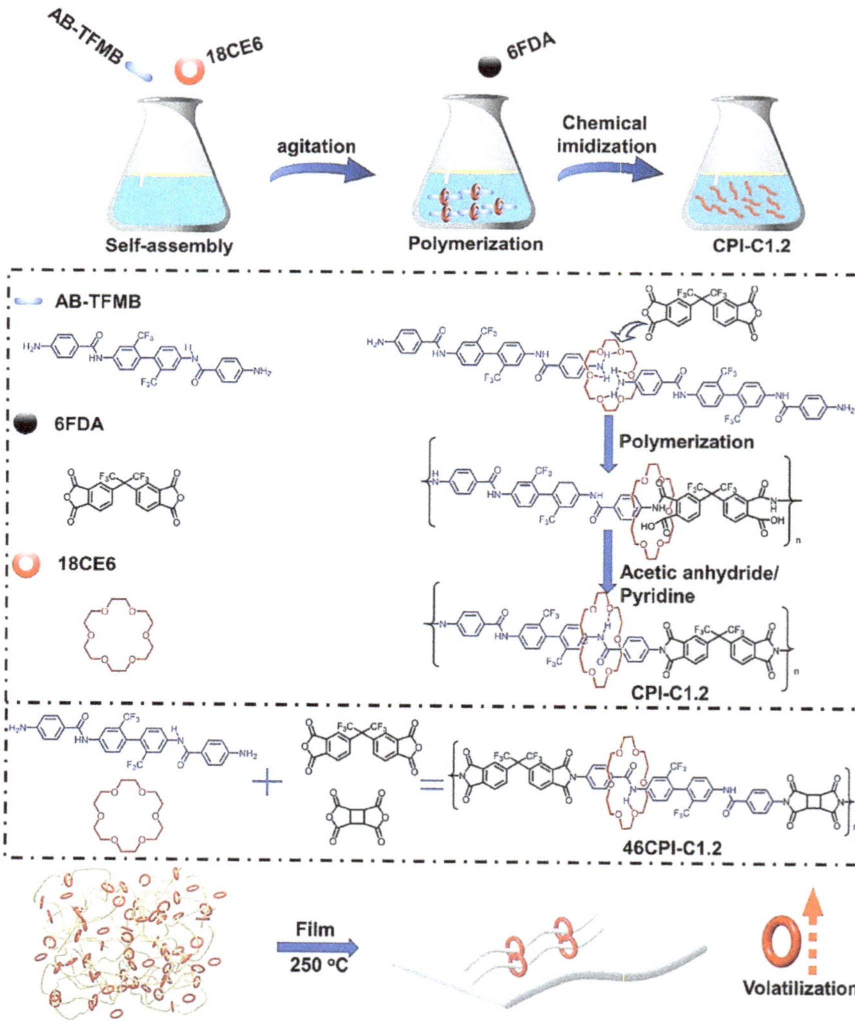

Figure 8.9 Synthesis process of CPI/18 CE6 composite films [109].

the PTFE filler content reaches 30 wt%, the dielectric constant of PI/PTFE-30 composite films decreases to 2.42 [107].

Incorporation of Low-Dielectric Constant Materials: Embedding materials with low dielectric constants, such as PTFE or crown ethers (Figure 8.9) [108–110], disrupts the dense packing of PI chains, increases free volume, and reduces interchain polarization. Furthermore, the interfaces between different materials in composites can scatter light, enhancing transparency and reducing the dielectric constant. For example, Kim et al. developed a ternary composite of hyperbranched and linear PI with SiO_2, achieving both a low dielectric constant and high transparency [111]. The introduction of nanofillers such as graphene oxide (GO) can further optimize the dielectric properties of PI.

Thermal Decomposition of Additives: The thermal decomposition of additives like polyethylene glycol (PEG) creates defined pores while maintaining the mechanical integrity of PI films, which is crucial for maintaining structural and functional stability under operational conditions [106]. For example, the thermal decomposition of PEG in the PI matrix can significantly lower the dielectric constant, even down to 1.85 [112].

Molecular Necking Strategy: using strategies like molecular necking and inserting fluorinated graphene can significantly improve the dielectric constant of PI films without sacrificing mechanical properties. Transparent PI films prepared using the molecular necking strategy exhibit excellent mechanical strength and toughness while maintaining high transparency and a low dielectric constant [113].

These studies indicate that optimizing the structure of the PI matrix and incorporating porous or composite materials can significantly enhance the dielectric properties of PI, making it more suitable for advanced electronic devices and high-frequency circuits. These improvements not only help lower the dielectric constant but also enhance the transparency and thermal stability of the material, offering new possibilities for PI applications in modern electronic products.

8.4.7 Polyimide Research Driven by Molecular Simulation and Data Science

In recent years, molecular simulation and data science techniques (Figure 8.10) have significantly advanced the development of low ε and transparent PIs [114, 115]. These methods provide a more detailed and accurate understanding of the relationship between molecular structure and material properties, which is critical for optimizing the dielectric performance and transparency of PIs. The integration of molecular simulation and data science offers several advantages over traditional experimental methods. Simulations can systematically vary molecular parameters and predict their impact on dielectric properties, instead of time-consuming and challenging experiments. These methods eliminate variability and potential errors associated with different experimental conditions, ensuring more consistent and reproducible data. Additionally, simulations can explore a wider range of molecular

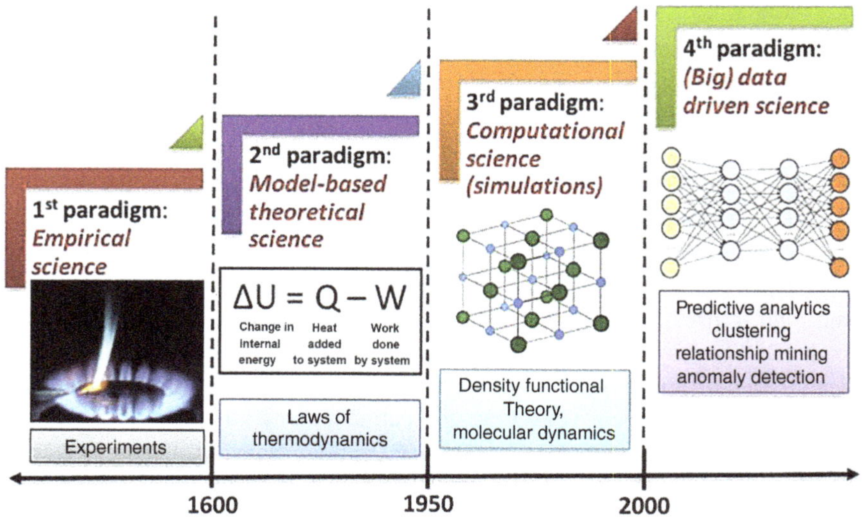

Figure 8.10 The four paradigms of science: empirical, theoretical, computational, and data-driven [115].

structures and compositions than is feasible experimentally. This broad exploration accelerates the discovery of new materials with optimized properties, reducing the time and cost associated with experimental trials.

For example, the dielectric constant of PIs can be accurately predicted using density functional theory (DFT), with results closely matching experimental values [116]. Molecular dynamics (MD) simulations provide insights into how fluorine substituents affect molecular polarization, fractional free volume (FFV), and interactions with water molecules, all of which are crucial for lowering the dielectric constant [117]. These simulations have shown that introducing fluorine atoms not only reduces the dielectric constant but also enhances optical transparency by disrupting CTCs and reducing crystallinity [118]. This dual improvement is essential for applications in high-frequency communication technologies and optoelectronic devices. Furthermore, data science techniques, particularly machine learning, have enhanced the understanding and development of PIs. Machine learning models can predict the properties of PIs by analyzing large datasets of molecular structures and their corresponding properties. For instance, Ramprasad et al. [119] developed a machine learning-based model capable of instantaneously and accurately predicting the frequency-dependent permittivity of polymers across 15 orders of magnitude. This model was trained on a dataset consisting of experimental measurements of ε values at 1210 different frequencies, using an advanced polymer fingerprinting scheme and a Gaussian process regression algorithm. The developed ML model was employed to predict the ε values of 11,000 potential polymers within the frequency range of 60–10^15 Hz (Figure 8.11). Tao et al. [120] created a large dataset containing 8 million hypothetical PIs and used machine learning models to identify the top three candidates with high glass transition temperature (Tg), Young's modulus (E), and yield strength (σ_y). Zhang et al. [121] developed a rapid

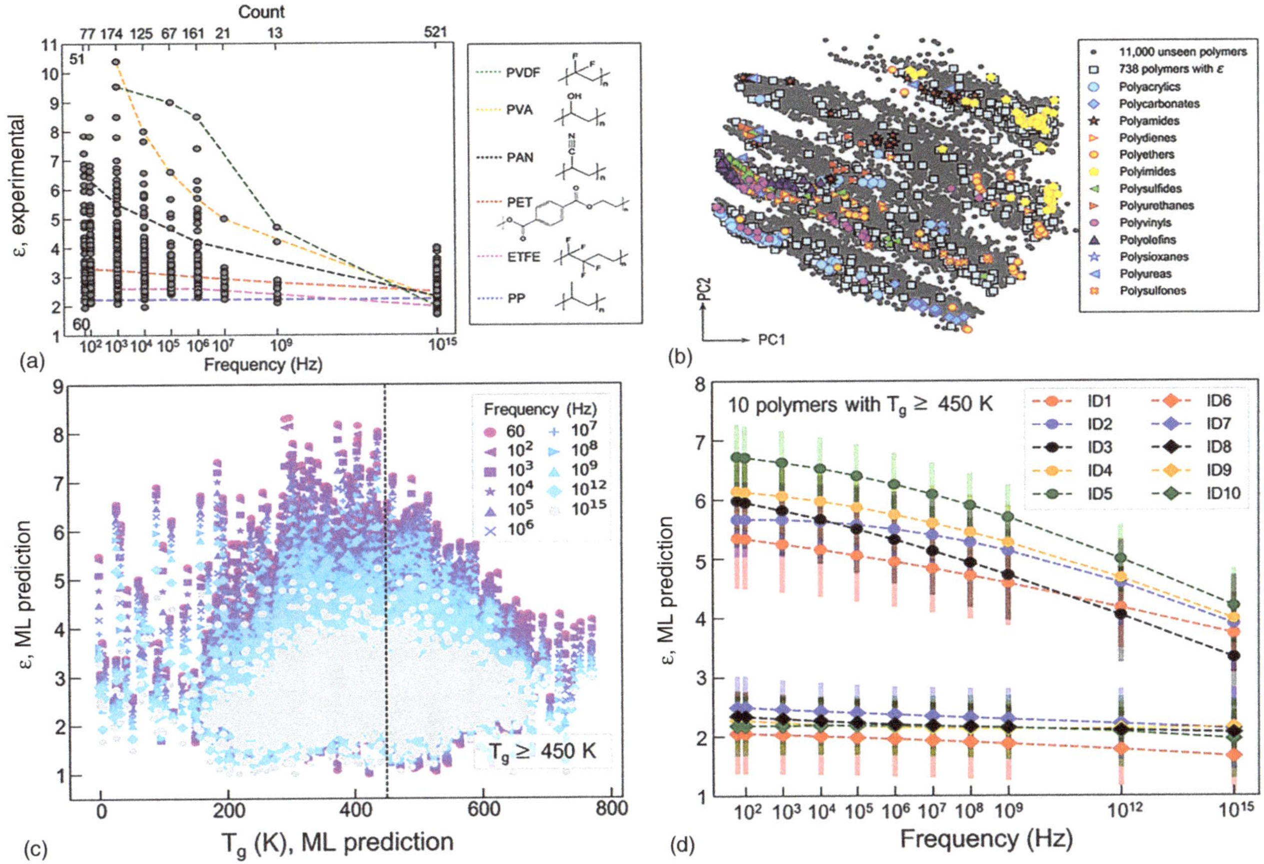

Figure 8.11 (a) Experimental ε as a function of the frequency, along with the data count at each frequency. The trends in ε values of six representative polymers are also shown using dashed lines. (b) Chemical space of the training set (738 polymers) considered this work (light blue squares), with respect to a larger unseen dataset of 11,000 polymers (gray circles), illustrated using the first two principal components (PC1 and PC2). (c) ML predicted ε at various frequencies for 11,000 unseen polymers, along with their ML predicted T_g values. (d) Ten representative polymers with high T_g (≥450 K) selected from a, such that five polymers (ID 1–5) have high ε (≥5), and remaining five (ID 6–10) have low ε (2–2.5) [119].

and accurate machine learning method to design three PI candidates with desired Tg values and experimentally validated the model's predictive accuracy.

These studies demonstrate that machine learning techniques have tremendous potential in accelerating the discovery and optimization of high-performance PIs. They not only improve accurate and effective prediction but also provide important chemical insights and design guidance.

8.5 Summary and Outlook

In this review, we discussed the design and synthesis of high-transparency, low-dielectric PIs, emphasizing the balance between dielectric and optical properties. Key findings include the importance of structural design in achieving desirable properties and the role of process parameters in optimizing performance. Specifically, the incorporation of fluorinated groups (such as trifluoromethyl, partially fluorinated, and fully fluorinated structures), alicyclic structures, bulky pendant groups (such as triphenyl, triphenylmethyl, and naphthyl), non-coplanar structures (such as spiro and cardo), and silicon-based particles have been widely applied in the development of high-transparency, low-dielectric PIs. These structural designs significantly reduce the polarization effects between molecules, increase free volume, and decrease CTCs, thereby lowering the dielectric constant and enhancing the transparency and dielectric performance of PIs.

Despite significant progress, challenges remain in terms of material stability, production cost, and achieving the ideal balance of properties. The introduction of fluorinated monomers, while effective in reducing dielectric constants, often comes with high costs. Cycloaliphatic structures can improve transparency but may suffer from lower thermal stability and reactivity. The inclusion of inorganic particles can enhance properties but often at the expense of mechanical flexibility. Future research should focus on further optimizing material properties and exploring new application areas. The difference in the position of diamine substitutions, where ortho-substituents favor high T_g transparent PIs and meta-substituents favor low-dielectric transparent PIs, needs deeper investigation. Comparative analysis of different structural design methods is necessary, considering the pros and cons of each approach, such as the high cost of fluorinated monomers and the reduced thermal stability of cycloaliphatic monomers.

Moreover, the role of molecular simulation and machine learning in the development of these materials is expected to grow, offering new avenues for rapid material optimization and property prediction. Integrating computational methods with experimental validation could become a critical strategy in developing next-generation PIs with tailored properties. For example, molecular dynamics simulations and DFT have shown significant efficacy in predicting the dielectric properties of PIs, providing deeper insights into how fluorine substituents affect molecular polarization and FFV. Machine learning models can predict the dielectric properties of PIs by analyzing large datasets of molecular structures and their corresponding properties. These models are trained on data obtained from both experimental results and molecular simulations, enabling rapid screening of new PI formulations. This combination offers powerful tools for the rational design of PIs with customized properties.

In conclusion, the continued development of high-transparency, low-dielectric PIs will rely on innovative design strategies, advanced characterization techniques, and the integration of computational tools to meet the evolving demands of electronic and optoelectronic applications.

References

1 Li, S., Tian, M., Gao, Q. et al. (2019). Nanometre-thin indium tin oxide for advanced high-performance electronics. *Nature Materials* 18 (10): 1091–1097.

2 Chang, J.-H. (2020). Equibiaxially stretchable colorless and transparent polyimides for flexible display substrates. *Reviews on Advanced Materials Science* 59 (1): 1–9.

3 Ren, X., He, Z., Wang, Z. et al. (2023). Design, synthesis and properties of semi-alicyclic colorless and transparent polyimide films with high glass transition temperatures and low retardation for potential applications in flexible electronics. *Polymers* 15 (16): 3408.

4 Fang, Y., He, X., Kang, J.-C. et al. (2024). Colorless transparent and thermally stable terphenyl polyimides with various small side groups for substrate application. *European Polymer Journal* 202: 112640.

5 Li, H., Bao, F., Lan, X. et al. (2023). Fluorinated polyimide with triphenyl pyridine structure for 5G communications: low dielectric, highly hydrophobic, and highly transparent. *European Polymer Journal* 197: 112327.

6 Zuo, H., Qian, G., Li, H.-B. et al. (2022). Reduced coefficient of linear thermal expansion for colorless and transparent polyimide by introducing rigid-rod amide units: synthesis and properties. *Polymer Chemistry* 13 (20): 2999–3008.

7 Yin, X., Feng, Y., Zhao, Q. et al. (2018). Highly transparent, strong, and flexible fluorographene/fluorinated polyimide nanocomposite films with low dielectric constant. *Journal of Materials Chemistry C* 6 (24): 6378.

8 Lian, M., Zhao, F., Liu, J. et al. (2023). The pivotal role of benzimidazole in improving the thermal and dielectric performance of upilex-type polyimide. *Polymers* 15 (10): 2343.

9 Tapaswi, P.K. and Ha, C.-S. (2019). Recent trends on transparent colorless polyimides with balanced thermal and optical properties: design and synthesis. *Macromolecular Chemistry and Physics* 220 (3): 1800313.

10 Liu, Y., Wang, Y.-K., and Wu, D.-Y. (2022). Synthetic strategies for highly transparent and colorless polyimide film. *Journal of Applied Polymer Science* 139: e52604.

11 Nagella, S.R. and Ha, C.-S. (2023). Structural designs of transparent polyimide films with low dielectric properties and low water absorption: a review. *Nanomaterials* 13 (14): 2090.

12 Zha, J.-W., Tian, Y., Zheng, M.-S. et al. (2023). High-temperature energy storage polyimide dielectric materials: polymer multiple-structure design. *Materials Today Energy* 31: 101217.

13 Zhang, D., Li, L., Wang, Y. et al. (2023). Methods and strategies to decrease the dielectric properties of polyimide films: a review. *Journal of Sol-Gel Science and Technology* 108 (1): 1–12.

14 Zhang, M., Wang, L., Xu, H. et al. (2023). Polyimides as promising materials for lithium-ion batteries: a review. *Nano-Micro Letters* 15 (1): 135.

15 Wang, Y., Zhou, J., Hou, J. et al. (2019). High-performance polyimides with high T_g and excellent dimensional stability at high temperature prepared via a cooperative action of hydrogen-bond interaction and cross-linking reaction. *ACS Applied Polymer Materials* 1 (8): 2099–2107.

16 Liaw, D.-J., Huang, C.-C., and Chen, W.-H. (2006). Optically transparency and light color of novel highly organosoluble alicyclic polyimides with 4-*tert*-butylcyclohexyl group. *Macromolecular Chemistry and Physics* 207 (4): 434–443.

17 Xu, Z., Croft, Z.L., Guo, D. et al. (2021). Recent development of polyimides: synthesis, processing, and application in gas separation. *Journal of Polymer Science* 59 (11): 943–962.

18 Liu, Y., Zhao, X.-Y., Sun, Y.-G. et al. (2023). Synthesis and applications of low dielectric polyimide. *Resources Chemicals and Materials* 2 (1): 49–62.

19 Ni, H.-J., Liu, J.-G., Wang, Z.-H., and Yang, S.-Y. (2015). A review on colorless and optically transparent polyimide films: chemistry, process and engineering applications. *Journal of Industrial and Engineering Chemistry* 28: 16–27.

20 Kim, Y.N., Lee, J., Kim, Y.-O. et al. (2021). Colorless polyimides with excellent optical transparency and self-healing properties based on multi-exchange dynamic network. *Applied Materials Today* 25: 101226.

21 Liu, J., Zhang, D., Yu, D. et al. (2021). Machine learning powered ellipsometry. *Light: Science & Applications* 10 (1): 55.

22 Zuo, H.-T., Gan, F., Dong, J. et al. (2021). Highly transparent and colorless polyimide film with low dielectric constant by introducing meta-substituted structure and trifluoromethyl groups. *Chinese Journal of Polymer Science* 39 (4): 455–464.

23 Sawada, R. and Ando, S. (2024). Polarization analysis and humidity dependence of dielectric properties of aromatic and semialicyclic polyimides measured at 10 GHz. *The Journal of Physical Chemistry C* 128 (16): 6979–6990.

24 Liang, Y., Sun, W., Li, T. et al. (2023). All organic polymer dielectrics for high-temperature energy storage from the classification of heat-resistant insulation grades. *Journal of Polymer Science* 61 (22): 2777–2795.

25 Morsalin, S., Phung, T.B., Danikas, M., and Mawad, D. (2019). Diagnostic challenges in dielectric loss assessment and interpretation: a review. *IET Science, Measurement & Technology* 13 (6): 767–782.

26 Ning, Z., Yan, Y., Wang, J. et al. (2024). High T_g, low dielectric, and colorless transparent polyimides based on rigid 9,9-spirobixanthene diamines. *ACS Applied Polymer Materials* 6 (9): 5434–5442.

27 Liaw, D.-J., Wang, K.-L., Huang, Y.-C. et al. (2012). Advanced polyimide materials: syntheses, physical properties and applications. *Progress in Polymer Science* 37 (7): 907–974.

28 Mo, X., Wang, C.-Y., Li, G., and Jiang, J.-M. (2012). High optical transparency and low dielectric constant polyimides containing trifluoromethyl and cyclohexane groups. *Journal of Macromolecular Science, Part B* 51 (7): 1370–1383.

29 Zhang, T., Pan, Y., Song, C. et al. (2020). Synthesis, characterization and comparison of properties of novel fluorinated polyimides derived from bisphenol–A-di(phthaleic anhydride). *Polymer Bulletin* 77 (8): 4077.

30 Ghosh, A. and Banerjee, S. (2008). Synthesis, characterization and comparison of properties of novel fluorinated polyimides derived from bisphenol–A-di(phthaleic anhydride). *Journal of Macromolecular Science, Part A* 45 (7): 578–584.

31 Hu, K., Ye, Q., Fan, Y. et al. (2021). Preparation and characterization of organic soluble polyimides with low dielectric constant containing trifluoromethyl for optoelectronic application. *European Polymer Journal* 157: 110566.

32 Chen, Z., Zhang, S., Feng, Q. et al. (2020). Improvement in mechanical and thermal properties of transparent semi-aromatic polyimide by crosslinking. *Macromolecular Chemistry and Physics* 221 (12): 2000085.

33 Yi, L., Li, C., Huang, W., and Yan, D. (2016). Soluble and transparent polyimides with high T_g from a new diamine containing tert-butyl and fluorene units. *Journal of Polymer Science Part A: Polymer Chemistry* 54 (7): 976–984.

34 Wang, C., Chen, W., Chen, Y. et al. (2014). Synthesis and properties of new fluorene-based polyimides containing trifluoromethyl and isopropyl substituents. *Materials Chemistry and Physics* 144 (3): 553–559.

35 Xu, J., He, C., Wei, T.S., and Chung, T.S. (2002). Low relative dielectric permittivities of polyimides and copolyimides derived from non-coplanar diamines and 4,4′-(hexafluoroisopropyl)diphthalic anhydride. *Plastics, Rubber and Composites* 31 (7): 295–299.

36 Liu, C., Pei, X., Huang, X. et al. (2015). Novel non-coplanar and tertbutyl-substituted polyimides: solubility, optical, thermal and dielectric properties. *Chinese Journal of Chemistry* 33 (2): 277–284.

37 Li, L., Xu, Y., Che, J. et al. (2018). Preparation, characterization and degradation kinetics of transparent fluorinated polyimides with low dielectric constants and excellent hydrophobic properties. *Polymer Bulletin* 75 (12): 5777–5793.

38 Zhao, X., Liu, J., Yang, H. et al. (2008). Novel polyfluorinated polyimides derived from α,α-bis(4-amino-3,5-difluorophenyl)phenylmethane and aromatic dianhydrides: synthesis and characterization. *European Polymer Journal* 44 (3): 808–820.

39 Hsiao, S.-H., Guo, W., Chung, C.-L., and Chen, W.-T. (2010). Synthesis and characterization of novel fluorinated polyimides derived from 1,3-bis(4-amino-2-trifluoromethylphenoxy)naphthalene and aromatic dianhydrides. *European Polymer Journal* 46 (9): 1878–1890.

40 Yang, C.-P., Hsiao, S.-H., and Chung, C.-L. (2005). Organosoluble, low-dielectric-constant fluorinated polyimides based on 2,6-bis(4-amino-2-trifluoromethylphenoxy)naphthalene. *Polymer International* 54 (4): 716–724.

41 Chung, C.-L., Tzu, T.-W., and Hsiao, S.-H. (2006). Organosoluble, low-colored fluorinated polyimides based on 1,1-bis[4-(4-amino-2-trifluoromethyl-phenoxy)phenyl]-1-phenyl-2,2,2-trifluoroethane. *Journal of Polymer Research* 13 (6): 495–506.

42 Hsiao, S.-H., Yang, C.-P., and Chung, C.-L. (2003). Synthesis and characterization of novel fluorinated polyimides based on 2,7-bis(4-amino-2-trifluoromethylphenoxy)naphthalene. *Journal of Polymer Science Part A: Polymer Chemistry* 41 (13): 2001–2018.

43 Chung, C.-L. and Hsiao, S.-H. (2008). Novel organosoluble fluorinated polyimides derived from 1,6-bis(4-amino-2-trifluoromethylphenoxy)naphthalene and aromatic dianhydrides. *Polymer* 49 (10): 2476–2485.

44 Yang, C.-P., Hsiao, S.-H., Tsai, C.-Y., and Liou, G.-S. (2004). Novel, organosoluble, light-colored fluorinated polyimides based on 2,2′-bis(4-amino-2-trifluoromethylphenoxy)biphenyl or 2,2′-bis(4-amino-2-trifluoromethylphenoxy)-1,1′-binaphthyl. *Journal of Polymer Science Part A: Polymer Chemistry* 42 (10): 2416–2431.

45 Wu, F., Zhou, X., and Yu, X. (2017). Synthesis and characterization of novel star-branched polyimides derived from 2,2-bis[4-(2,4-diaminophenoxy) phenyl]hexafluoropropane. *RSC Advances* 7 (57): 35786–35794.

46 Kim, K.-E., Jee, M.H., and Baik, D.H. (2023). Effect of terephthaloyl chloride comonomer on thermal and optical properties of colorless and transparent polyimide films based on 4, 4′-(hexafluoroisopropylidene) diphthalic anhydride and bis(3-aminophenyl) sulfone. *Fibers and Polymers* 24 (7): 2275–2282.

47 Banerjee, S., Madhra, M.K., Salunke, A.K., and Maier, G. (2002). Synthesis and properties of fluorinated polyimides. 1. Derived from novel 4,4″-bis(aminophenoxy)-3,3″-trifluoromethyl terphenyl. *Journal of Polymer Science Part A: Polymer Chemistry* 40 (8): 1016–1027.

48 Bao, F., Qiu, L., Zou, B. et al. (2024). Development of fluorinated colorless polyimides of restricted dihedral rotation toward flexible substrates with thermal robustness. *Macromolecules* 57 (8): 3568–3579.

49 Jia, M., Li, Y., He, C., and Huang, X. (2016). Soluble perfluorocyclobutyl aryl ether-based polyimide for high-performance dielectric material. *ACS Applied Materials & Interfaces* 8 (39): 26352–26358.

50 Jang, W., Lee, H.S., Lee, S. et al. (2007). The optical and dielectric characterization of light-colored fluorinated polyimides based on 1,3-bis(4-amino-2-trifluoromethylphenoxy)benzene. *Materials Chemistry and Physics* 104 (2): 342–349.

51 Liaw, D.-J. and Chang, F.-C. (2004). Highly organosoluble and flexible polyimides with color lightness and transparency based on 2,2-bis[4-(2-trifluoromethyl-4-aminophenoxy)-3,5-dimethylphenyl]propane. *Journal of Polymer Science Part A: Polymer Chemistry* 42 (22): 5766–5774.

52 Mi, Z., Liu, Z., Tian, C. et al. (2017). Soluble polyimides containing 1,4:3,6-dianhydro-d-glucidol and fluorinated units: preparation, characterization, optical, and dielectric properties. *Journal of Polymer Science Part A: Polymer Chemistry* 55 (19): 3253–3265.

53 Wu, Q., Ma, X., Zheng, F. et al. (2019). High performance transparent polyimides by controlling steric hindrance of methyl side groups. *European Polymer Journal* 120: 109235.

54 Guan, Y., Wang, D., Song, G. et al. (2014). Novel soluble polyimides derived from 2,2′-bis[4-(5-amino-2-pyridinoxy)phenyl]hexafluoropropane: preparation, characterization, and optical, dielectric properties. *Polymer* 55 (16): 3634–3641.

55 Wang, Y., Zhou, X., Feng, X. et al. (2022). Polyimide films containing trifluoromethoxy groups with high comprehensive performance for flexible circuitry substrates. *ACS Applied Polymer Materials* 4 (8): 5831–5839.

56 Kim, J.W. and Chang, J.-H. (2020). Syntheses of colorless and transparent polyimide membranes for microfiltration. *Polymers* 12 (7): 1610.

57 Liu, B., Hu, W., Matsumoto, T. et al. (2005). Synthesis and characterization of organosoluble ditrifluoromethylated aromatic polyimides. *Journal of Polymer Science Part A: Polymer Chemistry* 43 (14): 3018–3029.

58 Yu, H.-C., Jung, J.-W., Choi, J.-Y. et al. (2017). Structure-property relationship study of partially aliphatic copolyimides for preparation of flexible and transparent polyimide films. *Journal of Macromolecular Science, Part A* 54 (2): 97–104.

59 Liu, H., Zhai, L., Bai, L. et al. (2019). Synthesis and characterization of optically transparent semi-aromatic polyimide films with low fluorine content. *Polymer* 163: 106–114.

60 Dautel, O.J., Wantz, G., Flot, D. et al. (2005). Confined photoactive substructures on a chiral scaffold: the design of an electroluminescent polyimide as material for PLED. *Journal of Materials Chemistry* 15 (41): 4446–4452.

61 Oishi, Y., Onodera, S., Oravec, J. et al. (2003). Synthesis of fluorine-containing wholly alicyclic polyimides by in situ silylation method. *Journal of Photopolymer Science and Technology* 16 (2): 263–266.

62 Toshihiko, M., Daisuke, M., Takahiro, H. et al. (2009). Alicyclic polyimides – a colorless and thermally stable polymer for opto-electronic devices. *Journal of Physics: Conference Series* 187 (1): 012005.

63 Hasegawa, M. and Horii, S. (2007). Low-CTE polyimides derived from 2, 3, 6, 7-naphthalenetetracarboxylic dianhydride. *Polymer Journal* 39 (6): 610–621.

64 Kumar, S.V., Yu, H.-C., Choi, J. et al. (2011). Structure–property relationships for partially aliphatic polyimides. *Journal of Polymer Research* 18 (5): 1111–1117.

65 van Velthoven, J.L.J., Gootjes, L., Noordover, B.A.J., and Meuldijk, J. (2015). Bio-based, amorphous polyamides with tunable thermal properties. *European Polymer Journal* 66: 57–66.

66 Nagao, R., Marumo, F., Saito, Y., and Asahara, T. (1971). The crystal structure of butane-1, 2, 3, 4-tetracarboxylic dianhydride. *Acta Crystallographica, Section B: Structural Crystallography and Crystal Chemistry* 27 (3): 569–572.

67 Shpil'man, N., Lyapunov, V., Dubitskii, Y., and Zhubanov, B. (1991). Interpretation of the carbon-13 NMR spectra of alicyclic dianhydrides using model compounds. *Zhurnal Prikladnoi Spektroskopii* 55: 683.

68 Hasegawa, M., Horiuchi, M., and Wada, Y. (2007). Polyimides containing trans-1, 4-cyclohexane unit (II). Low-K and Low-CTE semi-and wholly cycloaliphatic polyimides. *High Performance Polymers* 19 (2): 175–193.

69 Hu, X., Yan, J., Wang, Y. et al. (2017). Colorless polyimides derived from 2 R, 5 R, 7 S, 10 S-naphthanetetracarboxylic dianhydride. *Polymer Chemistry* 8 (39): 6165–6172.

70 Yi, M.H., Huang, W., Lee, B.J., and Choi, K.Y. (1999). Synthesis and characterization of soluble polyimides from 2, 2-bis (4-aminophenyl) cycloalkane derivatives. *Journal of Polymer Science Part A: Polymer Chemistry* 37 (17): 3449–3454.

71 Shiotani, A., Shimazaki, H., and Matsuo, M. (2001). Preparation of transparent polyimides derived from cis-and trans-dicyclohexyl-3, 3′, 4, 4′-tetracarboxylic dianhydrides. *Macromolecular Materials and Engineering* 286 (7): 434–441.

72 Carta, M., Malpass-Evans, R., Croad, M. et al. (2013). An efficient polymer molecular sieve for membrane gas separations. *Science* 339 (6117): 303–307.

73 Matsumoto, T., Ishiguro, E., Nakagawa, S., and Kimura, R. (2013). Alicyclic polyimides derived from alkanone bis-spironorbornanetetracarboxylic dianhydrides. *Journal of Photopolymer Science and Technology* 26 (3): 361–365.

74 Wang, H.G., Yuan, S., Ma, D.L. et al. (2014). Tailored aromatic carbonyl derivative polyimides for high-power and long-cycle sodium-organic batteries. *Advanced Energy Materials* 4 (7): 1301651.

75 Hasegawa, M., Kasamatsu, K., and Koseki, K. (2012). Colorless poly (ester imide) s derived from hydrogenated trimellitic anhydride. *European Polymer Journal* 48 (3): 483–498.

76 Popovici, D., Barzic, A.I., Barzic, R.F. et al. (2016). Semi-alicyclic polyimide precursors: structural, optical and biointerface evaluations. *Polymer Bulletin* 73: 331–344.

77 Ahn, B.-H., Lee, D.-W., and Lee, J.-K. (2001). Synthesis of newel positive type photosensitive polyimide. *Macromolecular Research* 9 (5): 247–252.

78 Abdulhamid, M.A., Ma, X., Miao, X., and Pinnau, I. (2017). Synthesis and characterization of a microporous 6FDA-polyimide made from a novel carbocyclic pseudo Tröger's base diamine: effect of bicyclic bridge on gas transport properties. *Polymer* 130: 182–190.

79 Zhang, M., Liu, W., Gao, X. et al. (2020). Preparation and characterization of semi-alicyclic polyimides containing trifluoromethyl groups for optoelectronic application. *Polymers* 12 (7): 1532.

80 Kwac, L.K., Kim, H.G., and Chang, J.-H. (2021). Comparison of properties of colorless and transparent polyimide nanocomposites containing different diamine monomers. *ACS Omega* 6 (29): 19006–19016.

81 Zhuang, Y., Orita, R., Fujiwara, E. et al. (2019). Colorless partially alicyclic polyimides based on Tröger's base exhibiting good solubility and dual fluorescence/phosphorescence emission. *Macromolecules* 52 (10): 3813–3284.

82 Barzic, A.I., Albu, R.M., Ioanid, E.G., and Hulubei, C. (2018). Molecular design of some semi-alicyclic polyimides as a route to improve refraction and dielectric properties for liquid crystal display applications. *High Performance Polymers* 30 (7): 776–786.

83 Hasegawa, M., Horiuchi, M., and Wada, Y. (2007). Polyimides containing trans-1,4-cyclohexane unit (II). Low-K and Low-CTE semi- and wholly cycloaliphatic polyimides. *High Performance Polymers* 19 (2): 175–193.

84 Zhou, D., Peng, Q., Kong, L.-L. et al. (2023). Synthesis of highly thermostable and transparent colorless polyimides based on a semi-aromatic tetracarboxylic anhydride. *ChemistrySelect* 8 (9): e202204830.

85 Yao, J., Wang, C., Zhao, X. et al. (2018). Highly transparent and soluble polyimides with synergistic effects of pyridine and cyclohexane. *High Performance Polymers* 30 (4): 418–426.

86 Mi, Z., Liu, Z., Wang, C. et al. (2018). Transparent and soluble polyimide films containing 4,4′-isopropylidenedicyclohexanol (Cis-HBPA) units: preparation, characterization, thermal, mechanical, and dielectric properties. *Journal of Polymer Science Part A: Polymer Chemistry* 56 (18): 2115–2128.

87 Bei, R., Qian, C., Zhang, Y. et al. (2017). Intrinsic low dielectric constant polyimides: relationship between molecular structure and dielectric properties. *Journal of Materials Chemistry C* 5 (48): 12807–12815.

88 He, J., Wu, X., and Cheng, Y. (2023). Low dielectric post-cured benzocyclobutene-functionalized fluorine-containing polyimide material. *European Polymer Journal* 196: 112334.

89 Zheng, H., Wang, C., Ma, Y. et al. (2021). High thermal stability and low dielectric constant of soluble polyimides containing asymmetric bulky pendant groups. *Journal of Macromolecular Science, Part A* 58 (12): 880–889.

90 Wang, X., Li, Y., Gong, C. et al. (2007). Synthesis and characterization of novel soluble pyridine-containing polyimides based on 4-phenyl-2,6-bis[4-(4-aminophenoxy)phenyl]-pyridine and various aromatic dianhydrides. *Journal of Applied Polymer Science* 104 (1): 212–219.

91 Hsiao, S.-H., Wang, H.-M., Chou, J.-S. et al. (2011). Triptycene poly(ether-imide)s with high solubility and optical transparency. *Journal of Polymer Research* 19 (1): 9757.

92 Huang, X., Mei, M., Liu, C. et al. (2015). Synthesis and characterization of novel highly soluble and optical transparent polyimides containing tert-butyl and morpholinyl moieties. *Journal of Polymer Research* 22 (9): 169.

93 Sheng, S., Li, D., Lai, T. et al. (2011). Organosoluble, low-dielectric-constant fluorinated polyimides based on 9,9-bis[4-(4-amino-2-trifluoromethyl-phenoxy)phenyl]xanthene. *Polymer International* 60 (8): 1185–1193.

94 Xiao, P., He, X., Zheng, F., and Lu, Q. (2022). Super-heat resistant, transparent and low dielectric polyimides based on spirocyclic bisbenzoxazole diamines with $T_g > 450$ °C. *Polymer Chemistry* 13 (24): 3660–3669.

95 Zhang, S.-J., Bu, Q.-Q., Li, Y.-F. et al. (2011). High organosolubility and optical transparency of novel polyimides derived from 2′,7′-bis(4-amino-2-

trifluoromethylphenoxy)-spiro (fluorene-9,9′-xanthene). *Materials Chemistry and Physics* 128 (3): 392–399.

96 Chern, Y.-T. and Shiue, H.-C. (1997). Low dielectric constants of soluble polyimides based on adamantane. *Macromolecules* 30 (16): 4646–4651.

97 Chern, Y.T. (1998). Low dielectric constant polyimides derived from novel 1,6-Bis[4-(4-aminophenoxy)phenyl]diamantane. *Macromolecules* 31: 5837–5844.

98 Lee, Y.-H., Lin, J.-W., Huang, Y.-C. et al. (2023). High thermally stable anthracene-based polyimides with low dielectric constants and dissipation factors. *ACS Applied Polymer Materials* 5 (12): 10196–10204.

99 Mathews, A.S., Kim, D., Kim, Y. et al. (2008). Synthesis and characterization of soluble polyimides functionalized with carbazole moieties. *Journal of Polymer Science Part A: Polymer Chemistry* 46 (24): 8117–8130.

100 Wu, Y., Ji, J., Huang, H. et al. (2021). Facile synthesis of acyloxy-containing fluorene-based Cardo polyimides with high optical transparency, fluorescence and low dielectric constant. *Reactive and Functional Polymers* 166: 104979.

101 Qu, C., Shan, L., Zhang, G., and Sun, R. (2023). Preparation and dielectric properties research of a novel kind of intrinsic silane-containing polyimide. *Polymer* 285: 126361.

102 Chen, Y., Liu, Y., and Min, Y. (2022). Synthesis and properties comparison of low dielectric silicon containing polyimides. *Materials* 15 (8): 2755.

103 Kurinchyselvan, S., Hariharan, A., Prabunathan, P. et al. (2019). Fluorinated polyimide nanocomposites for low K dielectric applications. *Journal of Polymer Research* 26 (9): 207.

104 Li, H., Wang, X., Gong, Y. et al. (2023). APDS modified several bisphenol A polyimides with low dielectric constant under high frequency. *Journal of Polymer Research* 30 (11): 407.

105 Kim, S., Ando, S., and Wang, X. (2015). Highly dispersible ternary composites with high transparency and ultra low dielectric constants based on hyper-branched polyimide with organosilane termini and cross-linked polyimide with silica. *RSC Advances* 5 (119): 98419–97428.

106 Chen, Z., Zhu, D., Tong, F. et al. (2019). Low dielectric constant polyimide hybrid films prepared by in situ blow-balloon method. *ACS Applied Polymer Materials* 1 (8): 2189–2196.

107 Zhang, W., Jiang, H., Nie, Y. et al. (2022). Composite films with low dielectric constant and dielectric loss factor at high frequency prepared from polyimide and polytetrafluoroethylene. *Polymer Engineering & Science* 62 (12): 4226–4234.

108 Li, H., Wang, X., Gong, Y. et al. (2023). Polyimide/crown ether composite film with low dielectric constant and low dielectric loss for high signal transmission. *RSC Advances* 13 (11): 7585–7596.

109 Xia, X., Zhang, S., He, X. et al. (2022). Molecular necklace strategy for enhancing modulus and toughness of colorless transparent polyimides for cover window application. *Polymer* 259: 125358.

110 Qiu, G., Ma, W., and Wu, L. (2020). Low dielectric constant polyimide mixtures fabricated by polyimide matrix and polyimide microsphere fillers. *Polymer International* 69 (5): 485–491.

111 Kim, J.W., Lee, S.J., Choi, M.Y., and Chang, J.-H. (2021). Colorless and transparent polyimide microporous film with excellent physicochemical property. *Polymers* 13 (8): 1298.

112 Lv, P., Dong, Z., Dai, X. et al. (2018). Synthesis and properties of ultralow dielectric porous polyimide films containing adamantane. *Journal of Polymer Science Part A: Polymer Chemistry* 56 (5): 549–559.

113 Zhang, P., Zhang, L., Zhang, K. et al. (2021). Preparation of polyimide films with ultra-low dielectric constant by phase inversion. *Crystals* 11 (11): 1383.

114 Peng, Z., Ye, A., Zhang, L. et al. (2024). Micro-crosslinked polyimide nanocomposites with low dielectric constant and low dielectric loss for microwave antenna with molecular dynamics. *Composites Communications* 46: 101804.

115 Agrawal, A. and Choudhary, A. (2016). Perspective: materials informatics and big data: realization of the "fourth paradigm" of science in materials science. *APL Materials* 4 (5): 053208.

116 Park, H., Choi, H., Kim, J. et al. (2024). Density functional theory-based approach for dielectric constant estimation of soluble polyimide insulators. *The Journal of Physical Chemistry B* 128 (10): 2528–2536.

117 He, X., Zhang, S., Zhou, Y. et al. (2022). The "fluorine impact" on dielectric constant of polyimides: a molecular simulation study. *Polymer* 254: 125073.

118 Peng, W.-F., Lei, H.-Y., Zhang, X.-X. et al. (2022). Fluorine substitution effect on the material properties in transparent aromatic polyimides. *Chinese Journal of Polymer Science* 40 (7): 781–788.

119 Chen, L., Kim, C., Batra, R. et al. (2020). Frequency-dependent dielectric constant prediction of polymers using machine learning. *npj Computational Materials* 6 (1): 61.

120 Tao, L., He, J., Munyaneza, N.E. et al. (2023). Discovery of multi-functional polyimides through high-throughput screening using explainable machine learning. *Chemical Engineering Journal* 465: 142949.

121 Zhang, S., He, X., Xia, X. et al. (2023). Machine-learning-enabled framework in engineering plastics discovery: a case study of designing polyimides with desired glass-transition temperature. *ACS Applied Materials & Interfaces* 15 (31): 37893.

9

Polyimide Dielectrics in Extreme Environment

Yuhuai Wang and Jin Li

Tianjin University, School of Electrical and Information Engineering, No. 92 Weijin Road, Nankai District, Tianjin 300072, China

Space technology has become an important method of daily communication, weather observation, and universe exploration. The development of space technology cannot be separated from higher requirements for the quality and service life of spacecraft. Among these, near-Earth orbit (NEO) spacecraft, located at an altitude of about 200–1200 km above the ground, such as meteorological satellites, communications satellites, manned spacecraft, and space stations, are numerous, multifunctional, and important. Most LEO spacecraft carry the task of operating in the NEO environment for extended periods, making the long-term stability of their materials even more critical. Polyimide (PI), a class of polymers containing an imide ring in the main chain, is one of the organic polymer materials with the best overall performance. Polyimide has good mechanical properties, a low coefficient of thermal expansion, and good dielectric properties, and can withstand extremely low temperatures. Due to these superior properties, polyimide has been widely used in satellites, airships, probes, and launch vehicles, among other spacecraft structural components. The environmental extremes of low-Earth orbit (LEO) pose challenging conditions for the stable operation of materials. Atomic oxygen (AO) is the residual gaseous environment in the form of atomic oxygen in low-Earth orbit (200–700 km). At this orbital altitude, the environmental components are N_2, O_2, Ar, He, H, O, etc., and the corresponding particle density is about 10^6–10^9/cm^3. The AO flux density can reach 10^{13}–10^{16}/cm^3. The AO erosion of the space materials mainly comes from the AO with high relative collision energy, and when the AO and high-speed vehicle collide, the relative impact velocity is as high as 8 km/s, which leads to the advective energy of the AO collision to reach about 5 eV. When the AO impacts with a high-speed vehicle, the relative impact velocity is as high as 8 km/s, resulting in the translational kinetic energy of the AO collision reaching about 5 eV. These AOs with high energy and high chemical reactivity are sufficient to break the chemical bonds of spacecraft polymer materials and induce oxidative decomposition [1].

Functional Polyimide Dielectrics: Structure, Properties, and Applications, First Edition.
Edited by Jun-Wei Zha.

9.1 Atomic Oxygen Erosion Mechanism of Polyimide

9.1.1 Atomic Oxygen Formation and Characteristics

In the microgravity environment of low-Earth orbit (LEO), atomic oxygen (AO) is the most abundant component [2]. Unlike the atmospheric environment on the surface of the Earth, the gas in LEO consists primarily of 80% AO and 20% N_2, meaning 99% of the oxygen exists in the form of AO. AO is generated by the decomposition of oxygen molecules in LEO due to solar ultraviolet ($\lambda \leq 243$ nm) radiation. Its generation mechanism primarily includes photodissociation, complex dissociation of oxygen molecules, and thermoelectronic collisions [3].

The erosion of space materials by atomic oxygen (AO) primarily results from the high relative collision energy of AO. Although AO is static, when it collides with high-speed vehicles, the relative impact velocity can reach up to 8 km/s. This extremely high-impact velocity causes the translational kinetic energy of the AO collision to reach approximately 5 eV. These AOs, with their high energy and chemical reactivity, are sufficient to break the chemical bonds of spacecraft polymer materials and induce oxidative decomposition [4]. In contrast, the average thermal velocity of other gas molecules in LEO is only about 0.4 km/s, resulting in very low collision energy, which is insufficient to induce any surface chemical reactions.

9.1.2 Atomic Oxygen Erosion Mechanism of Polyimide

The erosion effect of atomic oxygen (AO) primarily involves the high-speed collision of AO with the surface of spacecraft materials, resulting in the formation of corresponding oxides through oxidation reactions. AO erosion can lead to changes in the surface morphology of the materials and is accompanied by the degradation of various physical properties. These include accelerated outgassing, increased mass loss rate, decreased mechanical strength, and deterioration of optical and electrical properties. Additionally, volatile gases and non-adherent oxides produced by the interaction of AO with materials can contaminate spacecraft surfaces, seriously affecting the thermal and optical properties of optical systems, temperature-controlled coatings, and other components and materials. This contamination can lead to changes in the solar absorptivity of coatings and reductions in the power output of batteries. After an in-depth study of the role and mechanism of a single environmental factor on the degradation behavior of materials in the space environment, researchers found that the erosion between AO and materials results from the synergistic action of various effects. These effects include the chemical reaction caused by strong oxidative properties, changes in polymer structure due to the chemical reaction, and the loss of surface material caused by atomic sputtering. The process of AO's erosion on space materials involves a complex interplay of chemical reactions, high-speed bombardment by AO, and mechanical damage processes, as shown in Figure 9.1.

Vacuum ultraviolet (VUV) radiation is a significant environmental factor in low-Earth orbit (LEO) that greatly impacts material properties. VUV refers to ultraviolet light with wavelengths ranging from 0 to 200 nm, which exists in LEO

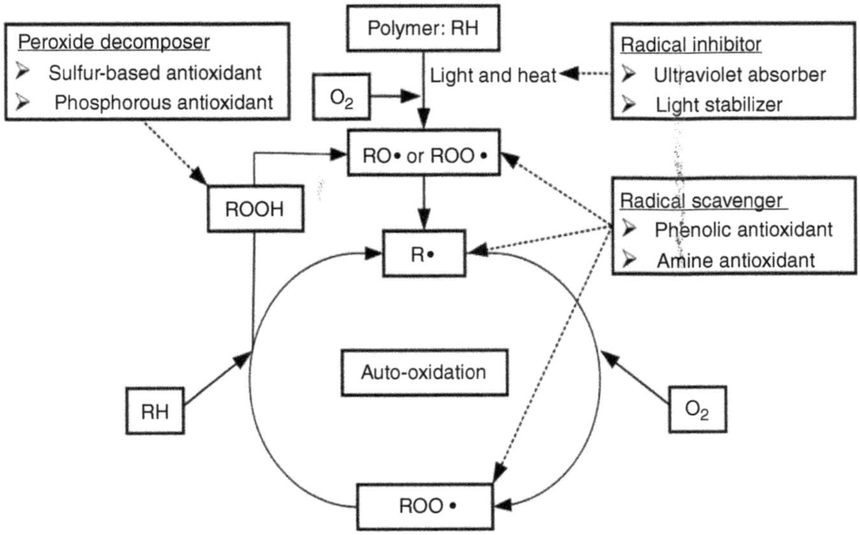

Figure 9.1 Atomic oxygen erosion processes.

and can cause the breaking of chemical bonds in some functional groups within polymer materials. Although VUV alone does not cause atoms to leave the material matrix directly, the synergistic effect of AO and VUV can further accelerate material erosion. There is a strong coupling effect between AO and VUV. For example, consider the commonly used Kapton material in an experiment. Figure 9.2a shows that after 6 hours of atomic oxygen irradiation, the original Kapton surface was severely eroded, exhibiting large areas of oxidative erosion with concave surfaces, typical characteristics of polyimide materials in an AO environment [5]. In contrast, Figure 9.2b shows a sample of SiO_2 nanorods coated on the Kapton surface. After AO irradiation, there is no trace of erosion on the surface of the sample, and the microspheres display a columnar distribution in a certain direction on the Kapton surface. This columnar distribution is related to the tensile orientation during Kapton film

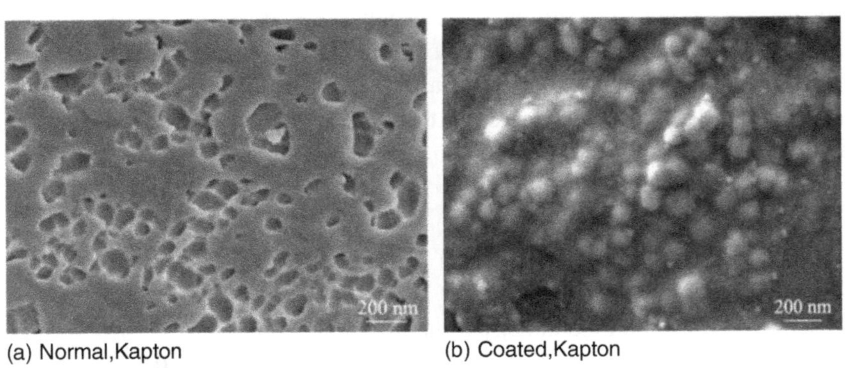

Figure 9.2 Surface SEM image of Kapton sample after atomic oxygen irradiation test. Source: Ref. [5]/with permission of American Chemical Society.

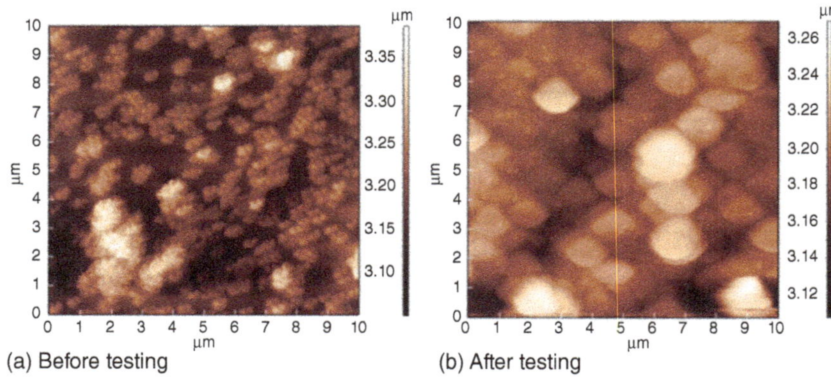

Figure 9.3 Surface AFM image of Kapton sample after atomic oxygen irradiation test. Source: Ref. [5]/with permission of American Chemical Society.

preparation and the lift-off coating method used [6]. Compared to the original morphology of SiO_2 nanomicrospheres in Figure 9.2a, the particle aggregates appear to be dissolved, the diameter of the aggregates has increased, and some small nanoparticles have detached from the aggregates and dispersed in the sol–gel coating. It was analyzed by AFM atomic force microscopy to observe the changes in the surface of the coating before and after AO irradiation, as shown in Figure 9.3.

Figure 9.3a shows the morphology of a Kapton sample coated with SiO_2 before AO irradiation. The modification improved the wettability of the Kapton surface, resulting in a uniformly extended coating. The SiO_2 nanorods prepared by the Stöber method exhibited an aggregated distribution. After AO irradiation, the surface of the coating underwent significant changes (Figure 9.3b). The oxidation by AO caused deep oxidation of the organosilicon materials that were not fully hydrolyzed in the coating, and the surface became covered with micron-sized particles of varying sizes and continuity. The white dots in these micron-sized oxidized particles are likely SiO_2 nanorods prepared by the Stöber method. Theoretically, this coating of silicon oxide particles is more resistant to AO erosion, which is beneficial for the long-term use of Kapton materials in the space environment. Atomic oxygen erosion causes varying degrees of damage to different spacecraft materials. Inhibiting AO erosion can greatly extend the service life of spacecraft and improve their safety. Therefore, to enhance the anti-AO erosion performance of spacecraft materials and ensure their safety and durability, it is necessary to research and develop various types of protective coatings.

9.2 Atomic Oxygen Protection Technology for Polyimide

At present, AO protection technology is mainly carried out in the following three directions: (i) Body material modification protection. This method introduces specific element groups such as P and Si or specific element nanoparticles into the molecular structure of the polymer material, so that the polymer material generates

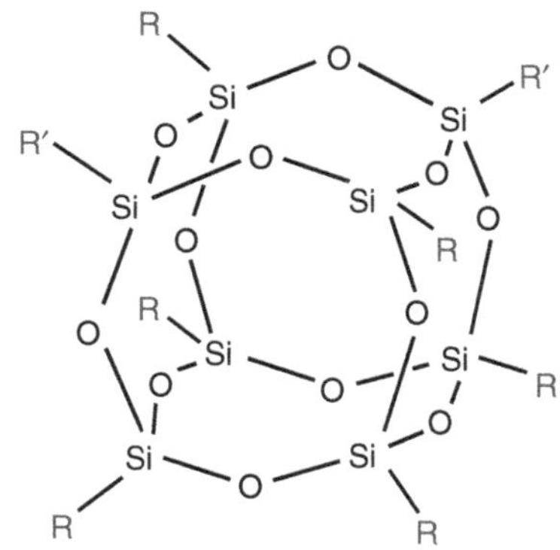

Figure 9.4 Sesquicarbazone molecular structure formula (POSS).

an inert layer of AO in situ when it suffers from the impact of AO, preventing AO from further eroding the polymer matrix. (ii) Surface modification protection. This method is to prepare an AO protective layer or modified layer on the surface of the polymer material for AO protection. (iii) Surface coating protection. This method involves spraying or brushing coatings with AO protection on the surface of the polymer material. The role of the protective coating is to protect the substrate material, which is usually achieved by isolating the substrate material from the environment.AO protective coatings need to have the following conditions: good resistance to AO erosion, a certain degree of flexibility, thin thickness, lightweight, strong bonding with the substrate, smooth and uniform coating surface, no air holes, and a certain degree of radiation resistance. In addition, the AO protective coating is also required to have excellent electrical conductivity to prevent the discharge phenomenon generated by the uneven potential of the spacecraft surface.

9.2.1 Material Modification Protection

Introducing specific element groups or nanoparticles into the molecular structure of polymer materials can generate an inert layer of AO in situ when the polymer material is subjected to AO impact, thereby preventing further erosion of the polymer matrix. This method of protection is known as body material modification protection. The introduced elements include P, Zr, Si, and others. Since the oxidation product of Si is SiO_x, which has a very high AO protection performance, polymers with Si-based body modification protection have gained more attention. The representative modification groups are caged sesquisiloxanes and polysiloxanes. Polyhedral oligomeric silsesquioxane (POSS) nano-matrix exhibits an organic and inorganic hybrid structure. Its molecular structure (Figure 9.4) consists of an inorganic framework composed of silicone resin and oxygen, which enhances the AO protection ability of the polymer. The outer part of the polymer is encapsulated by a variety of functionalized nano-compounds with organic groups, facilitating the introduction of POSS into polymers for the protection of Si-bearing body materials. Introducing POSS into the molecular structure of polymers is highly favorable for AO protection [7–10].

9.2.2 Surface Protection

Surface protection generates an AO protective layer on the surface of the polyimide film of the flexible substrate by physical or chemical means. Depending on the timing of the protection, the technology route can be divided into two types, namely, to assemble the flexible substrate product after the protection treatment on the surface of the polyimide film, and to produce the flexible substrate product and then carry out the surface protection treatment. According to the composition and nature of the coating material, it can be divided into inorganic coatings, organic coatings, and organic/inorganic composite modified layers.

Inorganic protective layer is the use of magnetron sputtering and other physical methods on the surface of polyimide coated with SiO_x, Al_2O_3, and other inorganic/metal oxides as a representative of the protective layer, the structure of this type of protective layer is dense, can effectively prevent the erosion of AO. However, it is easy to crack and peel off after transport, use and high and low temperatures [11–13].

Organic protective coatings usually choose polysiloxanes, polysilicon aluminum nitrides, polyfluorides and polysilazanes as the protective groups, among which, polysilazanes can be transformed into SiN/SiC with better protection against AO under thermal decomposition, which has received more attention. Compared with inorganic coatings, organic protective coatings have better flexibility, and the combination with polymer matrix is also better than inorganic coatings, but limited by the nature of the organic matter itself, facing the space environment of UV irradiation aging phenomenon.

After the surface silanization of PI, there are a large number of hydrocarbon groups in the newly generated surface structure, which contribute to the generation of siloxyl groups in the reaction with AO, and indirectly enhance the AO protection ability of polymer materials. Domestic researchers generally use wet chemical method to modify the surface activation of PI. Using a certain concentration of alkali for surface chemical modification can remove the passivation layer on the surface of PI and increase the hydroxyl content on the surface [14]; using silane coupling agent to assist the surface modification of silylation can improve the interfacial state of the modified layer and the polymer matrix. Liu et al. used different kinds of silane coupling agent to modify the surface of PI in acidic and alkaline environments, and found that the surface modification with silane coupling agent APTES in acidic environment had better AO protection performance compared with other silane coupling agents [14, 15].

In general, the organic/inorganic composite protective layer on the polymer surface combines the advantages of body material modification and general protective coating, and has a good combination with the substrate and a certain degree of flexibility, with good comprehensive protective performance. However, at present, the synergistic protective mechanism between the components of the composite system and AO is not clear, and there is a lack of research on the mechanism of the composite protective system and AO. At the same time, due to the lack of mechanism, there is no AO protection performance evaluation standard applicable to the composite protection system, and there is a long test cycle and high test cost defects. In addition, polymer composite anti-AO materials can be produced by using the new material, graphitic dilution. Xu Bo et al. prepared a series of graphene oxide/polyimide (GO/PI) composite films with good anti-AO erosion properties by simple mechanical blending (ball-mill blending) and high-temperature pressing. The results show that the introduction of trace amount (0.5%) of GO can improve the AO erosion resistance of GO/PI composite films by 17.9%, and the composite films have good thermal stability and mechanical properties.

9.3 Example of Atomic Oxygen Erosion Protection

Atomic oxygen erodes aging spacecraft mainly in low-Earth and low-Earth orbits, including damaging spacecraft control coatings, causing structural deterioration and corrosion, and endangering the safe operation of spacecraft. Taking the International Space Station (ISS) and the Chinese "Tianhe" space station as examples, the application of anti-atomic oxygen erosion technology is introduced and its practical effects are compared.

9.3.1 Example of Protection Against Atomic Oxygen Erosion on the International Space Station (ISS)

The International Space Station (ISS), which was transferred to full use in 2010, is the largest space platform operating in orbit in LEO and is important for the study of protective erosion of AO. ISS is based on SiO_x-PTFE-based protective coatings. NASA's Low-Orbit Materials Selection Guide reports [16] that the mass loss rate (including protective coatings) of SiO_x–PTFE composite protective coatings after the action of 2.5×10^{22} atoms/cm^2 AO is less than 7.24×10^{-4} g/cm^2, which meets the ISS AO protection needs within 15 years, even after 15 years in orbit. The SiO_x–PTFE composite protective coating was applied to the Aurora functional cargo module of the ISS, which was launched and commissioned in November 1998, and has been in operation for more than 15 years until 2023, during which the SiO_x–PTFE coating has contributed to the operation of the module for more than 15 years. PTFE coating provided better AO protection for the solar wing of this cargo bay during this period, as shown in Figure 9.5.

9.3.2 Example of Protection Against Atomic Oxygen Erosion in the Core Module of the Chinese Space Station Tianhe

The Tianhe Core Module, with a design life of 15 years and commissioned in 2021, is the first segment of the Chinese space station to be launched into orbit. The Tianhe Core Module operates in low-Earth orbit and is subject to significant AO erosion. The Tianhe Core Module operates in low-Earth orbit (LEO) and is subject to significant erosion by AO. The cumulative flux of AO on the windward side of the core module is estimated to be 7.83×10^{22} atoms/cm^2, so AO protection is necessary to

Figure 9.5 International space station PI usage.

Figure 9.6 Interplate cables with silicone protective coating deposited on the surface. Source: China Manned Space Engineering Office/Wikimedia Commons.

ensure the normal performance and expected life of the surface materials. The AO protection technology is based on the plasma-enhanced chemical vapor deposition (PECVD) technology, with the inner structure dominated by the ($—CH_3$) group, and the surface structure dominated by the —[—Si—O—Si—]—, which is produced by the Lanzhou Institute of Space Physics (Lanzhou, Gansu Province) used a coating made of siloxane to protect the cables between the solar panels, as shown in Figure 9.6. The thickness of the coating is about 400 nm, which ensures good AO protection and toughness at the same time. The mass loss rate of the coating after 2.5×10^{22} atoms/cm^{2}AO in a ground simulation was only 3.3×10^{-4} g/cm^2, which is only 1/2 of the mass loss rate of the SiO_x–PTFE composite protective coating on the surface of the solar arrays on the former ISS under the same conditions.

9.4 Summary

This chapter introduces the application of PI in spacecraft equipment insulation as well as the generation and basic characteristics of atomic oxygen in LEO space, overviews the erosion mechanism and erosion effect of atomic oxygen on polyimide insulation in microgravity environment, and focuses on atomic oxygen protection technology and application examples such as ontogeny and surface modification with the main conclusions as follows.

(1) Atomic oxygen is generated by the decomposition of oxygen molecules in LEO space by solar ultraviolet radiation, which is sufficient to break the chemical bonds of spacecraft polymer materials and induce oxidative decomposition of the materials. Experimental results show that the erosion of atomic oxygen, i.e., the chemical reaction of space materials under the action of high-speed bombardment, involves a complex process of chemical and physical action as well as mechanical damage at the same time.
(2) The function of protective coatings is to provide protection for the substrate material, which is usually achieved by isolating the substrate material from the environment. Atomic oxygen protection technology at home and abroad is mainly divided into three types of methods: body material modification protection, surface modification protection, and surface coating protection.
(3) Application of SiO_x-PTFE composite protective coatings against atomic oxygen attack on the International Space Station (ISS). An example of the application of atomic oxygen on the core module of China's "Tianhe," based on plasma-enhanced chemical vapor deposition technology, with the application of (—CH_3) groups in the inner structure and —[—Si—O—Si—]— groups in the surface structure to resist proton-oxygen erosion, with a doubling of mass-loss rate compared with ISS.

References

1 Cui, Z.Y., Wang, L.S., Qian, Y.H. et al. (2021). Atomic oxygen erosion-corrosion to polyimide in low earth orbit and its protective measures. *Corrosion & Protection* 42 (11): 12–19.

2 Li, G., Liu, X., and Li, T. (2013). Effects of low earth orbit environments on atomic oxygen under cutting of space craft polymer films. *Composites Part B: Engineering* 44 (1): 60–66.

3 Skurat, V.E., Barbashev, E.A., Dorofeev, Y.I. et al. (1996). Simulation of polymer film and surface behaviour in a space environment. *Applied Surface Science* 92: 441–446.

4 Liu, Y., Liu, X., Li, G. et al. (2010). Numerical investigation on atomic oxygen under cutting of the protective polymer film using Monte Carlo approach. *Applied Surface Science* 256 (20): 6096–6106.

5 Grossman, E. and Gouzman, I. (2003). Space environment effects on polymers in low earth orbit. *Nuclear Instruments and Methods in Physics Research Section B: Beam Interactions with Materials and Atoms* 208: 48–57.

6 Ding, M.X. (2006). *Polyimides: Relationship Between Chemistry, Structure and Properties and Materials[M]*. Beijing: Science Press.

7 Li, G., Wang, L., Ni, H. et al. (2001). Polyhedral oligomeric silsesquioxane (POSS) polymers and copolymers: a review. *Journal of Inorganic and Organometallic Polymers* 11 (3): 123–154.

8 Minton, T.K., Wright, M.E., Tomczak, S.J. et al. (2012). Atomic oxygen effects on POSS polyimides in low earth orbit. *ACS Applied Materials & Interfaces* 4 (2): 492–502.

9 Lei, X.F., Qiao, M.T., Tian, L.D. et al. (2015). Improved space survivability of polyhedral oligomeric silsesquioxane (POSS) polyimides fabricated via novel POSS-diamine. *Corrosion Science* 90: 223–238.

10 Tomczak, S.J., Marchant, D., Svejda, S. et al. (2004). Properties and improved space survivability of POSS (polyhedral oligomeric silsesquioxane) polyimides. *Materials Research Society Symposium Proceedings* 851: 395-406.

11 Duo, S.W., Song, M.M., Liu, T.Z. et al. (2012). SiO_2 Coatings Prepared by Sol-Gel Process Protecting Silver from Atomic Oxygen Erosion. *Applied Mechanics and Materials* 121–126: 3044–3047.

12 Heimann, R.B., Kleiman, J.I., Litocsky, E. et al. (2014). High-pressure cold gas dynamic (CGD)-sprayed alumina-reinforced aluminum coatings for potential application as space construction material. *Surface and Coating Technology* 252: 113–119.

13 Kleiman, J.I. (2010). Surface modification technologies for durable space polymers. *MRS Bulletin* 35 (1): 55–65.

14 Shu, M., Li, Z., Man, Y. et al. (2016). Surface modification of poly (4, 4-oxydiphenylene pyromellitimide) (Kapton) by alkali solution and its applications to atomic oxygen protective coating. *Corrosion Science* 112: 418–425.

15 Liu, K., Mu, H., Shu, M. et al. (2017). Improved adhesion between SnO_2/SiO_2 coating and polyimide film and its applications to atomic oxygen protection. *Colloids and Surfaces, A: Physicochemical and Engineering Aspects* 529: 356–362.

16 Edward, M.S. (1995). Space Environmental Effects on Spacecraft: LEO Materials Selection Guide TRW Space & Electronic Group: 14–19.

10

Smart Polyimide Dielectrics

Deyan Kong, Yu Kang, Shuting Jiang, Hongxia Xie, Pengjie Wu, Ying Jiang, Yueheng Qiao, and Xinli Xiao

Harbin Institute of Technology, Department of Applied Chemistry, School of the Chemistry and Chemical Engineering, 92 Xidazhi Street, Nangang District, Harbin City, Heilongjiang Province 150001, P. R. China

10.1 Introduction

The twenty-first century is the era of materials, and in the new era, materials with diverse functions and excellent performance can be better applied to various aspects of production and life. Nowadays, the intelligence of materials has become a research hotspot in the field of science.

Smart materials have seven major functions, namely sensing function, feedback function, information recognition and accumulation function, response function, self-diagnosis ability, self-healing ability, and adaptive ability [1]. Smart materials have excellent performance and can better adapt to and meet the multiple needs of modern society, which is the cornerstone of high-tech development. Smart materials enrich the connotations of functional materials and structural materials, enabling their cross-fusion and achieving material diversification. As scientists predict, smart materials will lead the trend of material science development.

Shape memory polymer as an emerging smart material, compared with the traditional shape memory alloy, has the advantages of low cost, light weight, large shape variable (up to 800%), easy processing, a variety of response ways, wide range of stimulus response, and good chemical stability and biocompatibility; some materials also have a certain transparency and adjustable biodegradation rate. Therefore, the deformable structure based on this smart material shows great application potential in many fields such as biomedicine and aerospace.

Among numerous smart materials, shape memory polyimide (SMPI) is a polymer material with excellent intelligence. SMPIs can be activated through various methods such as heat, magnetic field, electricity, light, sound, chemical solvents and humidity [2–5]. At the same time, SMPIs have the advantages of low density, large recoverable deformation, good biocompatibility [6], chemical corrosion resistance [7], easy control of deformation temperature, and easy recycling and reuse. They

Functional Polyimide Dielectrics: Structure, Properties, and Applications, First Edition.
Edited by Jun-Wei Zha.

have potential applications in textile weaving, biomedical equipment, aerospace components, and space satellites, and have received widespread attention from the scientific and industrial communities.

10.2 Shape Memory Polymer

10.2.1 Introduction of Shape Memory Polymer

As a typical representative of stimulus-responsive deformable polymer materials, shape memory polymer deformation includes two steps: shape programming and recovery. Shape programming refers to the process by which the material deforms and fixes its temporary shape under certain conditions, while shape recovery refers to the process by which the material returns to its initial shape under heat, light, pH, and other stimuli [8] (Figure 10.1). The property by which SMP completes the shape programming and return cycle is called the shape memory effect (SME) of the material.

There are many ways to classify SMP, and they tend to be categorized according to a particular orientation. Common classifications are as follows:

(A) According to different stimulus drive methods, SMP can be divided into: (i) Heat-driven type: this is the most common type of SMP. Through direct or indirect heating (such as photothermal effect and electric/magnetothermal effect), the material temperature is changed, which causes the material to produce shape memory effect. This method is relatively easy to control and has a rapid response. (ii) Electric-driven type: It is usually composed of heat-driven SMP and conductive material, using electric field to heat the material, so that the material is subjected to thermal stimulation and return to shape. This material not only has good electrical conductivity, but also is easy to operate, has higher controllability, and is widely used in the field of electronics. (iii) Magnetic-driven type: The heat-driven SMP is combined with a ferromagnet or soft magnetic material to generate heat under the action of a magnetic field, so that the material has a shape memory process. (iv) Light-driven type: it refers to the shape memory effect of the material under the stimulation of the external light source. Light-driven SMP can be divided into two types of polymers – with glass transition temperature (T_g) or melting temperature (T_m) and containing photoactivity factors. Such materials are widely used in printing and other fields because they are fast, precise, and can be controlled remotely. (v) Solution-driven type: It is the shape memory effect caused by the hydrogen bond between the material and the solution or the glass transition temperature change of the material in the solution, and its operating conditions are mild. (vi) pH-driven type: reversible cross-linking can occur to changes in pH value, resulting in shape memory behavior.

(B) According to the number and direction of shape changes, SMP can be divided into: (i) unidirectional dual SMP, which is the most common and most basic class, composed of a cross-linked network and a reversible switch, can be

Figure 10.1 Deformation process of SMP: Shape programming and shape recovery.

programmed from a permanent shape to a temporary shape, and eventually return to a permanent shape, but each shape change needs to be reprogrammed; (ii) unidirectional multiple SMP, which is composed of a cross-linked network and a number of reversible transition switches, can be programmed from permanent shape to two or more temporary shapes, and gradually return to the initial permanent shape; (iii) bidirectional SMP, which is considered to be a kind of SMP with great potential for future application, can realize reversible transformation between two or more shapes when subjected to opposite stimuli without secondary programming.

(C) According to the reprocessability of the polymer, SMP can be divided into: (i) thermoplastic SMP, usually refers to physical cross-linking. The SMP has excellent extensibility and re-processability, but the irreversible segment slip will lead to its poor shape memory effect; (ii) thermosetting SMP, with relatively greater shape recovery stress, is more suitable for the preparation of the drive, but cannot be reprocessed; (iii) thermal adaptive SMP, both the advantages of the above two types of SMP, in order to ensure the excellent shape memory effect at the same time, but also through reversible dynamic bond to give the material permanent shape solid-state reshaping, self-repair, reprocessing, and other multiple advantages, has become one of the research hotspots in recent years.

10.2.2 The Shape Memory Molecular Mechanism

The memory properties of shape memory polymers are mainly due to the two-component structure of the material: the fixed phase and the reversible phase (otherwise known as the intersection point and the molecular switch). Among them, the fixed phase mainly plays a role in forming the skeleton of the original shape of the memory material and provides the driving force for the temporary shape to become the original shape. It can be a chemical cross-linking point (thermosetting) or a molecular entanglement, van der Waals force, etc. as a physical cross-linking point (thermoplasticity). The reversible phase, on the other hand, is used to produce a temporary shape of the material when the polymer is deformed, usually crystalline or amorphous. The reversible phase freezes or softens near the transition temperature of the material, resulting in a change in the overall mechanical properties of the material and in a change in the shape [9]. Therefore, the switching temperature of the shape memory polymer, T_{trans}, can be either the melting temperature, T_m, or the glass transition temperature, T_g.

The typical unidirectional shape memory process is shown in Figure 10.2a [10]. First, the SMP is heated above the shape memory transition temperature (T_{trans}) to give a temporary shape under the action of an external force, after which

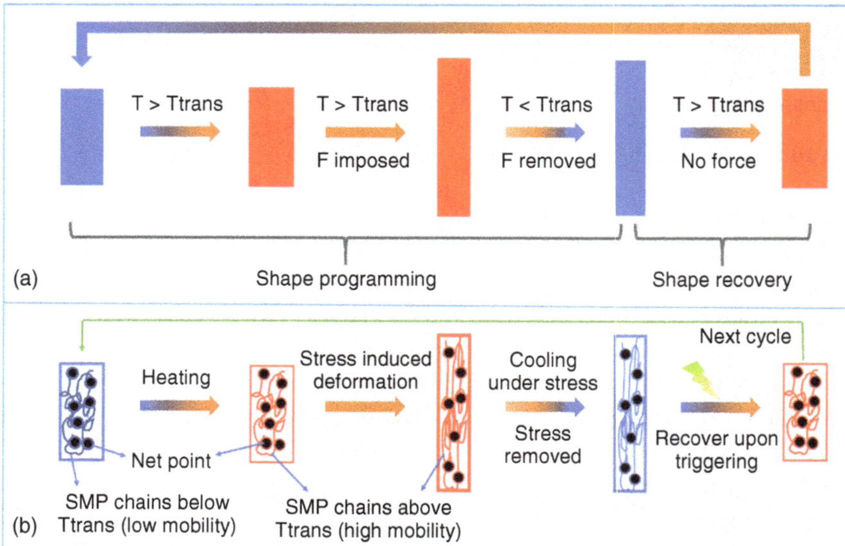

Figure 10.2 Shape memory process of unidirectional SMP: (a) The shape memory process of one-way SMP; (b) the molecular mechanism of the shape memory process of SMP. Source: Ref. [10]/with permission of Elsevier.

the maintenance of the external force cools the SMP below the T_{trans}, and the SMP can still maintain the temporary shape after the external force is removed. When reheated above T_{trans}, the SMP reverts to its original shape. In terms of the microscopic molecular mechanism, the shape memory effect results from the "activation" and "freezing" of molecular segments caused by external stimuli. Figure 10.2b shows that, at the microscopic scale, the molecular chain of SMP can change from the frozen state (blue) to the activated state (red) when heated to T_{trans}. The "black dots" are the network cross-linking points in SMP that are used to remember permanent shapes. The initial shape of the SMP is heated above T_{trans}, at which point the reversible phase changes from a "frozen" state to a highly elastic state, and the molecular segments between the stationary phases appear to be freely curled. Then, when the temperature drops below T_{trans}, the reversible phase changes from a highly elastic state to a "frozen" state, and the internal stress is also stored in the "frozen" molecular chain. When SMP is heated to T_{trans} again, the "frozen" molecular chain segment is transformed into an active molecular chain segment, and the internal stress is released to drive the SMP shape recovery, thus completing a thermodynamic cycle process.

10.2.3 Shape Memory Performance Test

The shape memory properties were measured by DMA under the tension mode by the following process: (i) the sample with a constant force was stretched at an elevated temperature ($T_{high} = T_g + 50\,°C$), (ii) the sample was then cooled to a low temperature ($T_{low} = 20\,°C$) at a rate of 5 °C/min, and held at that temperature for

5 minutes, (iii) the external load was then removed and the sample was held for another 5 minutes, and finally (iv) the sample was reheated to T_{high} at 5 °C/min and held for more than 1 hour to observe the shape recovery process. The shape memory properties were then evaluated by shape fixation (R_f) and shape recovery ratio (R_r) according to Eqs. (10.1) and (10.2). R_f represents the capability of fixing the temporary shape, whereas R_r represents the capability of memorizing the original shape. In the equations, ε_x represents the strain of each shape obtained from the strain curve obtained from DMA.

$$R_f = (\varepsilon_B - \varepsilon_A) / (\varepsilon_{B,load} - \varepsilon_A) \tag{10.1}$$

$$R_r = (\varepsilon_B - \varepsilon_{A,rec}) / (\varepsilon_B - \varepsilon_A) \tag{10.2}$$

10.3 Heat-driven SMPI

In 2012, Yoonessi from NASA and others first reported on polyimides with shape memory effect. Researchers introduced graphene into the polyimide matrix and successfully prepared graphene/polyimide composite materials. The test results showed that the mechanical and thermal mechanical properties of the composite materials were improved, and the shape recovery rate of the composite material was greatly improved through experiments, proving that it has better shape memory performance. Figure 10.3 shows its shape memory curve at 230 °C. This study is considered the starting point of SMPI research, opening the door to SMPI research. Since then, domestic and foreign researchers have followed closely and started studying SMPI.

In 2015, Xiao et al. used 4,4′- diaminodiphenyl ether (ODA) and hexafluorodianhydride (6FDA) for polymerization to prepare SMPI thin films with flexible units in the main chain. By adjusting the proportion of diamine, the molecular weight (Mn) of thermoplastic SMPI was controlled, and T_g could be adjusted from 299 °C to 325.1 °C. The shape memory recovery process is shown in Figure 10.4. Xiao found that due to low-density covalent cross-linking, thermosetting SMPI exhibits higher

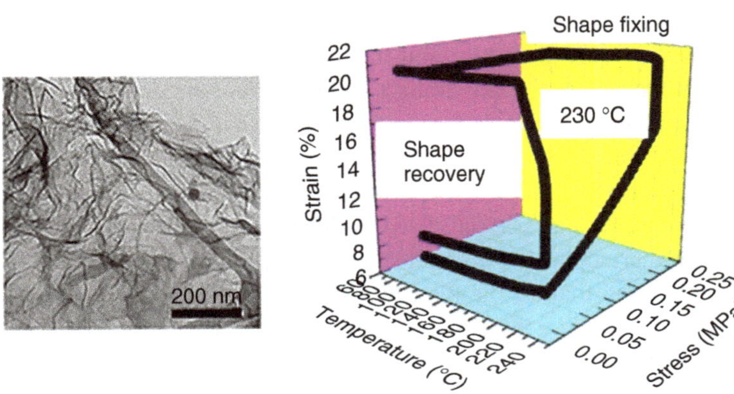

Figure 10.3 Shape memory effect curve of graphene-reinforced polyimide. Source: Ref. [11]/American Chemical Society.

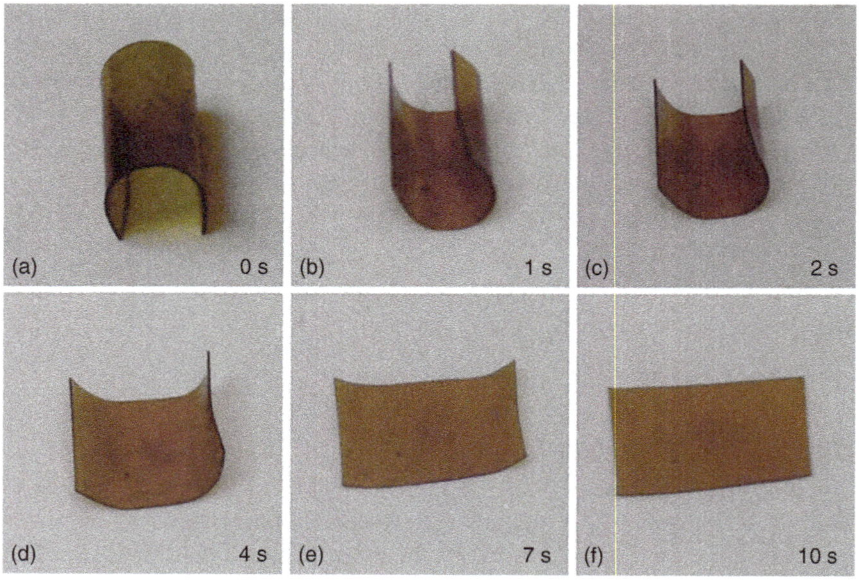

Figure 10.4 Shape recovery process of shape memory polyimide. Source: Ref. [12]/American Chemical Society.

T_g and better shape memory performance than SMPI. This study explores the shape memory effect mechanism of high-temperature SMPI based on molecular chain flexibility, molecular weight, and cross-linking density, providing a theoretical and experimental basis for the study of high-temperature SMPs.

T_g is a key factor determining the application of high-temperature shape memory polymer materials, but the preparation conditions of high T_g polymer materials are generally complex and cumbersome, and the economic cost is high. In 2017, Xiao et al. obtained a high-temperature shape memory polymer with controllable T_g by copolymerizing BPADA with diamine combinations of BAB and BAPB. The T_g of the SMPI produced varies between 183 and 230 °C, and the relationship between T_g and its components conforms to the Fox equation. The significant difference in the storage modulus between the glass state and the high elastic state of the material gives it a high shape fixation rate, while strong $\pi - \pi$ interactions and a large number of molecular chain entanglements give the material a good shape recovery rate. The preparation of SMPI through simple copolymerization reaction is an effective way to obtain high-temperature shape memory polymers with ideal properties.

In 2015, Wang synthesized a series of PIs with different molecular structures to check whether neat thermoplastic PIs could show high shape memory properties. Interestingly, Wang found that all the PIs could show shape memory properties, shown in Figure 10.5. The influences of chain structure and cycle time on shape memory properties were also carefully investigated to confirm the crucial factor of shape memory properties for PIs. Wang also found that the addition of nanosilica to the PI networks could further improve the shape memory properties due to the introduction of hydrogen bonds.

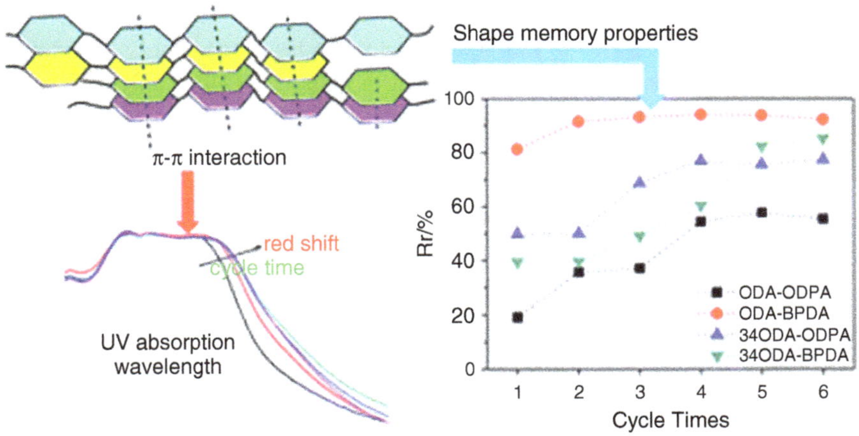

Figure 10.5 Shape memory properties of PIs during different cycle times.

10.4 Remote-driven SMPI

In early stages, SMP is typically referred to a kind of thermal-responsive material. As a stimulation approach, thermal field exhibits the advantages of wide source, simple operation, and easy implementation [13, 14]. However, thermal-actuated SMPs cannot be remotely controlled, which limits their further applications in wireless-controlled smart systems [15]. Remote-actuated SMPs have fast, accurate, wireless-controlled shape-morphing properties [16], and have become a current hot topic due to their great prospects in many fields. Electrical and light-responsive shape memory polyimides can be driven remotely.

10.4.1 Electrical-driven SMPI

Like common polyimides, SMPIs are electrically insulated, which limits their applications. Electrically driven shape memory polyimide is usually composed of thermally driven SMP combined with conductive material, using electric field to heat the material, so that the material is heat stimulated and restore the shape. This material consists of conductive fillers combined with the SMP matrix. At the same time, compatibility between filler and matrix is required. The electrical response mechanism is to apply a certain alternating voltage to the material, causing the conductive layer inside the material to generate current and ensure its transfer to the substrate layer. After the electric thermal conversion is completed, the material is endowed with heat energy. When the glass transition temperature (T_g) is reached, its shape memory behavior can be activated. The conductivity and thermal conductivity of composite materials are key influencing factors for electrically driven shape memory polymers. Various conductive fillers such as carbon nanotubes, carbon fibers, graphene, conductive polymer material, and silver wires have been introduced into polymers for the synthesis of electrically driven SMP materials [17–19].

Transparent metal mesh is used to replace the conductive fillers to construct a conductive network for electrical actuation for the first time. The self-cracking aluminum mesh is embedded in the transparent shape memory polyimide film (Alm@TSMPI) by solution casting. As a new type of flexible and transparent heater, the Alm@TSMPI has advantages of fast response and high steady-state temperature. As an electric actuator, the Alm@TSMPI is active and deformable depending on its variable stiffness characteristics, and reverts to its original shape within 13 seconds under electric stimulation. Figure 10.6 shows the shape memory recovery process of the Alm@TSMPI film under electrical stimulation. This is the first report on an electro-active transparent shape memory polymer while maintaining light transmission [20].

Electroactive shape memory polyimide (EASMPI) is synthesized by incorporating ground carbon fibers (GCFs) into a polyimide matrix to form a compact conductive network, and percolation threshold of $0.312\,S\,m^{-1}$ meets the demands of both electro-active and shape memory effects. The glass transition temperature of 302 °C for EASMPI is much higher than that of other electroactive SMPs, and its shape recovery is activated effectively by a voltage of 15.87 V. Recovery stress is crucial for shape memory materials to output work, and EASMPI produces high recovery stress of 40.1 MPa. EASMPI shows vast potential in practical applications with its electroactive shape memory effect and high recovery stress at high temperature .

10.4.2 Optically-driven SMPI

The optically driven shape memory polymers are materials that exhibit reversible deformation or recovery in response to light. The light source for this process can include sunlight, infrared laser, or ultraviolet light. Based on its structural composition, it can generally be categorized into two types: the incorporation of photoisomerized chemical groups in the polymer or the addition of fillers with photothermal effects. Common chemical groups include cinnamic acid groups, azophenyl groups, and spiral pyranes. When exposed to light, these groups undergo photochemical reactions, or photoisomerization reactions, in order to fix or restore the shape of the material. Fillers with photothermal effects include polydopamine, metal nanoparticles, carbon-based materials, conjugated polymers, and dye molecules.

A chemical network with azobenzene as the side chain creates a polyimide that can integrate multiple shape controls. Trifunctional cross-linkers serving as net points of the chemical network render polyimide thermal-induced shape memory effects, which enable shape transformation. (Figure 10.7) Azobenzene as a photoresponsive group is employed to achieve the photofixity and reversible photodeformability. Such photo-sensitive behaviors are independent of molecular prealignment and remain available after thermally shaping and fixing. As a result, these noninterfering performances induced by heat and light allow us to arbitrarily combine them to meet different needs. This study provides a new approach for the preparation of multifunctional actuators and has potential applications in the field of intelligent drivers [21].

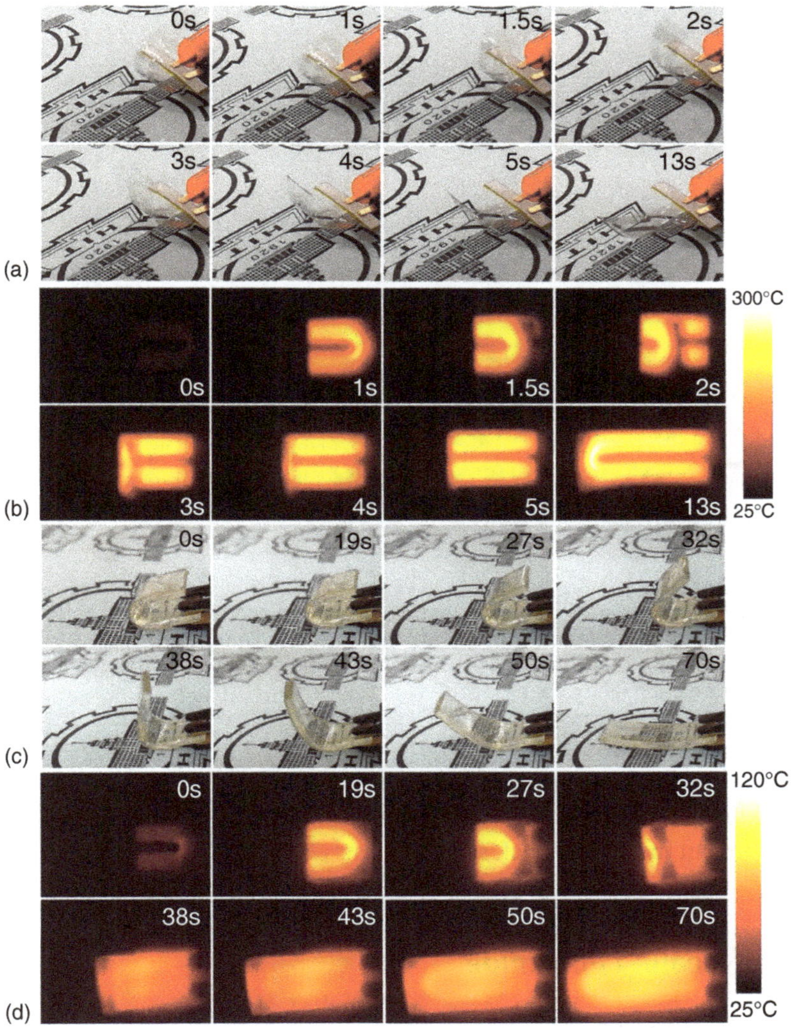

Figure 10.6 (a) Digital photos and (b) infrared thermal images of shape memory recovery process of the Alm@TSMPI at 12 V DC, (c) digital photos and (d) infrared thermal images of shape memory recovery process of transparent SMPS at 6 V DC. Source: Ref. [20]/with permission of Elsevier.

A new 3D printed polyimide (PI)/paraffin wax (PW)/carbon nanotube (CNT) NIR-responsive photoactuators with controlled deformation mode and ultrasensitive light response has been proposed (Figure 10.8). The process involves printing a CNT slurry according to a predetermined path, followed by lamination with PW and PI film to create PI/PW/CNT (PPC) NIR responsive photoactuators. Then, the effects of the CNTs pattern (line width and line spacing) on the deformation efficiency were studied and optimized. Upon optimization, the deformation time can be reduced to 3 seconds and the deformation angle can be raised to 70°. Moreover, the PPC actuator is highly sensitive to temperature and

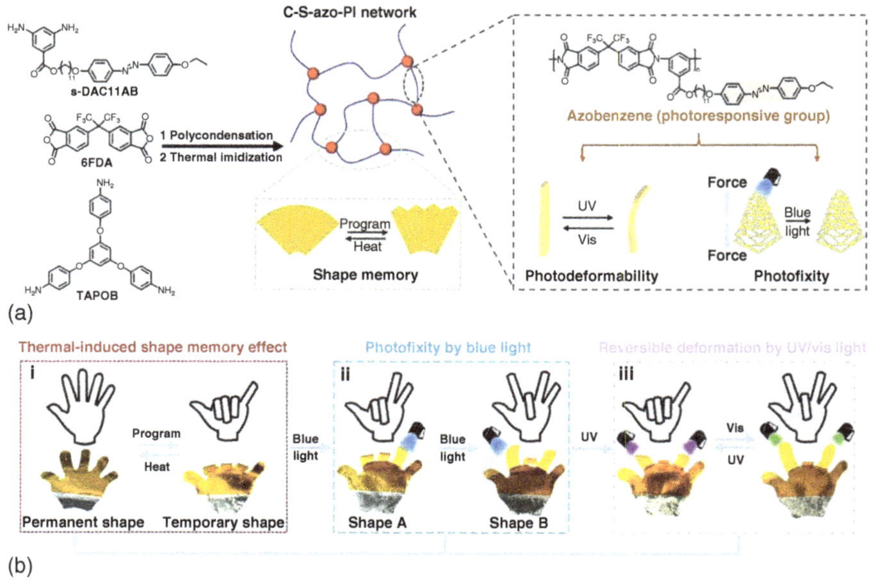

Figure 10.7 (a) The chemical structures of reactive monomers and cross-linkers, as well as the schematic network of the resulting polyimide with integrated multiple shape manipulation (c-s-azo-PI). (b) Photographs showing the shape programming (i), photofixity (ii), and photodeformation (iii) of the c-s-azo-PI film. Source: Ref. [21]/with permission of Elsevier.

can be deformed within a temperature difference of about 2–3 °C. Finally, various PPC photoactuators including flower-shape and soft mechanical hand have been prepared to present the different deformation forms [22].

10.5 Self-healing Polyimide

10.5.1 Introduction of Self-healing

With the development of modern social sciences, polymer materials are increasingly widely used in various aspects of industry and life. However, during the use of polymer materials, macroscopic damage and microcracks are inevitable due to the influence of external environment, which affects the normal use of the material and leads to material fracture, shortening the service life of the material. For externally visible cracks, traditional repair techniques such as mechanical connections or adhesive bonding are generally chosen for repair, while more invisible small cracks cannot be repaired using these traditional methods. Therefore, there is an urgent need for a new type of repair method to repair these small cracks.

Self-healing polymer materials are generally divided into two types: external self-healing polymer materials and intrinsic self-healing polymer materials [23]. External self-healing polymer materials store repair agents in one-dimensional,

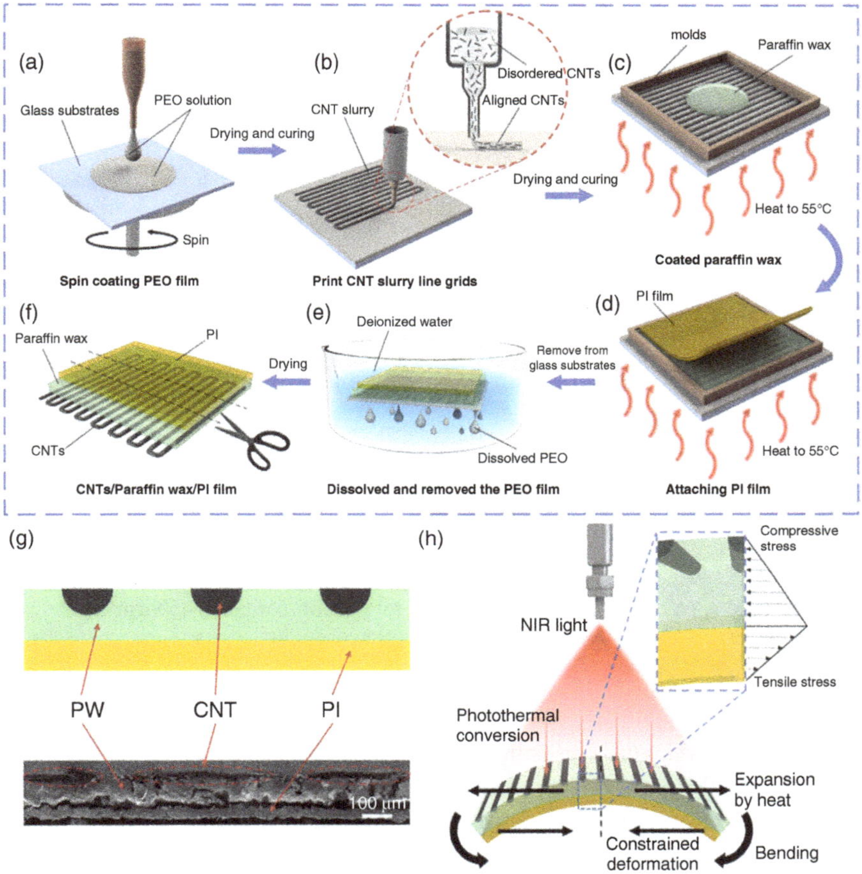

Figure 10.8 (a–f) Preparation process. (g) Composition structure of prepared PPC photoactuators. (h) Schematic diagram of deformation principle of PPC photoactuators. Source: Ref. [22]/with permission of Elsevier.

two-dimensional, or three-dimensional repair units (such as capsules, laminates and fiber networks), and then add repair units to the matrix material. When cracks and damage in the material extend to the repair unit, the repair agent flows out to heal the wound, as shown in Figure 10.9.

External self-healing polymer materials are repaired through the chemical reaction of repair agents, thus, having the advantages of fast response and repair speed. But they have problems such as difficulty in repeated repair, short lifespan of repair agents, and difficulty in thermoplastic processing. The self-healing mechanism of intrinsic self-healing polymer materials is to achieve self-healing of polymer cross-linked networks through dynamic bonds breaking, recombination, or dynamic exchange in the main or side chains of molecular chains under external stimuli. Unlike externally assisted self-healing, intrinsic self-healing polymer materials do not require the addition of repair agents. At the same time, the stability of dynamic bonds is much higher than that of reactive repair agents, and they are not

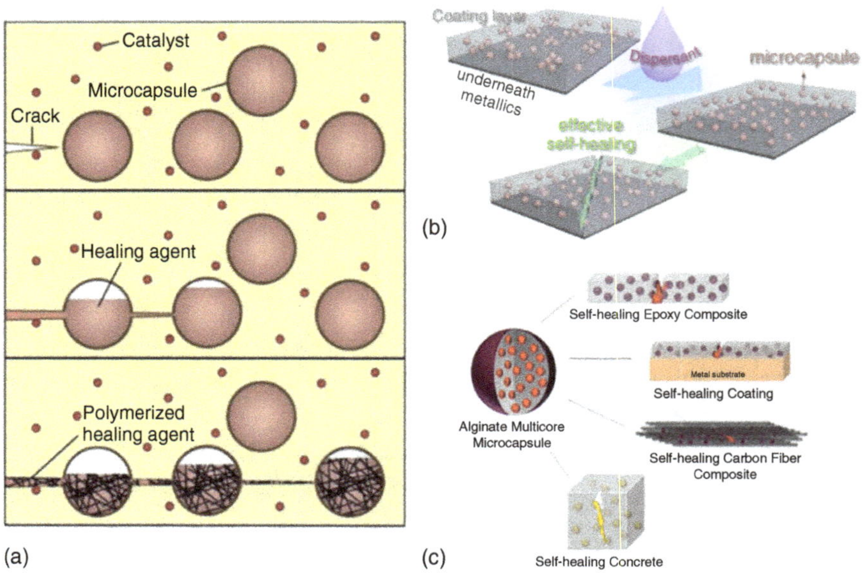

Figure 10.9 (a) Repair process of microcapsules; (b) the anti-corrosion application of microcapsule self-healing resin; and (c) the application of sodium alginate microcapsules. Source: Ref. [23]/Springer Nature.

easily lost during the repair process. Therefore, theoretically, intrinsic self-healing polymer materials can undergo infinite repairs.

According to the types of reversible chemical reactions, they can be divided into two categories: non-dynamic covalent bond self-healing (physical type) and dynamic covalent bond self-healing (chemical type), as shown in Figure 10.10. The upper row is reversible non-covalent bond self-healing, mainly relying on intermolecular interactions, while the lower row is reversible covalent bonds [24].

Reversible non-covalent bonds mainly include hydrophobic interactions, host guest interactions (metal ligand interactions), hydrogen bonding, ionic bonding, crystallization, polymer nanoparticle interactions, and multiple intermolecular interactions. Dynamic covalent bonds generally include disulfide bonds, boron ester bonds, ester bonds, imine bonds, acylhydrazone bonds, and Diels Alder (D–A) cycloaddition reactions. Compared with dynamic noncovalent bonds, dynamic covalent bonds have higher energy and are more advantageous for the mechanical functions of polymer materials. Meanwhile, dynamic covalent bonding can endow traditional thermosetting polymer materials with the machinability of thermoplastic materials, which is of great significance for the recycling and processing of thermosetting polymer materials.

Dynamic covalent bonds can be divided into two categories: ordinary reversible reactions and dynamic reversible reactions. Most of the dynamic covalent bonds belong to the category of dynamic reversible reactions, such as disulfide bonds, boron ester bonds, ester bonds, imine bonds, and acylhydrazone bonds. The characteristic of the reaction is that in different external environments, the forward polymerization rate constant and the reverse depolymerization rate constant always

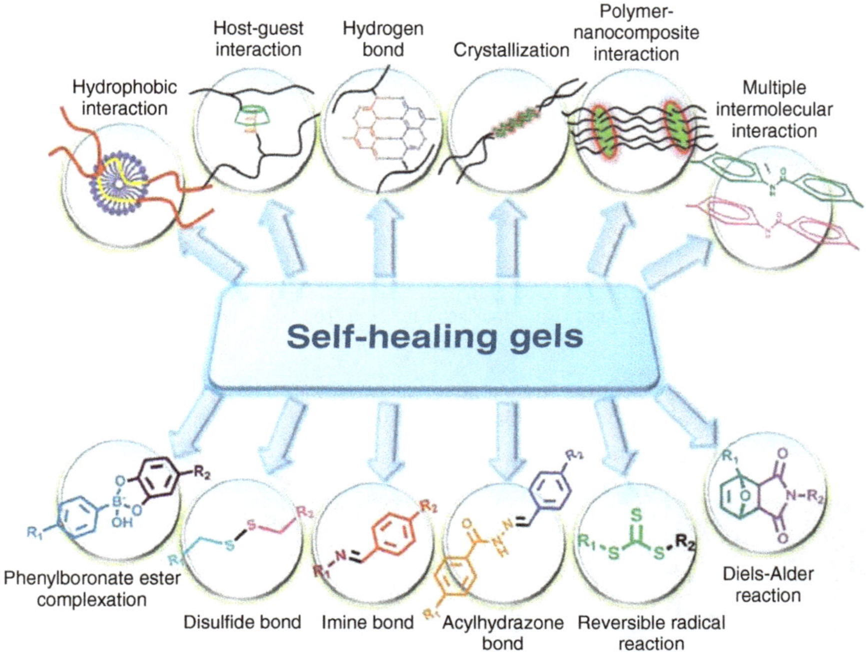

Figure 10.10 Classification of intrinsic self-healing materials. Source: Adapted from Ref. [24].

remain the same, and the formation and breaking rates of dynamic covalent bonds are the same. Overall, the number of dynamic covalent bonds remains unchanged. The D–A reaction and the hydrolysis/dehydration of boron ester bonds belong to ordinary reversible reactions, with different polymerization rate constants and reverse depolymerization rate constants.

10.5.2 Development Status of Self-healing Polyimide

Compared with the self-healing of elastomers and gel, polyimide has been considered as a kind of material unable to achieve self-healing in the past due to its difficult molecular structure to realize dissociation and polymerization and its high T_g. Until recent years, there have been reports on the self-healing performance of polyimide and its composite materials, and their research is still in its early stages.

10.5.2.1 External Self-healing Polyimide

In 2016, Susa replaced aromatic diamines with long-chain aliphatic branched fatty dimers and achieved self-healing by relying on branched van der Waals forces [25]. The resulting material exhibited high self-healing efficiency, as shown in Figure 10.11; however, it had low tensile strength (<6 MPa) and T_g (20 °C), showing significant differences in performance compared to traditional polyimides and no advantages compared to other self-healing polymer materials.

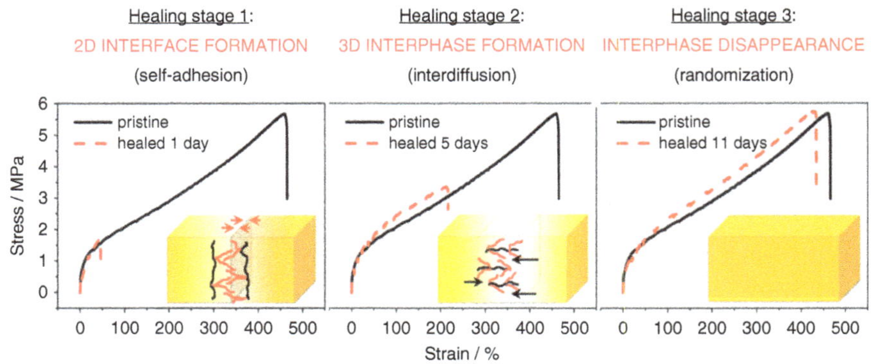

Figure 10.11 Effect of self-healing time on repair performance of semi aromatic polyimide. Source: Ref. [25]/American Chemical Society.

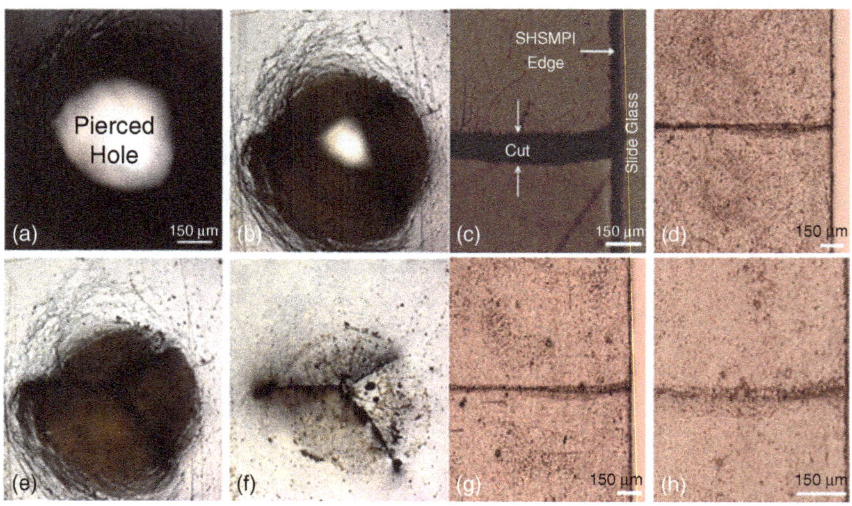

Figure 10.12 SHSMPI perforation and scratch self-healing process diagram. Source: Ref. [2]/Elsevier.

In 2019, Kong prepared SHSMPI, which can repair various damages at higher temperatures, by adding 8% weight fraction of fusible thermoplastic polystyrene (PS) to the SMPI matrix [2]. The results indicate that SHSMPI has good shape memory performance and thermal mechanical properties, and can effectively self-healing damage such as cracks, perforation, and cutting at 243 °C, as shown in Figure 10.12.

In 2019, Li blended traditional polyurethane (PU) with functional polyimide to prepare a self-healing polymer (PUPI) system with high mechanical properties [26], as shown in Figure 10.13. The excellent self-healing and mechanical properties are attributed to a unique intermolecular network, which is generated by strong supramolecular interactions between the amino ester groups in polyurethane and the imide groups in polyimide. Therefore, introducing intrinsic self-healing

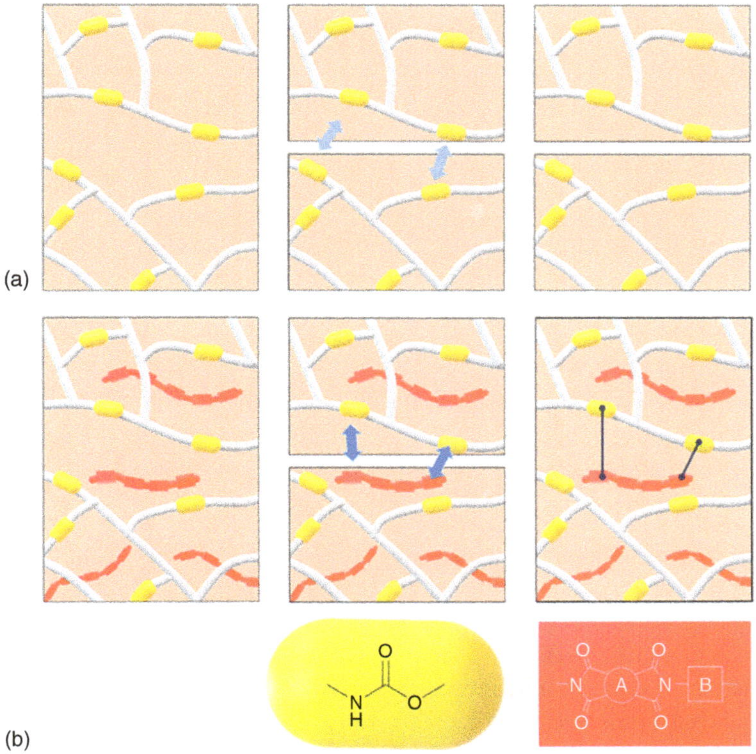

Figure 10.13 (a) Schematic diagram of self-healing mechanism of PU and (b) PUPI. Source: [26]/American Chemical Society.

polymer materials into polyimide systems to achieve self-healing ability is of certain significance for the development of self-healing materials.

In 2017, Lei utilized the mechanism of dynamic reversible exchange of imine bonds through trace amounts of primary amines to prepare polyimide imide thermosetting hybrid polymers, which exhibited excellent mechanical properties (tensile strength up to 94.1 MPa, tensile modulus up to 1.96 GPa), high glass transition temperature (274 °C), and good thermal stability (thermal weight loss temperature up to 429 °C) [27]. Subsequently, in 2019, Lei Xingfeng prepared thermosetting polyimide films based on imine bonds using the same principle [28]. Two materials can self-heal at temperatures below T_g. However, Lei only tested the self-healing of the adhesive state and did not test the self-healing of scratches. At the same time, Lei needed to add a drop of extremely dilute primary amine solution and hot press for 2 hours to the fitting area of the spline during repair, which does not have practical application value.

10.5.2.2 Intrinsic Self-healing Polyimide

Under normal circumstances, dynamic reversible exchange of imine bonds can only occur at temperatures higher than the T_g of the polymer. However, if trace amounts of primary amine are added during the repair reaction, the imine bonds

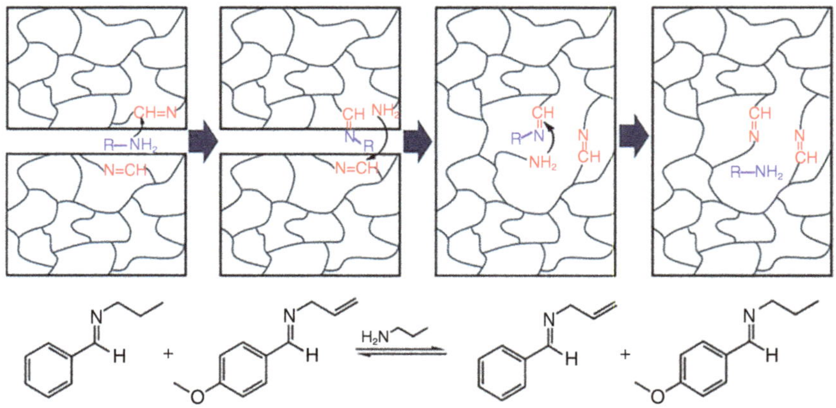

Figure 10.14 Schematic diagram and reaction equation of the healing mechanism of imine exchange reaction promoted by primary amine. Source: Ref. [28]/Oriprobe Information Services, Inc.

in the polyimide can undergo rapid dynamic exchange even under mild conditions below T_g, causing the polyimide molecular chains to recombine and achieve self-healing effects. At temperatures below T_g, the dynamic imine bonds in rigid polyimide molecular chains cannot undergo thermodynamically driven reversible exchange reactions. However, when trace amounts of primary amine are added, the kinetic-induced imine exchange reaction can still occur and the molecular chains can be reorganized. Therefore, polyimide based on imine bonds achieved self-healing at temperatures below T_g under the action of trace amounts of primary amine (Figure 10.14). This research approach provides some inspiration for other dynamic covalent self-healing polymer materials and provides new guidance for the research and development of dynamic covalent self-healing materials.

In 2021, Kim introduced 4,4′- diaminodiphenyl disulfide (DTDA) into polyimide to prepare a self-healing polyimide film with good optical properties [29]. It can quickly and easily recover any damage caused by external stress at $T_g + 30\,°C$ to achieve self-healing. In August 2022, Wan synthesized a self-healing polyimide film using DTDA, 1,3-bis (4-aminophenoxy) benzene (BAPB) and bisphenol A type diethyl dianhydride (BPADA) as monomers [30]. The film has good Young's modulus ($E > 4$ GPa), high T_g ($T_g > 190\,°C$), and excellent insulation performance (breakdown strength $E_b > 300\,kV\,mm^{-1}$). In addition, Wang also focused on studying the damage mechanism and self-healing effect of the film after corona damage. However, the film cannot fully self-heal under individual heating conditions and can only be repaired under hot pressing conditions.

Besides, Wan designed a low-molecular-weight polyimide gene unit for cross-linking with polyimide ligases to prepare smart films [31], as shown in Figure 10.15. Due to the variability of gene units and ligase combinations, polyimide films that balance flexibility and rigidity are designed into three forms through a Mimosa-like biomimetic strategy to adapt to different application scenarios. At the same time, the film can self-heal through hydrogen bonding and imine, giving it good self-healing efficiency and excellent recyclability, making it reusable.

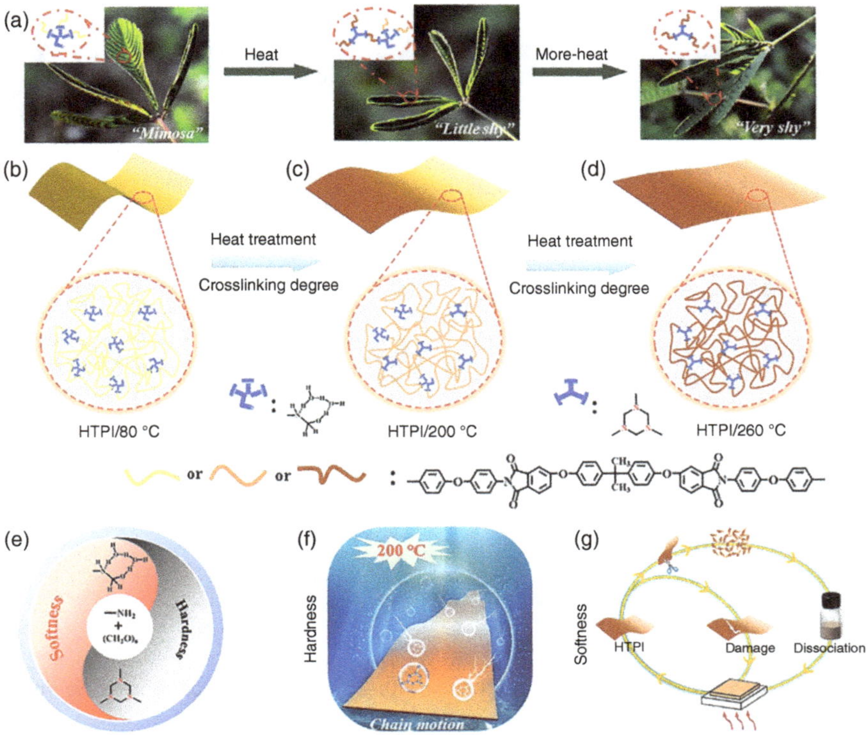

Figure 10.15 Schematic diagram of Mimosa smart thin-film principles. Source: [31]/John Wiley & Sons.

From the perspective of Mimosa biomimetics, the design concept of preparing high-strength dynamic polyimides through simple polyimide gene units and ligase cross-linking provides a solid foundation and clear example for the sustainable development of electrical and electronic polyimides.

In December 2022, Wan successfully prepared a dynamic cross-linked polyimide insulation film with fully closed-loop recyclability [32]. The film has good self-healing ability after mechanical/electrical damage due to the presence of imine bonds, as shown in Figures 10.16. The recycled and self-healing polyimide film still maintains its good mechanical properties, high thermal stability, excellent insulation performance, and breakdown strength ($E_b > 358\,\text{kV}\,\text{mm}^{-1}$), and is a very promising low-energy consumption and high-temperature-resistant insulation material. The method of utilizing high T_g structured polyimide through reversible exchange of imine bonds has promoted the wider application of this sustainable and recyclable material in the fields of electrical and microelectronics.

10.6 SMPI-based Flexible Optical/Electronic Devices

As we all know, shape memory polyimide has been used extensively on flexible electronics due to their good optical transmittance, high thermal and chemical

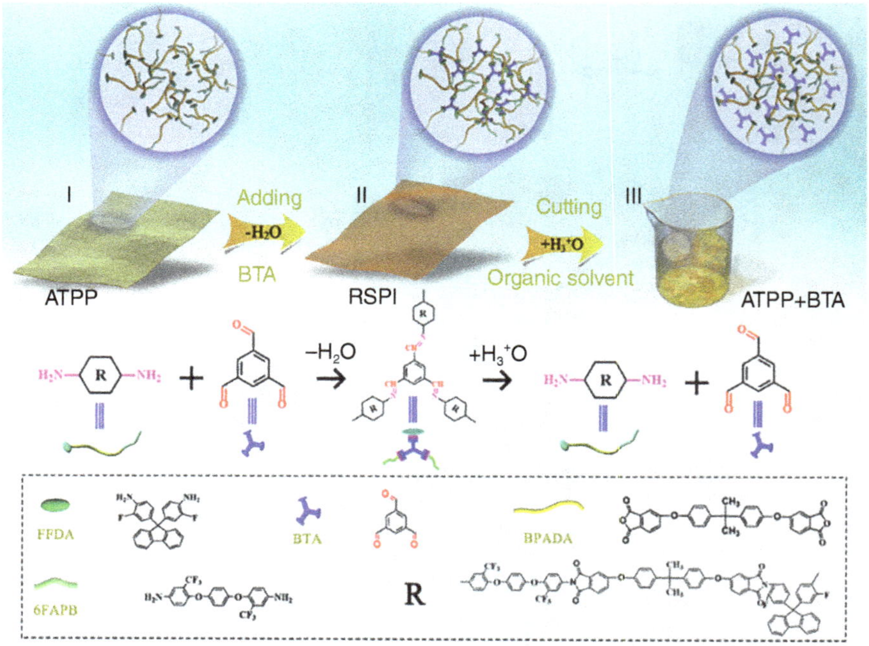

Figure 10.16 Schematic diagram of RSPI film synthesis and dissociation process. Source: Ref. [32]/John Wiley & Sons.

resistance, and high mechanical properties, especially in thin-film solar cells, organic thin-film transistors (OTFTs), organic light-emitting displays (OLED), sensors, and so on. Also, polyimide-based deployable structures and devices can be used in lunar base, deployable solar batteries, solar sail, sun shield of next-generation space telescope, and so on.

Smart, deformable, and transparent electrodes are a significant part of flexible optoelectronic devices. Huang et al. [33] proposed a novel approach for making highly transparent, smooth, and conductive shape memory polyimide hybrids. Colorless shape memory polyimide (CSMPI) with high optical transparency and high heat resistance is served as the substrate for flexible electronic devices for the first time, as shown in Figure 10.17. A hybrid (Au/Ag) metal grid electrode embedded in CSMPI (BMG/CSMPI) is first fabricated via self-cracking template and solution coating, the advantages of which include ultra-smooth surface, superior mechanical flexibility and durability, strong surface adhesion, and excellent chemical stability due to the unique embedded hybrid structure. The resulting white polymer light-emitting diodes (WPLEDs) based on BMG/CSMPI with shape memory effect are active and deformable, and are converted from 2D devices into 3D devices depending on their variable stiffness characteristics. Furthermore, ultra-thin and flexible 3D optoelectronic devices fabricated using shape memory polymers can promote the development of advanced optoelectronic applications in the future.

The current electroactive shape memory polymer composites (SMPCs) show low electric heating efficiency, and the transparency of the matrix is seriously damaged

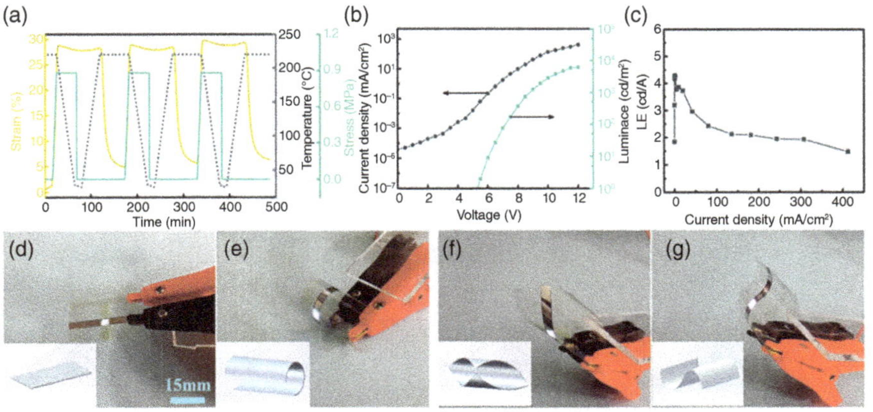

Figure 10.17 (a) Stretchable shape memory cycles of CSMPI substrate. (b) Current density–voltage luminance characteristics of the WPLEDs fabricated with BMG4/CSMPI. (c) Luminance efficiency–current density characteristics of the WPLEDs. Photo of WPLEDs lit at 8 V: (d) initial 2D planar device, (e, f) 3D cylindrical device after the shape is fixed; and (g) 3D waved device with fixed shape. Source: Ref. [33]/American Chemical Society.

by the conductive fillers. Huang et al. [34] used transparent metal mesh to replace the conductive fillers for the first time and to construct a conductive network for electrical actuation. The self-cracking aluminum mesh is embedded in the transparent shape memory polyimide film (Alm@TSMPI) by solution casting, as shown in Figure 10.18. As a new type of flexible and transparent heater, the Alm@TSMPI has advantages of fast response and high steady-state temperature. As an electric actuator, the Alm@TSMPI is active and deformable depending on its variable stiffness characteristics, and reverts to its original shape within 13 s under electric stimulation. This is the first report on an electroactive transparent shape memory polymer while maintaining light transmission.

10.7 Smart Polyimide Foam

In addition to polyimide film, polyimide foam plays an important supporting role in the development of national economy and industry due to its many excellent properties (Figure 10.19). Usually, polymer foams can be divided into open-cell foams and closed-cell foams, which correspond to thermoplastic foams with flexibility and thermosetting foams with rigidness, respectively. With regard to polyimide (PI) foams, they were first invented by DuPont and Monsanto Company in the 1960s, and then they gained interest from domestic and foreign academic communities as well as industrial circles for their excellent performance [35–37].

Polyimide foams, as one kind of polymer foam with imide rings in the main chains, are mainly composed of polyimide resin and internal open or closed pore structures with different sizes [38, 39]. In particular, polyimide foam materials have excellent thermal properties, splendid mechanical properties, preeminent environmental protection functions, which are superior compared with traditional polymer foams

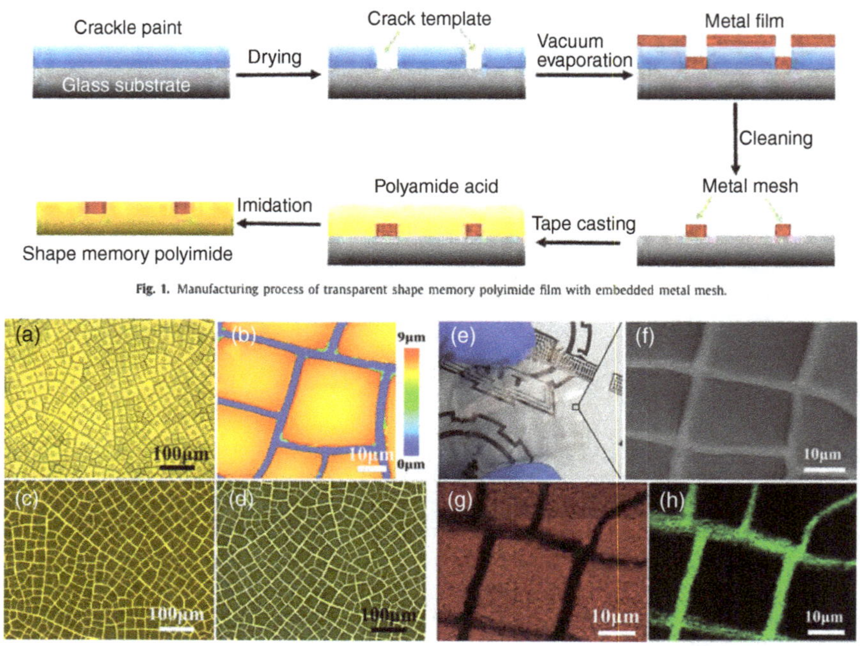

Figure 10.18 (a) Optical microscope and (b) laser confocal microscope image of micro-crack template. Optical microscope image of metal mesh (c) on glass substrate and (d) embedded in polyimide substrate. (e) Digital photo of TSMPI film with patterned transparent conductive circuit. (f) SEM image of metal mesh embedded in polyimide substrate. EDS mapping of (g) C element and (h) Al element on the Alm@TSMPI. Source: [34]/Elsevier.

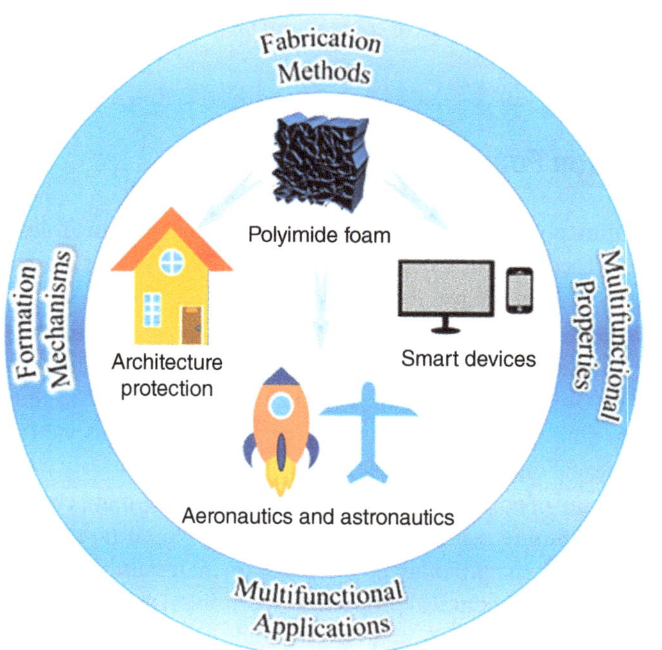

Figure 10.19 Application of smart polyimide foam.

(such as polyurethane foams, polyethylene foam, and polystyrene foam). Therefore, polyimide foams can not only act as structural materials but also play the role of functional materials. [40] Integrating the PI foams with functional materials such as Ag, rGO, and CoFe alloys can bring about more functions like sensing and electromagnetic shielding properties. [41] Therefore, polyimide-based foam materials belong to advanced functional materials, which are increasingly used as key materials for thermal insulation, vibration, and noise reduction in aerospace, aviation, transportation, military hardware, and other high-tech areas [42–44].

10.7.1 Electromagnetic Interference Shielding Performance of Polyimide-Based Foams

The polymer foams are ideal candidates for applications in the fields of thermal management, flexible sensors, electromagnetic interference (EMI) shielding due to their three-dimensional (3D) pore structure, high porosity, and light weight. Polyimide foams are a class of aromatic heterocyclic porous materials with imide and benzene rings on the macromolecular chain backbone, which endows PIFs with high mechanical strength, excellent thermal stability, flame retardancy, and thermal insulation properties.

Yang et al. [45] prepared lightweight PIFs with 3D interconnected framework using polyester ammonium salt (PEAS) precursors as the starting derivatives (Figure 10.20). Then, PI-based electrode with a highly efficient palladium nanocatalyst embellished 3D reduced graphene oxide/PI foam (Pd/3D rGO@PI foam) toward H_2O_2 electroreduction was fabricated by dip-drying and electrodeposition method. The Pd/3D rGO@PI foam electrode showed excellent electrocatalytic performance and good stability, and it was considered as a potential candidate for fuel cells. In addition, PIFs have been used as the starting derivatives to prepare carbon foams due to their high carbon yield under inert atmospheres.

Li et al. [46] carbonized rigid thermosetting PIFs to prepare lightweight carbon foams, and the carbon foams which were treated at 1200 °C presented excellent EMI

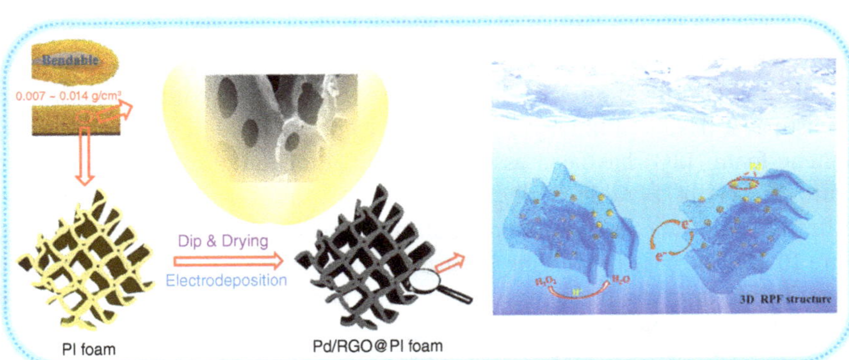

Figure 10.20 Polyimide-based electrode based on a highly efficient palladium nanocatalyst embellished three-dimensional reduced graphene oxide/polyimide foam. Source: Ref. [45]/American Chemical Society.

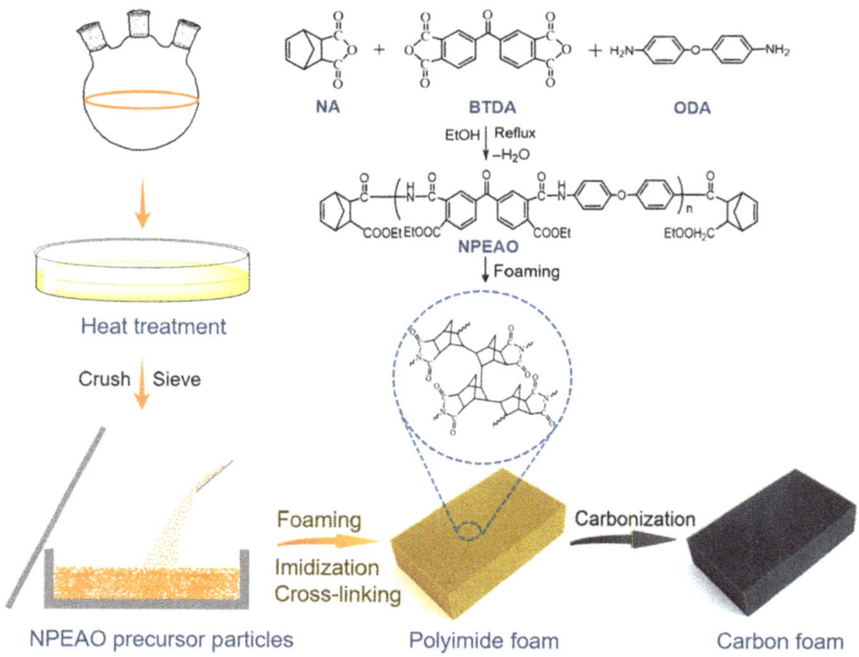

Figure 10.21 Schematic illustration for fabricating carbon foams derived from thermosetting PI foam. Source: Ref. [46]/with permission of Elsevier.

shielding performance of 54 dB at a sample thickness of 2.0 mm and a density of 91 kg/m^3 (Figure 10.21). These stiff CFs open up the possibility for the applications as structural–functional integrated materials in the fields of aerospace or telecommunication.

10.7.2 Sensing Performance of Polyimide-Based Foams

With the mechanical properties of the polyimide foam, researchers can produce sensors derived from polyimide composites. As previous study has proved that the structure and electronic properties can be observably altered with the adsorption of moisture, a flexible humidity sensor made of polyimide-derived laser-induced graphene foam (LIG) has been assembled to monitor the conductivity precisely (Figure 10.22a) [47]. Additionally, Figure 10.22b,c displays the LIG sensor on the flexible substrate and the intentional microgap for the sensor. As shown in Figure 10.22d, the peaks appearing in current curves indicate that the conductivity across the microgap can be increased, which can be attributed to the adsorption of moisture by polyimide when people breathe.

The schematic of a smart electronic skin (5 × 5 pixels) made of PI/CNT foam is exhibited in Figure 10.23a; wearable pressure sensors were patterned in electronic skin. [48] From Figure 10.23b,c,d, when giving press to the e-skin in different ways, the recorded 2D and 3D maps show mechanical stimuli position (the darker the

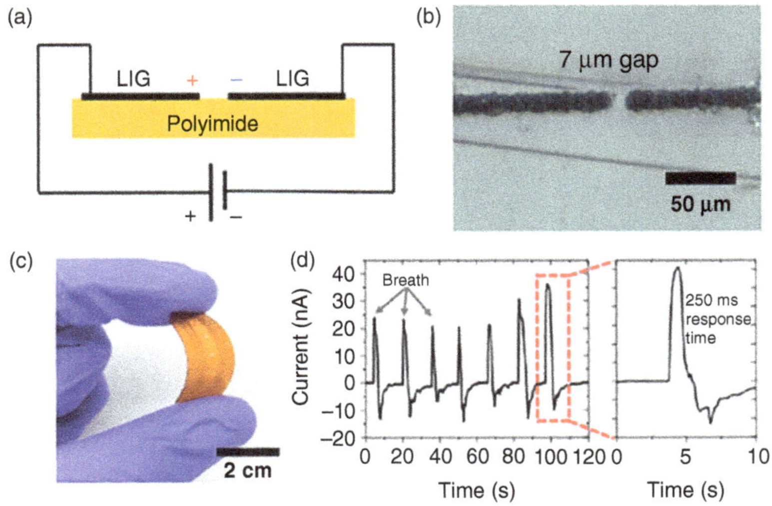

Figure 10.22 (a) Schematic diagram of polyimide-derived laser-induced graphene foam with microgap. (b) Image of the microgap for the flexible humidity sensor. (c) Digital image of the smart sensor. (d) Response current changes of the graphene sensor to human breath. Source: Ref. [47]/American Chemical Society.

color, the larger the resistance variation in 2D maps) as well as the conspicuous pressure distribution (more precise pressure position result in 3D maps).

10.8 Atomic-Oxygen-Resistant Polyimide Films with Shape Memory Function

Since the Soviet Union launched the first artificial satellite, Sputnik, in 1957, space technology has been developing rapidly, providing a reliable technological support for human observation of astronomy and meteorology, geological exploration, daily communication and navigation, and exploration of the mysteries of the universe 33. Improving spacecraft function and efficiency is an important way forward for spacecraft design on the premise of limited mass to delivery. Deployable structures and devices have excellent advantages in cost, storage bulk, weight, and reliability. They can achieve some special abilities which traditional structures and devices cannot [49]. Polymer film is an important part of such deployable structures and devices. It is usually directly exposed to a space environment, and therefore its performances are highly susceptible to space environmental factors, such as atomic oxygen (AO), vacuum ultraviolet irradiation (VUV), thermal cycling, space debris, ions, and electrons [50, 51]. AO, as one of the significant compositions in low-earth orbit, has major threat on polymers, leading to polymers lose their excellent performance and seriously reduce the service life of polymers and shorten the life time of the spacecraft [52, 53]. In the low-earth orbit, neutral gases (mainly AO) appear to be the most prominent among the various factors affecting the serviceability of spacecraft, with an impact factor of about 9. (Note: An impact factor of 0 means

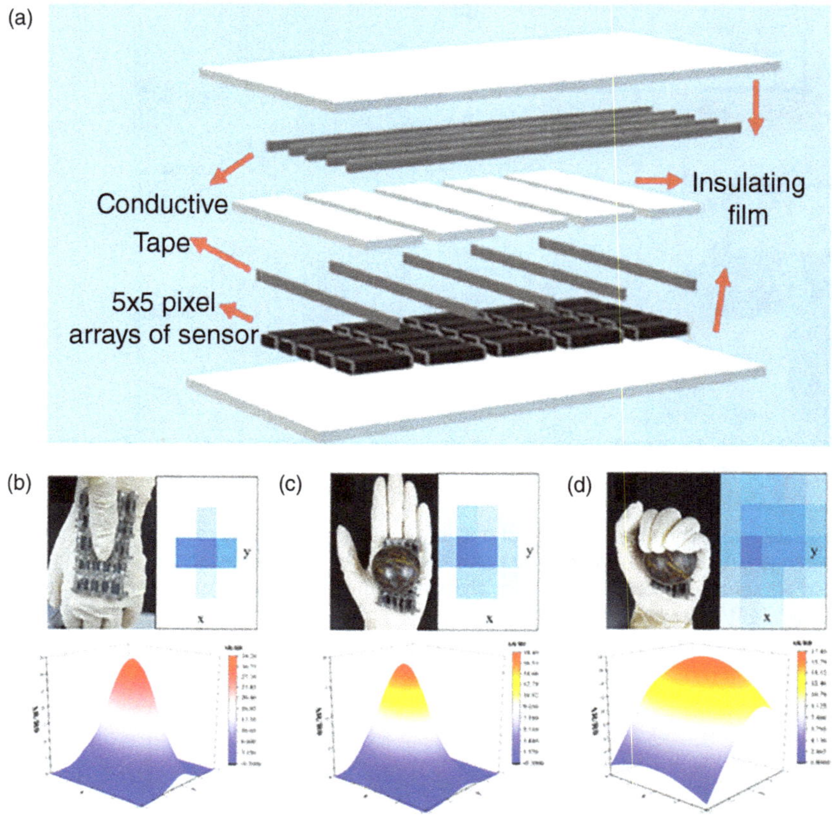

Figure 10.23 (a) Schematic of electronic skin made of PI/CNT foam. Optical photos of (b) a finger pressing the electronic skin, (c) a stone ball in the hand, and (d) a hand gripping the stone ball; 2D and 3D maps show the corresponding resistance variation, respectively. Source: Ref. [48]/American Chemical Society.

that the factor is not referenced, while an impact factor closing to 10 can lead to catastrophic consequences for the spacecraft.)

10.8.1 Mechanism of Atomic Oxygen Formation and Its Role in Polymers

Oxygen molecules by wavelengths less than 243 a.m. solar ultraviolet irradiation, photodecomposition generated by the AO occupies nearly 80% of the low-altitude orbital gas environment, although its density is not high, not more than 108 atoms/cm^3, its own energy is also relatively low, not more than 0.1 eV, but the spacecraft in orbit is a high speed (about 7.8 km/s) running. The relative kinetic energy of AO hitting the surface of the vehicle is as high as 4–5 eV; AO also has strong oxidizing and reactive activity, and it is very easy to interact with the material, and its reaction mechanism is mainly divided into the following steps: extraction, addition, elimination, embedding, and substitution (Figure 10.24) [42, 54]. In

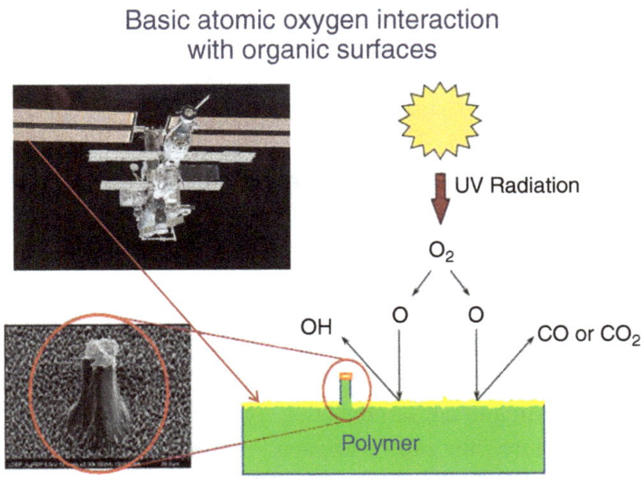

Figure 10.24 The production of AO and its erosive effect to polymer. Source: NASA/Public domain.

addition, once the erosion opens up and breaks through the surface barrier, it will lead to the "palliation" phenomenon, which not only destroys the polymer, but also the surface of the polymer. In addition, once the erosion opens and breaks through the surface barrier, it will trigger the phenomenon of "amalgamation," which not only destroys the surface structure of polymer PI, but also infiltrates into the lower layer of corrosion and causes cracks, embrittlement, loss of quality, and deformation of the material, resulting in the degradation of the thermal, optical, and mechanical properties of the PI and its gradual failure.

In the low-altitude orbit AO has obvious oxidative stripping effect on PI materials, such as biphenyl-type polyimide Upilex.S@ atomic oxygen erosion rate of $0.92 \times 10\text{–}24\,\text{cm}^3/\text{atom}$, while homobenzene-type polyimide Kapton@, PMR, and Ultem@ are relatively high at $2.8 \times 10\text{–}24\,\text{cm}^3/\text{atom}$, $3.0 \times 10\text{–}24\,\text{cm}^3/\text{atom}$, and $3.3 \times 10\text{–}24\,\text{cm}^3/\text{atom}$, respectively; although the erosion rates of polyimides with different structures are different, all of them can lead to serious degradation or even failure of material or device performance during long-term service.

10.8.2 Atomic Oxygen Resistance Modification of Polyimide

Among polymer films, polyimide film has been applied predominantly to thermal control films, attached where it is exposed directly to a space environment, and also has been used as a structural material for deployable structures due to its high specific strength and rigidity, high dimensional accuracy, shape memory function, and low rate of thermal expansion. Polyimide is usually used as the substrate of such deployable structures as solar sails and large flexible solar paddles for the International Space Station (ISS) or artificial satellites. Polyimide-based shape memory polymers (SMPs) not only add shape memory functions to deployable structures but can also expand their potential in space flexible electronics that can be durably used in harsh environments.

Nevertheless, traditional shape memory polyimide film does not have good capacity of resistance to high temperatures and tolerance against intense atomic oxygen (AO) and ultraviolet (UV) irradiation. Therefore, endowing shape memory polyimide films with atomic oxygen resistance is a key step to realize the further application of polyimide in the aerospace field. Nowadays, the techniques are broadly categorized into two types: surface modification techniques [43] and bulk phase modification techniques [44, 55].

10.8.2.1 Surface Modification Techniques

Physical phase modification technology generally involves the addition of AO-resistant components, such as silicone, ZnO, Al_2O_3, or fluorine or phosphorus doping, to raw materials such as polyamide acid used in the preparation of polyimides to form a very uniform and dense protective layer, so that the AOs are insulated from the substrate before they can be applied. Yang et al. [56] synthesized shape memory polyimide/silica (PI/silica) composites with atomic oxygen resistance, good thermal stability, and mechanical properties by co-condensation of poly(amic acid) terminated with (3-aminopropy)triethoxysilane (APTES) and tetraethoxysliane (TEOS) via a sol–gel method. The results show that the PI/silica composite films possess high thermal decomposition temperature ($T_d > 550\,°C$), high glass transition temperature ($T_g > 265\,°C$) as well as good shape fixation ratio ($R_f > 98\%$) and shape recovery ratio ($R_r > 90\%$). Compared to pristine PI, the mechanical properties, surface topography, and shape memory performances of PI-SiO_2-15 were less affected by AO exposure, which results from the silica protective layer formed on the composite film surface, indicating good AO-resistant ability of PI/silica composite films (Figure 10.25).

The bulk phase modification method has two outstanding advantages: first, some metal elements (such as Ti and Al) are added in the preparation process, which react with oxygen atoms to generate the corresponding metal oxides, and the coating of metal oxides can be stabilized, thus enhancing the anti-AO erosion performance of the material. The second is that the material after body phase modification has various excellent properties of the material itself, which makes the material fully play its own performance advantages, but also play the purpose of anti-AO. However, on the one hand, the technical requirements of this method are very high, and the cost of producing the material is also very large, so there will still be some bottlenecks in the wide-scale popularization and application; on the other hand, because the bulk phase modification does not change the fact that the material matrix is still organic structure, the erosion effect will still occur under the attack of atomic oxygen, and the anti-AO performance is difficult to guarantee (Figure 10.26).

10.8.2.2 Bulk Phase Modification Techniques

Compared to bulk phase modification, surface modification provides strong AO resistance to polyimide films while maintaining the excellent mechanical properties of the polyimide film itself. The AO resistance is achieved by plating a target protective layer on the substrate surface, which can be categorized into silicone, SiO_2, and metal oxides (e.g., ZnO, Al_2O_3, and indium tin oxide).

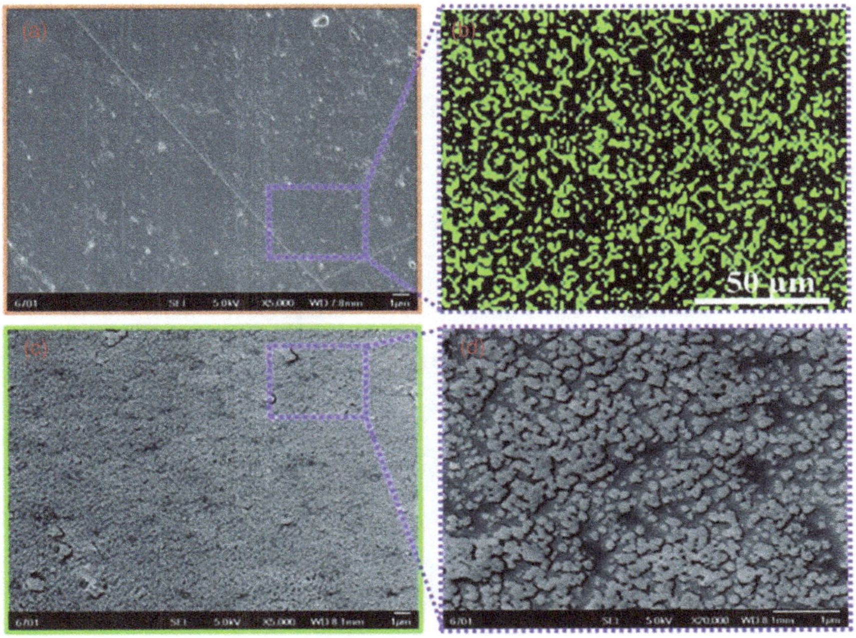

Figure 10.25 Surface morphology (a) and surface Si map (b) of PI-SiO_2-15; surface morphology of PI-SiO_2-15 after etching 24 hour with HF; and (c) the corresponding amplification image (d). Source: Ref. [56]/The Royal Society of Chemistry.

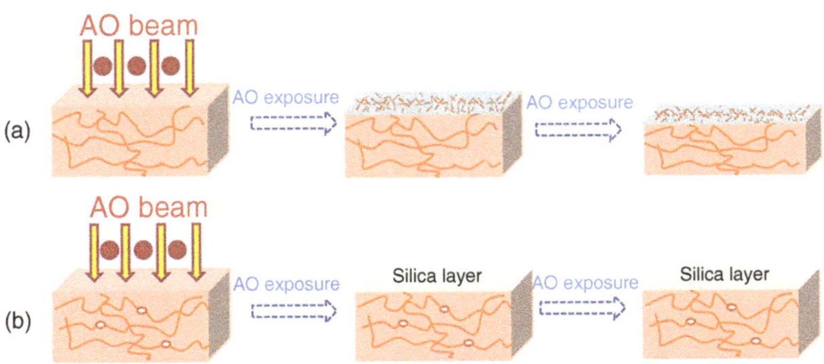

Figure 10.26 Schematic illustration of shape memory behaviors.

At present, the preparation of AO-resistant protective layers is mainly based on nanoparticle-doped silicone resins and other methods. Hu et al. [57] prepared a polysilazane coating on polyimide substrate by the polymeric precursor method to protect space materials from attack of atomic oxygen (AO) and vacuum ultraviolet radiation (VUV) in low-earth orbit (LEO) environment. Erosion kinetics, surface morphologies, and surface composition investigation indicate that this polysilazane coating possesses excellent AO resistance and displays low shrinkage tendency in AO exposure. The erosion yield of the prepared polysilazane coating is 3.5 × and 2.1 × 10 cm/atom during AO exposure and simultaneous AO and VUV exposure,

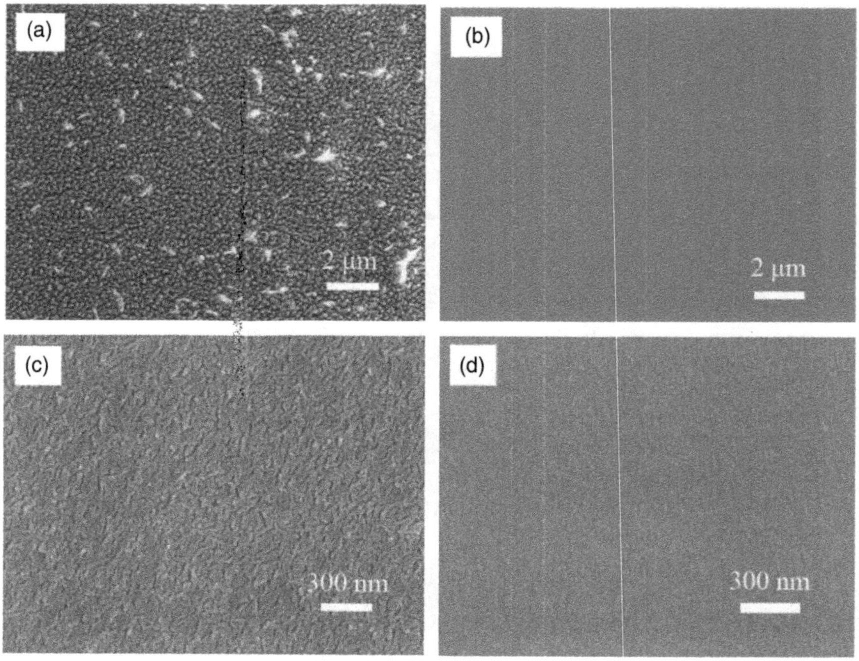

Figure 10.27 A typical SEM image of the AO-exposed surface (a) polyimide, (b) the PSZ coating and a magnified SEM image of the PSZ coating after exposure to AO alone, (c) AO + VUV, and (d) total AO fluence of 9.3×10^{20} atoms/cm^2. Source: Ref. [57]/with permission of Elsevier.

respectively. The low AO erosion yield of the polysilazane coating results from the formation of a SiO-rich layer on its top surface during exposure, and the VUV radiation is able to create free radical sites on the exposed surface, promoting the formation of a uniform SiO-rich surface layer (Figure 10.27).

However, the application of metal oxides in coatings is still relatively small. Compared to polymers, metal oxides are less susceptible to AO erosion, but metal oxides are hard and brittle materials, so it is difficult to produce a reliable bond with the substrate, but it is easy to stress changes and alternating temperature field, cracking, peeling, and other phenomena on the substrate, thus making the AO invasion; however, the problem is not solved and it is not easy to apply in the anti-AO coating. With the rise of integrated circuits at the end of the 1950s and its requirements for high density and high reliability, a large number of advanced surface modification methods and theories, such as epitaxial growth technology, came into being. Today, surface modification process has developed into a specialized discipline and field, from Ge, Si, GaAs, ZnSe, diamond to SiC, GaN, ZnO, all of which require advanced and mature surface modification process support. Nowadays, the more widely used surface modification methods are generally pulsed laser deposition, magnetron sputtering, chemical vapor deposition, molecular beam epitaxy, sol–gel method, vacuum evaporation, liquid-phase epitaxy, spray pyrolysis, and photo-silanation treatment.

Polyimide is widely used in aerospace applications because of its excellent mechanical properties and good chemical stability. However, AO in LEO space degrades polyimide films, leading to serious degradation of their properties. In order to further expand the application of polyimide films in the space environment, the research of AO-resistant polyimide films is essential.

References

1 Nakao, W., Osada, T., Nishiwaki, T. et al. (2021). Focus on self-healing materials: recent challenges and innovations. *Science and Technology of Advanced Materials* 22 (1): 243–234.

2 Kong, D.Y., Li, J., Guo, A.R. et al. (2019). Self-healing high temperature shape memory polymer. *European Polymer Journal* 120: 109279.

3 Zhang, Z.Y., Kong, D.Y., Li, X.F. et al. (2019). High temperature shape memory polymer with high wear resistance. *Smart Materials and Structures* 28 (10): 105005.

4 Yan, G.M., Wang, H., Li, D.S. et al. (2021). Design of recyclable, fast-responsive and high temperature shape memory semi-aromatic polyimide. *Polymer* 216: 123427.

5 Wang, F.F., Yuan, C., Wang, D. et al. (2020). A phase evolution based constitutive model for shape memory polymer and its application in 4D printing. *Smart Materials and Structures* 29 (5): 055016.

6 Jang, L.K., Fletcher, G.K., Monroe, M. et al. (2020). Biodegradable shape memory polymer foams with appropriate thermal properties for hemostatic applications. *Journal of Biomedical Materials Research Part A* 108 (6): 1281–1294.

7 Ma, S., Chu, W.L., Sun, X.L. et al. (2021). Effects of segmented layer suction and micro-vortex generator on a high-load compressor cascade performance. *Proceedings of the Institution of Mechanical Engineers Part C-Journal of Mechanical Engineering Science* 23 (21): 5309–5323.

8 Zhang, H., Xia, H., and Zhao, Y. (2012). Optically triggered and spatially controllable shape-memory polymer-gold nanoparticle composite materials. *Journal of Materials Chemistry* 22: 845–849.

9 Hu, J., Zhu, Y., Huang, H. et al. (2012). Recent advances in shape-memory polymers: structure, mechanism, functionality, modeling and applications. *Progress in Polymer Science* 37 (12): 1720–1763.

10 Zhao, Q., Qi, H.J., and Xie, T. (2015). Recent progress in shape memory polymer: new behavior, enabling materials, and mechanistic understanding. *Progress in Polymer Science* 49–50: 79–120.

11 Yoonessi, M., Shi, Y., Scheiman, D.A. et al. (2012). Graphene polyimide nanocomposites; thermal, mechanical, and high-temperature shape memory effects. *ACS Nano* 6 (9): 7644–7655.

12 Xiao, X.L., Kong, D.Y., Qiu, X. et al. (2015). Shape-memory polymers with adjustable high glass transition temperatures. *Macromolecules* 48 (11): 3582–3589.

13 Xia, Y., He, Y., Zhang, F. et al. (2021). *Advanced Materials* 33: 2000713.

14 Wang, W.X., Liu, Y.J., and Leng Coord, J.S. (2016). Recent developments in shape memory polymer nanocomposites: actuation methods and mechanisms. *Chemical Reviews* 320: 38–52. 10.1016/j.ccr.2016.03.007.
15 Chen, S.J., Hu, J.L., Zhuo, H.T. et al. (2010). Study on the thermal-induced shape memory effect of pyridine containing supramolecular polyurethane. *Polymer* 51: 240–248. 10.1016/j.polymer.2009.11.034.
16 Yu, Y.L. and Ikeda, T. (2005). Photodeformable polymers: a new kind of promising smart material for micro- and nano-applications. *Macromolecular Chemistry and Physics* 206: 1705–1708. 10.1002/macp.200500318.
17 Kun, W., Zhu, G.M., Tang, Y.S. et al. (2013). *Polymer Journal* 45 (6): 671–675.
18 Mohammad, S., Masoud, B., and Mohammadreza, R. (2017). *ACS Applied Materials & Interfaces* 9 (28): 24061–24070.
19 Zhou, J., Li, H., Tian, R. et al. (2017). *Scientific Reports* 7: 5535.
20 Huang, X.Z., Zhang, F.H., and Leng, J.S. (2020). *Applied Materialstoday* 21: 100797.
21 Sun, X.J., Zhang, M.H., Wei, J. et al. (2023). *ACS Macro Letters* 12 (7): 921–927.
22 Ma, L.M., Zhang, G.M., Hou, J.Q. et al. (2023). *Materials today Communications* 37: 107272.
23 Yu, H., Shi, Y., Yuan, B. et al. (2021). *Ionics* 27 (3): 907–923.
24 Hicyilmaz, A.S. and Bedeloglu, A.C. (2022). *SN Applied Sciences* 3 (3): 363.
25 Susa, A., Bose, R.K., Grande, A.M. et al. (2016). *ACS Applied Materials & Interfaces* 8 (49): 34068–34079.
26 Lee, S., Hong, P.H., Kim, J. et al. (2020). *Macromolecules* 53 (6): 2279–2286.
27 Lei, X.F., Jin, Y.H., Sun, H.L. et al. (2017). *Journal of Materials Chemistry A* 5 (40): 21140–21145.
28 Lei, X.F., Ru, H., Sun, Y. et al. (2019). *Chinese Polymer Bulletin* 10: 86–92.
29 Kim, Y.N., Lee, J., Kim, Y.O. et al. (2021). *Applied Materials Today* 25: 1012261.
30 Wan, B.Q., Dong, X.D., Yang, X. et al. (2022). *Journal of Materials Chemistry C* 10 (31): 11307–11315.
31 Wan, B.Q., Yang, X., Dong, X.D. et al. (2022). *Advanced Materials* 35 (2): 2207451.
32 Wan, B.Q., Dong, X.D., Yang, X. et al. (2022). *Energy & Environmental Materials* 6 (1): e12427.
33 Huang, X.-Z., Zhang, F.-H., Liu, Y.-J. et al. (2020). Active and deformable organic electronic devices based on conductive shape memory polyimide. *ACS Applied Materials & Interfaces* 12 (20): 23236–23243.
34 Huang, X.-Z., Zhang, F.-H., Leng, J.-S. et al. (2020). Metal mesh embedded in colorless shape memory polyimide for flexible transparent electric-heater and actuators. *Applied Materials Today* 21: 100797.
35 Xu, Y.H., Wang, L.L., Hu, A.J. et al. (2018). *Progress in Chemistry* 30 (5): 684–691.
36 Liu, H.T., Tian, H.F., Yao, Y.Y. et al. (2020). *New Journal of Chemistry* 44 (28): 12068–12078.
37 Miao, Z.N., Jia, Z.N., Yu, Z. et al. (2020). *Journal of Applied Polymer Science* 137: 49357.

38 Sun, G.H., Wang, W.P., Zhang, C. et al. (2017). *Journal of Applied Polymer Science* 134 (14): 44240.

39 Bakir, M., Meyer, J.L., Pang, S. et al. (2020). *Soft Matter* 16 (6): 1389–1140.

40 Zheng, Q., Kaur, S., Dames, C., and Prasher, R.S. (2020). *International Journal of Heat And Mass Transfer* 151: 119331.

41 Wei, Q., Pei, S., Qian, X. et al. (2020). *Advanced Materials* 32 (14): 1907411.

42 Fischer, H.R., Tempelaars, K., and Kerpershoek, A. (2010). *ACS Applied Materials & Interfaces* 2 (8): 22. 1 8.2225.

43 Shu, M., Li, Z., and Man, Y. (2016). *Corrosion Science* 112: 418–425.

44 Wang, P., Tang, Y., and Yu, Z. (2015). *ACS Applied Materials & Interfaces* 7 (36): 20144.

45 Yang, M., Zhang, C., Lv, Q. et al. (2020). *ACS Applied Materials & Interfaces* 12 (1): 934–944.

46 Li, J., Ding, Y., Yu, N. et al. (2020). *Carbon* 158: 45–54.

47 Stanford, M.G., Zhang, C., Fowlkes, J.D. et al. (2020). *ACS Applied Materials & Interfaces* 12 (9): 10902–10907.

48 Chen, X., Liu, H., Zheng, Y. et al. (2019). *ACS Applied Materials & Interfaces* 11 (45): 42594–42606.

49 Shen, Z.C., Mu, Y.Q., Ding, Y.G. et al. (1998). *SPIE* 9796: 97960U.

50 Hu, L.F., Li, M.S., Xu, C.H. et al. (2009). *Coating Technology* 203: 3338–33343.

51 Shimamura, H. and Nakamura, T. (2010). *Polymer Degradation and Stability* 95: 21–33.

52 Shimamura, H. and Nakamura, T. (2009). *Polymer Degradation and Stability* 94: 1389–1396.

53 Odian, G.G. and Odian, G. (2004). *Principles of Polymerization*. New York: Wiley-Interscience.

54 Wang, X., Li, Y., and Qian, Y. (2018). *Advanced Materials* 30 (36): 1, 803854.

55 Shivakumar, R., Bolker, A., Tsang, S.H. et al. (2020). *Polymer* 191: 122270.

56 Yang, Z.H., Wang, Q.H., Bai, Y.K., and Tm, W. (2015). *RSC Advances* 89: 72437–73326.

57 Hu, L.F., Li, M.S., Xu, C.H. et al. (2009). *Surface & Coatings Technology* 203: 3338–3343.

Index

Functional Polyimide Dielectrics: Structure, Properties, and Applications, First Edition.
Edited by Jun-Wei Zha.

www.ingramcontent.com/pod-product-compliance
Lightning Source LLC
Chambersburg PA
CBHW080932240926
51849CB00003BA/438
* 9 7 8 3 5 2 7 3 5 4 3 3 7 *